Lebesgue's Theory of Integration

Rahul Jain

Lebesgue's Theory of Integration

The Untouched Classic

Springer

Rahul Jain
Indian Police Service
Indian Institute of Science (IISc)
Bengaluru, Karnataka, India

Tata Institute of Fundamental Research (TIFR)
Bengaluru, Karnataka, India

ISBN 978-981-96-1171-3 ISBN 978-981-96-1169-0 (eBook)
https://doi.org/10.1007/978-981-96-1169-0

Mathematics Subject Classification: 26A42, 26A45, 26B30, 26A39

Translation from the French language edition: "Leçons sur l'intégration et la recherche des fonctions primitives, professées au Collège de France par Henri Lebesgue" by Rahul Jain, © Gauthier-Villars 1904. Published by Gauthier-Villars. All Rights Reserved.

This Springer imprint is published by the registered company Springer Nature Singapore Pte Ltd.
The registered company address is: 152 Beach Road, #21-01/04 Gateway East, Singapore 189721, Singapore

Preface to the Second Edition

Two years ago, when the house of Gauthier–Villars informed me that the first edition of these lessons was out of print, I was very perplexed. How can this book retain its character as a review of the main concepts of the integration and of the results acquired in the search of primitive functions, without incorporating in it the numerous works published on these subjects over 23 years?

I had to choose. I resolutely discarded anything that would not directly contribute to "making it more understandable". For example, if I discussed term–by–term integration of a series, it was because the possibility of this operation originates directly from those properties which characterise the integration and shed light on them. When one considers integration as the sum of an infinite number of indivisibles, one uses an extension of the notion of sum which, in some respect, is comparable to the one that gives the sum of a series. These two extensions are intimately linked. But I did not discuss the method of integration by parts and by substitutions, the second mean value theorem, the inequality of Schwarz, and its generalisations, which are nevertheless indispensable for the mathematical use of integration.

Generalisation is one of the best ways of "making people understand" mathematics. Whereas, in the particular case, one gets stuck into the observation of facts which are specific to this particular case. In the general case, there is nothing more to observe, than the facts on which one must reason. This approach leads to the same result as with axiomatic definitions, albeit in a logically less precise, but much more lively and suggestive manner!

However, I have not treated the functions of several variables, because the reader of this collection could refer to an excellent book by M. de la Vallée Poussin. Additionally, it has offered me a generalisation of a much broader integral: the integral of Stieltjes.

Indeed, it is almost a misconception to treat the integral of Stieltjes by restricting it to the functions of a single variable. Yet I thought I could do so. I contended myself by indicating the general physical meaning of the integral of Stieltjes.

When I shifted from the point of view of the quadrature and took up the perspective of primitive functions, I no longer had a choice. I had to talk about the work of M. Denjoy, which is fundamental and decisive enough, to have my undivided focus. While following the ideas of M. Denjoy, I have deviated conspicuously from his presentation in form and sometimes in substance. By doing so, I believe, that I have made this beautiful theory of functions of one real variable more accessible and contributed to its better comprehension.

The totalisation of M. Denjoy essentially uses the transfinite induction. Therefore, I had to use the transfinite induction more deliberately than I had done in the first edition.

Although this first edition has appeared, to some, audacious and purposely filled with outrageous novelties, it was the work of a timid man who, out of the seven chapters he had written, had devoted six to the exposition of the previous research before addressing the work that was considered revolutionary. If he did so, it was not out of propagandist skill seeking to recruit followers for revolution, but rather to reassure himself. He believed in fact, and still believes, that to do useful work it is necessary to walk on one of the paths opened by previous works. Otherwise, the risk is too great, of creating a science unrelated to the rest of mathematics. Therefore, he endeavoured to extract ideas that had consciously or unconsciously guided, the mathematicians in the study of integration, their ideals in this field, in the words of the late P. Boutroux, and to show that his personal ideas were closely connected with those of his predecessors.

It is with the same timidity that previously I discussed the transfinite numbers. I was able to proceed by allusions and affirmations because I was only using transformations of simply infinite series into more complex series provided by the method of chains of intervals. But, for M. Denjoy's totalisation, I had to develop the "Note" that I had devoted to transfinite numbers.

From this "Note" it follows, in particular, that I could have avoided the use of chains of intervals and, consequently, no longer relied on the transfinite numbers in many places of this book. I thought there would be drawbacks and some hypocrisy in doing so. Let me explain by analogy. The infinitesimals were once obscure entities that appeared in imprecise and inaccurate statements. Everything became clear, thanks to the notion of limit. We can, therefore, do away with the notion of infinitesimals. But on the other hand, there is no longer any obscurity in using them. And wouldn't it be somewhat hypocritical to discourage others from using the suggestive and convenient language of infinitesimals while continuing to use it oneself for reasoning? The chains of intervals are used quite naturally. The transfinite numbers are an excellent mathematical tool. It is advisable to get accustomed to using them.

To better understand the totalisation of M. Denjoy, I generalised it in a manner as Stieltjes generalised the ordinary integration. This led to the problems whose solutions are still awaited.

But what is the purpose of these studies? They would have been very useful even if their only effect had been of fixing our attention on integration and differentiation, sufficient for us to recognise this: integration is always an operation similar to the one required to calculate the amount of heat necessary to raise the temperature of a body by 1 degree, as a function of the masses of its part and their specific heats; the differentiation is the inverse operation. These operations connect two quantities attached with these bodies and a function attached to the points of these bodies.

"How, one could say, you did not know that?" Do not expect to get my confession so easily: "I knew it, I knew it very well." However, if I had known it in 1903 as perfectly as I do now, I would not have omitted to discuss the complete Stieltjes integration in the first edition of this book. And we must believe that this omission did not generally seem very serious because none of those who did me the honour of reviewing my book pointed it out.

I said integral of Stieltjes; shouldn't have I said integral of Cauchy? Cauchy, in fact, very clearly talked about the importance and the physical meaning of the new integration taken in all its generality, whereas Stieltjes primarily, logically defined the new operation, but only in case of one variable. I did not think it necessary to change the adopted terminology. If I had done so, would it have been necessary to take the name of the first inventor currently known or the name of the one who gave integral its widest definition, currently known? In any case, the attribution would have been inaccurate and unjust; it is better to stick to the established inaccuracies.

This provides me with an opportunity to apologise for the omissions and commissions. I only wanted to give some starting points for bibliographical research. I did not try to summarise by the history of intensive developments of the notion of integration during the past 20 years. Nor do I claim to succeed in listing everything I have borrowed from various recent works on the theory of functions of real variable; they have all been consistently helpful to me.

I thank M. Vasilesco who kindly helped me in the correction of the proofs.

Paris, France
New Delhi, India
December, 1926

Henri. Lebesgue
Translated by Rahul Jain

Preface by the Translator

This book presents an English translation of Lebesgue's renowned work in French, *Leçons sur lintégration et la recherche des fonctions primitives*, second edition, published in 1928 by *Gauthier–Villars*, [19]. The original work in French was an outcome of a course conducted by Mr. Henri Lebesgue at the Collège de France during the academic year 1902–1903. He was in charge of the course, which was sponsored by the Peccot family. Across 20 lessons, he imparted ideas that have been compiled into this book, emanating from the content covered in that course. The focus of the course lay in the development of the integration concept. However, rather than providing a comprehensive historical summary, only a selective account was offered during those twenty lessons. These results specifically concern the integration of functions involving only one real variable, particularly those which were easily generalisable. This translation is supplemented with a comprehensive historical summary, which may be essential for a reader of the 21st century.

In the year 1900 and around that time, the idea of generalisation held the utmost significance. Lebesgue explicitly stated a preference for following established paths rather than forging entirely new mathematical avenues. However, what started with a humble beginning, culminated ultimately in pioneering a new branch of mathematics. Initially, it was a humble beginning. He commenced by revisiting the "numbers of analysis" such as definite integral, lengths of curve, area of domains, etc., previously addressed in Camille Jordan's work.[1]

While Lebesgue did not aim to conduct a comprehensive historical survey of the development of functions, their integration, Fourier series convergence, etc., the translator endeavours to provide a historical perspective within this book to ensure its self–contained nature. This historical context and analysis, pertinent to the subsequent chapters, are detailed in the first chapter of this book.

It is important to take note of the mathematical backdrop against which Lebesgue laid the foundations of his seminal work. The developments commenced with the definition of

[1] *See*, Cours D'Analyse, [12] and [13].

the term "function" and gradually progressed to the integration of functions. Numerous mathematicians, including Euler, Descartes, Lagrange, Cauchy, D'Alembert, Bernoulli, Fourier, Darboux, Cantor, Dirichlet, Hankel, Riemann, Jordan, Borel, Lebesgue, Baire, Stieltjes, Denjoy, and others, contributed to these advancements. Initially, the term "function" was narrowly defined. It was perceived as a representation achievable through an equation with various terms linked by algebraic signs. Furthermore, these terms comprised distinct functions like radicals, exponents, trigonometric expressions, and so forth.

The primary focus was on determining which class of functions could be integrated and understanding the integration process. During Newton's era, integration was viewed as the inverse of differentiation, while in Cauchy's era, it evolved into a process involving a limit of sums. It was first established that continuous functions were integrable. However, the scope of integrable functions expanded over time, encompassing functions with a finite number of maxima, minima, and a finite number of discontinuity points. Subsequently, Riemann's integration emerged, encompassing functions with a countably infinite number of discontinuities distributed in a specific manner within a function. Although the set of discontinuities was extensive (comprising a countably infinite number), it remained "reducible" or of "measure zero". This signified that the length of intervals containing discontinuity points could be arbitrarily small.

These discussions culminated in the measure–theoretic estimation of the size of a set, leading to the development of integration within the realm of measure theory. This introduced the concepts of measurable and summable functions, signifying a departure from geometric quadrature and the Cauchy–Darboux type of sums and limit processes. Although the concept of limit of sums persisted in measure–theoretic integration, it differed from the Cauchy–Darboux sums, which were termed "integral by excess and defect".

It was D'Alembert and then Bernoulli who, while working on the wave equation, introduced the concept of representing a function through the summation of an infinite trigonometric series. Bernoulli's work was significantly supported by Fourier's theory presented in 1822[2].

These advancements led to the idea of representing a function via a trigonometric series, further advancing the abstraction of the concept of a function. This development brought forth various questions concerning the convergence of such trigonometric series, now known as Fourier series. Different types of convergence emerged, including point-wise convergence, uniform convergence, absolute convergence, almost everywhere convergence, and the convergence of square–integrable functions (Lennart Carleson 1966), among others. Accompanying this was the question of the term–by–term integrability of series, which revolved around whether one could interchange the integration and summation signs and, if so, under what conditions. This query was profound because properties such as continuity or integrability held by the terms of a series did not automatically transfer to the sum function; special conditions were necessary. It was demonstrated

[2] *See* [9].

that while the term functions of a series were Riemann integrable, the sum function was not. To establish the integrability of the sum function, prior assumptions were needed regarding the sum function (Osgood and Arzela). However, Lebesgue's definition of integration overcame this challenge. A sequence of bounded and Lebesgue integrable (summable) functions converged to bounded Lebesgue integrable (summable) functions. This way, Lebesgue expanded the definition of integrable functions, broadening the class beyond what Riemann had established.

It was observed that certain functions were differentiable and yielded bounded derivative functions upon differentiation. However, peculiarly, these derivative functions were not integrable according to Riemann's definition. This problem was resolved through Lebesgue's integration definition, ensuring that every bounded derivative became integrable. Consequently, this solution addressed the fundamental theorem of integral calculus. Thus, Lebesgue restored the integration process as the reverse of differentiation.

The development of the integration theory is intricately linked to the theory of measure and emerged due to a sheer dearth. Camille Jordan undertook the task of defining a set's extent, while Borel further elaborated on the concept of measure, later expanded by Lebesgue. As a precursor to measuring the set (an estimate of its size), set–theoretic approximations were made regarding the sizes of sets, including nowhere dense sets and everywhere dense sets within an interval. This estimation of size aimed to assess the size of a function's singularities within the set.

Lebesgue delved into the rectification of curves. An integral formula was devised to calculate the arc length of curves characterised by continuously varying Tangents-an approach known as the analytic method. Another method, without presumptions about the curve's tangents, emerged to measure the arc length. This method approximated curve lengths by inscribing polygons along the curve, attaining a limit when the polygon perfectly matched the curve. This process yielded valuable deductions, such as the approximation of continuous functions with polynomials and the concept of functions with bounded variation. Functions of bounded variation proved instrumental in solving the fundamental theorem of integral calculus and determining the primitive function of a bounded derivative function. It emerged as a necessary and sufficient condition for the existence of a primitive function derived from a bounded derivative function. Additionally, the concept of functions with bounded variation had implications in the Stieltjes integral, where the integrator function was considered a function of bounded variation.

All in all, this book is a valuable addition to the study of mathematics, as it presents the original work of the creator of the theory of integration and is now made accessible to English–speaking students worldwide. The theory of integration was in its infancy during the time when this work was written, and the author has provided references to the original work of several eminent mathematicians of that era. The relative simplicity of the theory at that time makes it easier to understand its intricacies without being overwhelmed by a multitude of tools. The process of building a theory from the ground up can be likened to constructing an intricate building. This book provides insight into the floor plan, the placement of different pillars, the location of stairs, the depth of the foundation, and the overall layout of the structure. This gives readers a glimpse into the mind of the author and how great minds worked during that era. The author's work strikes a delicate balance between straightforward ideas and intricate, sophisticated, abstract, esoteric, and specialised concepts.

Moreover, this translation preserves the philosophical and intuitive style of Lebesgue's original writing, which reflects the cultural and intellectual milieu of early 20th-century mathematics. While the verbose and detailed explanations may feel unfamiliar to modern readers accustomed to concise and formal texts, the historical and pedagogical value of this approach is undeniable. With time, readers will grow accustomed to this style and come to appreciate its depth and clarity, offering a unique window into the evolution of mathematical thought.

This book will be useful for students at all levels, including those taking their first course in analysis, advanced students, book collectors, professors, and general book lovers who appreciate the aesthetic value of great mathematical works. In conclusion, this book is a valuable resource for anyone interested in the theory of integration, its historical development, and the mind of its creator. The journey through the book promises to be rewarding, and I hope it will inspire and develop interest in the subject among readers.

New Delhi, India Rahul Jain

Acknowledgements

I am deeply grateful to several people whose contributions were instrumental in bringing this translation of Henri Lebesgue's 1928 masterpiece to fruition.

My deepest gratitude extends to my esteemed teacher, Prof. Adi Adimurthi (currently, working at the Indian Institute of Technology (IIT) Kanpur, retired from the Tata Institute of Fundamental Research, Centre for Applied Mathematics, Bengaluru). He kept the fire of my passion for this project burning brightly. Countless discussions on intricate mathematical concepts illuminated the path forward, and his guidance proved invaluable in navigating the journey of publishing this work with Springer Nature. Professor Adimurthy's subtle suggestions, born from his vast experience, were incredibly valuable. Despite his undoubtedly busy schedule, his prompt responses to my numerous queries provided unwavering support and direction.

Ms. Laurence Duquenne, a fellow at the University of Leeds, Leeds Institute of Rheumatic and Musculoskeletal Medicine, deserves my profound thanks for her meticulous proofreading. Her sharp eye meticulously combed through every detail, unearthing the finer subtleties of the French language and ensuring the translated text accurately reflects the original. Beyond her linguistic expertise, Ms. Duquenne offered unwavering encouragement and motivation throughout the process. Notably, her keen attention to detail uncovered missing paragraphs in my initial translation, a testament to her commitment to impeccable quality. I am deeply impressed by her ability to manage her own research alongside the added responsibility of proofreading such a significant volume. Her dedication has profoundly improved this work.

Connecting bridges: My heartfelt gratitude goes to Dr. S. B. Solanki, Consultant Orthopaedic and Joint Replacement Surgeon, Jaipur, and Dr. Professor (clinical) Hemant Pandit, the University of Leeds, Chapel Allerton Hospital, for playing crucial roles in connecting me with Ms. Duquenne. In my hour of need, they acted as silent and swift facilitators, bringing us together. Their timely assistance will forever be etched in my memory.

I owe a debt of gratitude to my friend, Manu Jaiswal (Professor, Department of Physics, IIT Madras), for his steady moral support. He readily lent a listening ear during countless conversations about the project, actively engaging in my deliberations and proposing insightful ideas. His resolute interest spurred me forward, and his suggestions played a crucial role in shaping the course of action. Notably, the very idea of translating complex mathematical concepts into a simpler, more accessible form for students, which served as the guiding light for this project, originated during our time together at the Indian Institute of Science in Bengaluru.

I am delighted to express my sincere gratitude to my esteemed colleagues, Mr. Shivansh Mehrotra and Ms. Shivani Mehta, both Ph.D. students at IIT Delhi, specialising in the fields of material science and computer science, respectively. Their invaluable collaboration brought immense joy to the process of developing this work. Their enthusiasm and technical expertise, particularly in the LaTex aspect of this book, significantly enhanced its quality and depth. Working alongside such dedicated and knowledgeable individuals has been an enriching experience, and I am grateful for their contribution to this project.

Finally, I would be remiss not to acknowledge the historical inspiration for this project. The legacy of His Highness Sawai Jai Singh Ji, the erstwhile King of Jaipur, Rajasthan, India and his esteemed courtiers, Pt. Kewal Ram Ji and Pt Govind Narayan Chaurasia, serves as a beacon of intellectual exchange and collaboration. The erstwhile King's profound interest in mathematical works led him to invite scholars from across the globe, including France and Portugal, and facilitate the translation of European mathematical and astronomical treatises into the Sanskrit language. The magnificent astronomical observatory Jantar Mantar in Jaipur stands as a testament to this dedication to cross-cultural knowledge. Their story fuels my own passion for bridging linguistic and cultural divides through translation, making complex mathematical concepts accessible to a wider audience.

Behind every successful endeavour, there is a family cheering you on. I would not have accomplished this project without the unwavering support of my Mother Smt. Manju Jain, my Father Sh. Ashok Jain, My Wife Ms. Preeti, and especially my 10 year-old son, Shrey. His unbounded inquisitiveness to learn served to keep my spirits high throughout this journey. He, indeed, served as a guiding light on this arduous path.

I am also thankful to my colleague at my office, Mr. Ganga Shankar, who was a constant motivation to me during my conversations about this work. He took a keen interest in the affair and despite his very busy schedule never missed an occasion to listen to my deliberations.

To each of these individuals, I express my heartfelt gratitude. Their contributions were essential in bringing this translation to light, and I deeply appreciate their support and guidance.

Contents

1 Historical Perspective ... 1
1.1 Introduction ... 1
1.2 Development of Concept of Function ... 5
1.2.1 Euler's Concept of Function ... 5
1.2.2 Work of D'Alembert and Solution of Wave Equation ... 6
1.2.3 Advent of Daniel Bernoulli ... 6
1.2.4 Fourier's Ideas in His Analytic Theory of Heat ... 7
1.2.5 Work of Dirichlet ... 9
1.3 Riemann's Ideas of Integration ... 10
1.3.1 Hankel's Work: Oscillation of f, Measure Theoretic and Topological Size of Sets ... 11
1.3.2 Darboux Upper and Lower Sums ... 12
1.3.3 Ulisse Dini and Others: Counterexample ... 16
1.3.4 Criticism of Riemann's Definition of Integration ... 16
1.4 Work of Camille Jordan ... 18
1.4.1 Notion of Extent of a Set ... 19
1.4.2 Integrable and Bounded Functions ... 20
1.5 Work of Borel ... 21
1.5.1 Synthesis of Borel Sets: Two Examples ... 22
1.5.2 Definition of Measurable Set ... 26
1.6 Work of Lebesgue ... 27
1.6.1 Problem of Measure ... 28
1.6.2 Exterior and Interior Measure ... 29
1.6.3 Measurable Sets ... 29
1.6.4 Borel Measurable Sets: A Procedure ... 30
1.6.5 Relation Between Jordan and Lebesgue Measurable Sets ... 31
1.6.6 Relation Between Borel Measurable Set and Jordan Measurable Set ... 31

1.6.7 Comparison Between Borel and Lebesgue Measurable Sets ... 32
1.7 Integration in Lebesgue's Sense ... 33
1.7.1 Definite Integral ... 33
1.7.2 Indefinite Integral ... 39
1.8 Work of Stieltjes ... 44
1.9 Work of Denjoy ... 45
1.10 Summary ... 48
References ... 50

2 The Integration Before Riemann ... 53
2.1 The Integration of Continuous Functions ... 53
2.2 Integration of the Discontinuous Functions ... 57

3 The Definition of Integral Given by Riemann ... 63
3.1 Properties Relating to Functions ... 63
3.2 Conditions of Integrability ... 69
3.3 Properties of Integral ... 74
3.4 Integration by Lower Sum and Upper Sum ... 76

4 Geometric Definition of the Integral ... 79
4.1 The Measure of the Sets ... 79
4.2 Definition of the Integral ... 86

5 The Functions of Bounded Variation ... 89
5.1 The Functions of Bounded Variation ... 89
5.2 The Rectifiable Curves ... 99

6 The Search for Primitive Functions ... 103
6.1 The Indefinite Integral ... 103
6.2 The Derivative Numbers ... 105
6.3 Functions Determined by One of Their Derivative Numbers ... 110
6.4 Search for Function Whose Derivative Number is Known ... 118
6.5 Riemannian Integration Considered as an Inverse Operation of the Differentiation ... 119

7 Definite Integration Using Primitive Functions ... 121
7.1 Direct Search for Primitive Functions ... 121
7.2 Properties of Derivative Functions ... 124
7.3 The Integral Deduced from the Primitive Functions ... 127

8 The Definite Integral of Summable Functions ... 131
8.1 The Problem of Integration ... 131
8.2 The Measure of the Sets ... 135
8.3 The Measurable Functions ... 142
8.4 Constructive Definition of the Integral ... 143
8.5 Other Forms of the Definition of Integral ... 153

9 The Indefinite Integral of Summable Functions ... 161
9.1 The Three Indefinite Integrals. The Additive Functions of a Set ... 161
9.2 The Absolutely Continuous Functions ... 174
9.3 The Singularities of Non-Absolutely Continuous Functions ... 177

10 The Search for Primitive Functions. The Existence of Derivatives ... 187
10.1 The Search for Primitive Functions ... 187
10.2 The Differentiation of Functions of Bounded Variation ... 195
10.3 The Rectification of Curves ... 206

11 The Totalisation ... 209
11.1 The Functions of the First Class ... 209
11.2 The Primitive Functions of Everywhere Finite Derivatives ... 214
11.3 The Primitive Functions of Everywhere Finite Derivative Numbers ... 222
11.4 The Totalisation ... 228

12 The Integral of Stieltjes ... 247
12.1 The Integral of Stieltjes Defined by the Theory of Summable Functions ... 248
12.2 The Linear Functionals ... 255
12.3 Direct Definition of Integral of Stieltjes ... 262
12.4 Physical Significance of Integral of Stieltjes ... 277
12.5 Function Primitives and Totalisation with Respect to a Function ... 281

Appendix: Note. On Transfinite Numbers ... 297

Historical Perspective

1

1.1 Introduction

In the social and cultural landscape of mathematics in which Lebesgue's[1] personality came to prominence on the world stage, the following mathematical developments were present:

1. Riemann's theory marked the pinnacle of generality in the realm of integration. His integrability criterion was the weakest under which the traditional definition, rooted in Cauchy sums, maintained its validity; it was so weak that the concept of integration could be extended to functions with a dense set of points of discontinuity—functions whose existence most mathematicians had not even considered. Therefore, until Cauchy's sums were the only approach to defining the integral, further generalisation of the integral was not conceivable. In this context, measure-theoretic concepts emerged as significant players, providing a novel foundation for defining Riemann's integral.
2. **Fourier's Problem of** 1822**:** Using a trigonometric series, Fourier sought to represent an arbitrary function. He had an insight that the coefficients of the series could be determined through term-by-term integration of the series against the $\cos kx$ and $\sin lx$ functions, using the orthogonality of trigonometric functions. His problem can be summarised as follows:

   M. If a trigonometric series can represent a bounded function f, is that series the Fourier series of f? A related problem is:
   N. When is the term-by-term integration of an infinite series of functions permissible? In other words, when is it true that

[1] Lebesgue: born in Beauvais, France, on 28 June 1875; died in Paris, France, on 26 July 1941.

R. Jain, *Lebesgue's Theory of Integration*,
https://doi.org/10.1007/978-981-96-1169-0_1

$$\int_{\alpha}^{\beta}\left(\sum_{n=1}^{\infty} f_n(x)\,dx\right)=\sum_{n=1}^{\infty}\int_{\alpha}^{\beta} f_n(x)\,dx?$$

Fourier had assumed that the answer to N was *Always* and had used N to prove that the answer to M was *Yes*. By the end of the 19th century, it was recognised that N was not always true even for uniformly bounded series precisely because $f(x)=\sum_{n=1}^{\infty} f_n(x)$ need not be Riemann integrable; moreover, the positive results that were obtained required extremely tedious efforts.

In 1829, Dirichlet was the first one to give a correct proof that for certain functions, the associated Fourier series converges to the function. He had to assume the function to be continuous and add a more stringent condition that the function had only a finite number of maxima and minima in the interval of periodicity.

Riemann took up this case further in 1854. He carefully analysed the results of Dirichlet. He suggested that different types of convergence may be taken into consideration. These developments, however, paved the way for Lebesgue's elegant proof that N is true for any uniformly bounded series of *Lebesgue integrable* functions $f_n(x)$. By applying this result to M, Lebesgue could affirm Fourier's belief that the answer is *Always.* Before the introduction of Lebesgue integration, when dealing with series of Riemann integrable functions, it was necessary to assume the uniform convergence of the series. In 1870, Eduard Heine stated the following.

> Up to the present time, it was believed that the integral of a convergent series, the terms of which remain finite between finite limits of integration, would be the sum of the integrals of the individual terms. M. Weierstrass observed first that the proof of this theorem demands that the series in the limits of integration not only converges but also converges uniformly.

So much so, even Cauchy fell into the error of assuming that the sum of a convergent series of continuous functions shared the common properties of its terms and, accordingly, was continuous and integrable (if the term functions were integrable).

3. **Fundamental Theorem of Integral Calculus:** the assertion

$$\int_{\alpha}^{\beta} f'\,dx = f(\beta)-f(\alpha),$$

is called the fundamental theorem of integral calculus. With Riemann's definition of integral, it was known that there exist non-integrable derivative functions. Thus, as defined by Riemann, integration does not always solve the fundamental theorem of integral calculus. Ulisse Dini and Vito Volterra showed that there exist functions with bounded, and non-integrable derivatives. Volterra constructed an example in *Giornale di Battaglini,* 1881 of such a function:

Example (**Volterra**)[2]**:** Let E be a perfect non-dense set which is not an integrable group.[3] Let (a, b) be an interval contiguous to E, and let us consider the function

$$\varphi(x, a) = (x - a)^2 \sin \frac{1}{x - a};$$

its derivative vanishes an infinite number of times between a and b. Let $a + c$ be the greatest value of x, $x \leq \frac{a+b}{2}$ at which φ' vanishes. With this in mind, let us define a function $F(x)$ by the following conditions: it is zero at the points of E. In every interval (a, b) contiguous to E, it is equal to $\varphi(x, a)$ from a to $a + c$; from $a + c$ to $b - c$, the function F is constant and equal to $\varphi(a + c, a)$; from $b - c$ to b, F is equal to $-\varphi(x, b)$. This function $F(x)$ is continuous, it has a derivative $F'(x)$, and the derivative F' is bounded. However, in the sense of Riemann, F' is not integrable, because at every point of E, the maximum of F' is $+1$, and its minimum is -1. Since this is the case at point $x = a$ for the function $\varphi'(x, a)$; however, by hypothesis E is not an integrable group. By the principle of condensation of singularities, many more derivative functions could be obtained, which were not integrable in the sense of Riemann, in any interval however small. M. Köpke constructed differentiable functions of bounded derivatives vanishing in every interval. These derivative functions are obviously not integrable.

4. **Length of curves.** For finding the length of a curve C, Jordan found two polygons S and S' without multiple points, one contained in the other, and inscribed the curve C between them. He then established that for any given ε, the polygons S and S' are such that each point in the annular space that separates them is at a distance less than ε from C. In this way, the length of curve C was approximated by the perimeter of the polygons, and limit was taken as described by the following procedure. Jordan considered the curve C defined by parametric equations

$$x = f(t), \quad y = \varphi(t), \quad t \in [t_0, T]$$

Then he divided the interval $[t_0, T]$ by points $t_0 < t_1 < t_2 < \cdots < t_n < T$ which corresponded to the points $(x_0, y_0), (x_1, y_1), \ldots (x_n, y_n), (X, Y)$ on the curve C. The perimeter of the polygon inscribed in the curve, of which the above points are vertices is given by

$$\sum \sqrt{(x_{k+1} - x_k)^2 + (y_{k+1} - y_k)^2}.$$

Jordan asserted that if this sum tended towards a definite and constant limit when the interval $T - t_0$ decreased indefinitely, then this limit would represent the *length of the curve's arc* corresponding to this interval.

---

[2] *See* [1], p. 100.

[3] A set of points of a straight line constitutes an integrable group, if the points of the set could be enclosed in a *finite* number of segments, the sum of lengths of which can be made as small as we want. *See* [1], P. 26.

Similarly, if the above curve has continuously varying tangents, Jordan gives a method of finding the length of the curve using the differential arc length of the curve, which is

$$ds = \sqrt{x'^2 + y'^2}\, dt;$$

and the length of the curve in the interval $[t_0, T]$ will be found by integrating the differential arc length between the two limits

$$s = \int_{t_0}^{T} \sqrt{x'^2 + y'^2}\, dt.$$

5. **Area of domains.** Jordan considered plane domains bounded by one or several curves with continuous derivatives except, possibly, at a finite number of points. He defined the curves parametrically as above.

$$x = f(t), \quad y = \varphi(t), \quad t \in [t_0, T].$$

Then, he defined the area of such domains by using double-integral

$$\iint dx\, dy,$$

extended to the interior of this domain. He also gave a procedure to reduce this double integral to a single integral using the formula $\int_{x0}^{X} y\, dx$ which yields the integral in parametric form as

$$\int_{t_0}^{T} \varphi(t) f'(t)\, dt.$$

Jordan discussed in his *Cours D'Analyse* [2], p. 132 and the following pages, calculation of the area of more general domains which were bounded by several rectifiable lines, each of which was decomposable in several partial arcs, such that in each of them x varies in the same direction

$$\int_{x_i}^{x_{i+1}} dx(-y_1 + y_2 - y_3 + \cdots),$$

where (x_i, x_{i+1}) is the interval where these partial arcs are taken. He further supposed that if along each of the arcs $A_1, A_2, \ldots, A_n$ there existed tangents whose direction varied continuously with the arc s then the integral depicting the area of the domain would take the form

$$\int y \cos\theta\, ds,$$

where θ is the angle between the y-axes and tangent to the arc. The above analysis was done for the curves without multiple points. After the advent of the measure theory of Borel (measurable Borel sets) and the subsequent work of Lebesgue, the area of the plane domains was defined as measure of the set of points constituting the plane domains.

1.2 Development of Concept of Function

1.2.1 Euler's Concept of Function

Although not originating with Euler, the conceptual groundwork for functions was found to have been remarkably elevated through his profound insights, transforming calculus into a formal theory of functions. In his seminal work, *Introductio in analysin infinitorum,* published in 1748, Euler defined a function of variable quantity as an *analytical expression* constituted by the variable itself and constants. This pioneering definition hinged upon the fundamental notion of an analytical expression, a concept Euler deemed fundamental to all known functions.

However, Euler's significant contribution went beyond a mere definition; he began a shift in perspective that eventually laid the foundation for the modern understanding of functions. Notably, his ground-breaking work in 1734 on partial differential equations marked the inception of *arbitrary functions* within integral solutions. This innovation didn't come without debate. When challenged by Jean D'Alembert, who argued that these arbitrary functions required representation by a single algebraic or transcendental equation for proper mathematical analysis, Euler countered. He asserted that the curves depicted by these arbitrary functions need not adhere to any predetermined law. Instead, they could possess irregularities and discontinuities, potentially composed of various curve fragments or traced freely across the plane.

Euler's terminology and conception of *discontinuity* were distinctive. He perceived a disruption in the analytical form of the functional relationship, allowing for functions that might be continuous in the modern sense yet discontinuous within Euler's framework. However, the notion that arbitrary functions could exhibit modern-style discontinuity at multiple points within a finite interval did not garner serious consideration during Euler's time. Instead, the focus lay primarily on the absence of unique determination by a single equation for arbitrary functions, rather than their properties as mappings $x \to f(x)$ between real numbers.

Euler's visionary perspective on functions laid the groundwork for a profound shift in mathematical thinking, setting the stage for the evolution of function theory as we understand it today.

This debate on the structure and definition of function further manifested itself in the solution of the wave equation, which D'Alembert proposed.

1.2.2 Work of D'Alembert and Solution of Wave Equation

The wave equation:

$$\frac{d^2y}{dt^2} = a^2 \frac{d^2y}{dx^2}. \tag{1.1}$$

D'Alembert proposed its solution as

$$y(x,t) = f(x+at) - g(x-at), \tag{1.2}$$

where the functions f and g possess sufficient *smoothness*, as interpreted in the modern terminology. However, it became evident that Euler did not consider this solution the most general solution of the wave equation. Euler argued that the string's position at any time (i.e., the progression of waves within the string) relied on the initial position (at $t = 0$) and the initial velocity (at $t = 0$). This initial shape of the string could take any form.

Euler's insight into the wave motion within the string highlighted what we now refer to as "initial data" or conditions concerning the position and velocity of the string. The initial shape of the string might manifest as a continuous curve, multiple continuous curves linked end-to-end, or even exhibit discontinuities at the ends. According to Euler, this array of possibilities couldn't be encompassed within just two functions f and g. He asserted that these functions alone could not accommodate the discontinuity in the 'analytic expression'. Hence to Euler, the solution of D'Alembert was a particular one. But D'Alembert contended that the functions should be *continuous*, meaning that a single equation for the function should suffice.

1.2.3 Advent of Daniel Bernoulli

As the discussions persisted, Daniel Bernoulli further complicated matters by introducing Fourier series, providing a solution for the vibrating string in the form of a trigonometric series:

$$y = \sum_{1}^{\infty} a_n \sin\frac{n\pi x}{l} \cos\frac{n\pi a t}{l}, \tag{1.3}$$

where l is the length of the string. Bernoulli claimed to have derived the most general solution for wave propagation along the string.

To determine the initial shape of the string, he substituted $t = 0$ into the equation, which resulted in:

$$y = \sum_{1}^{\infty} a_n \sin\frac{n\pi x}{l} \tag{1.4}$$

This raised a crucial question: could f and g from D'Alembert's solution be represented in trigonometric form, $\sum_{1}^{\infty} a_n \sin \frac{n\pi x}{l}$? In other words, does an arbitrary function allow such a trigonometric expansion? Is this true for any function?

Surprisingly, it was observed that even non-periodic functions f and g could be expanded into periodic trigonometric functions.

In this context, Euler found Bernoulli's solution more particular and of limited utility than D'Alembert's. Thus, D'Alembert and Lagrange joined Euler in contesting Bernoulli's solution. However, this controversy persisted without a definitive resolution until 1807. It was not until a quarter-century after Bernoulli's passing that Fourier introduced some of these concepts in his early communications to the French Academy, later expounded upon in his "Analytic Theory of Heat" in 1822.

1.2.4 Fourier's Ideas in His Analytic Theory of Heat

In this communication, Fourier introduced a revolutionary doctrine, stunning Lagrange with the concept that any given curve or function, regardless of its nature, could be expressed within any interval using a trigonometric series of the form

$$f = a_0 + (a_1 \cos x + b_1 \sin x) + (a_2 \cos 2x + b_2 \sin 2x) + \cdots . \tag{1.5}$$

Fourier did not seek a rigorous proof but offered concrete examples that validated the potency of his assertion. The precise constraints needed to make this claim entirely accurate remained, and to some extent, still remain for subsequent mathematicians to determine.

Fourier's findings not only upheld Daniel Bernoulli's claim regarding his series but also surpassed it by revealing the series' true potential. It swiftly dismantled Euler's and his contemporaries' belief that a mathematical function could be continuously extended in only one way beyond its defined interval. Furthermore, Fourier's examples indicated that the curve represented by his series could comprise disconnected segments—pieces devoid of logical or definitional linkage, not even connected sequentially at their ends.

Fourier's assertion made it evident that the power of representation through analytic expression was at least as formidable as the capacity for geometric visualisation.

Once it was understood that mathematical expression could adapt to vastly different and unrelated requirements, it became apparent that there was no logical endpoint until the definition we now accept for a function of a real variable, often referred to as the Dirichlet definition of a function. In this definition, if every value of x in an interval corresponds to a definite value of y, regardless of how fixed or determined, then y is termed a function of x. For example, $y = 1$ at all rational points in any interval, while being 0 at irrational points.

When it was established that an arbitrary function can be represented by a Fourier series of the form (1.5), the question arose of how to find the coefficients a_n and b_n of the series for a given function. Fourier evaluated these constants using Euler's method (Fourier himself admitted this!) and the *orthogonality* of trigonometric functions. He used the following integrals.

$$\begin{cases} a_n = \dfrac{1}{\pi} \displaystyle\int_{-\pi}^{+\pi} f(x) \cos nx \, dx, \\ b_n = \dfrac{1}{\pi} \displaystyle\int_{-\pi}^{+\pi} f(x) \sin nx \, dx. \end{cases} \tag{1.6}$$

Once a method of determining constants a_n and b_n was found, the next question that arose was the convergence of the Fourier series (1.5). Dirichlet later completed this task. It was Dirichlet who supplied the requisite sufficient conditions for the convergence of the Fourier series to the given function, and hence the truth of the theorem that hitherto remained to be ascertained. In his remarkable Memoir [3], known for its clarity and rigour, Dirichlet provides sufficient conditions for the convergence of Fourier series.

Fourier analysis elucidated the following:

1. Earlier, it was thought that with a given mathematical expression, a curve defined on an interval can be extended in a unique manner beyond the interval of definition. If it were done in another manner, it would disobey the mathematical expression which defined it in the interval of definition. But now, with the advent of Fourier analysis, the given curve could be extended beyond the interval of definition in any manner. There would be no violation of the given mathematical expression. Indeed, now there existed no definite mathematical expression, since all such arbitrary mathematical expressions were admissible which had Fourier expansion, and their coefficients could be found by the integration method.
2. The arbitrary curve representing the initial shape of the string could consist of separate pieces of any kind, which were independent of each other. They may not even be connected to each other at the end points. Such was the generality provided by Fourier analysis of the initial shape of the string. Thus, the arbitrariness of the function was justified. In other words, Fourier analysis widened the class of functions which could be admitted as the initial shape of the string and the solution of the wave equation. This accomplished the dual task of defining a *function* and widening the class of functions admissible as the solution of the wave equation.
3. This analysis freed the *functions* from the fetters of Analytic expressions using variables and constants and connecting symbols such as addition, subtraction, division, roots, exponents, etc. In this whole analysis one cannot miss to notice that the widening of the admissible class of functions is done through the integration present at that time. Indeed, such analysis pointed out that all those functions that are integrable over the interval of definition of the shape of string, against the trigonometric functions such as sin and cos, and yield a finite result, are admissible.

1.2.5 Work of Dirichlet

Fourier's announcements opened up the field of research in this direction. Many mathematicians of the time became engaged in researching on the convergence of the Fourier series. Particularly notable is the work of Dirichlet in his 1829 paper. To begin with, Dirichlet, in his analysis, admitted functions which were *bounded, piecewise monotonous, and piecewise continuous.* He further loosened the conditions and allowed the function to have infinite discontinuities even in a finite interval. He noted, if for any a, b in $[-\pi, +\pi]$ there exist some $r,\ s$ such that the function is continuous in the interval (r, s).[4]

> It is necessary that then the function $\varphi(x)$ be such that, if one denotes by a and b, any two quantities between $-\pi$ and $+\pi$, it is always possible to place between a and b, other quantities r and s close enough together so that the function remains continuous in the interval from r to s.

Then, for such functions, the Fourier series would converge. It was here that he constructed his famous example of *very discontinuous function* that could not be integrated in Riemann sense. It was constructed out of the sheer necessity of being able to integrate the function in the series.

$$f(x) = \begin{cases} c & \text{if } x \in \mathbb{Q}, \\ d & \text{if } x \in \mathbb{R} \setminus \mathbb{Q}. \end{cases} \tag{1.7}$$

Dirichlet states that such functions cannot be substituted into the series because the terms which are definite integrals would lose all significance.[5]

> Easily understood is the necessity of this restriction when considering that the different terms in the series are definite integrals, reverting to the fundamental notion of integrals. Then, it becomes clear that the integral of a function only holds meaning if the function satisfies the previously stated condition. An example of a function not meeting this condition would be if we supposed φ to be equal to a determined constant c when the variable x assumes a rational value, and equal to another constant d when the variable is irrational. The function defined in this manner has finite and determined values for any x; however, it cannot be substituted into the series because the different integrals involved in this series would lose all meaning in this scenario. The restriction I've specified, along with the requirement not to become infinite, are the only ones that apply to the function φ. Any cases not excluded by these restrictions can be reduced to those we have considered earlier.

Therefore, Dirichlet continued his work in this direction and tried to find out the extent to which discontinuities may be allowed in the function so that its integrability would not be compromised.

---

[4] *See* [3].

[5] *See* [3].

1.3 Riemann's Ideas of Integration

In 1831, Dirichlet assumed a professorship in Berlin, where Bernhard Riemann became one of his students. During this period, Riemann developed a keen interest in the Fourier series. In 1854, Riemann tried to find the necessary and sufficient conditions for a function such that, at any point x in $[-\pi, +\pi]$, the Fourier series for $f(x)$ should converge to $f(x)$. However, Riemann could prove only that if f is bounded and integrable over $[-\pi, +\pi]$, then the Fourier coefficients approach zero as n tends towards infinity. Riemann's integral definition was a leap forward in mathematics, allowing for the integration of discontinuous functions.

In his *Habilitation*, Riemann showed his ideas about the concept of integrals, providing a definition and a criterion for integrability. Riemann posed the fundamental question: What does the symbol $\int_a^b f(x)\,dx$ mean?

In a different approach from Cauchy, Riemann started with a partition $a = x_0 < \cdots < x_n = b$ of the interval $[a, b]$. He then constructed different sums

$$S = \sum_{i=1}^{n} \delta_i f(x_{i-1} + \delta_i \varepsilon_i)$$

where $\delta_i = (x_i - x_{i-1})$, $(i = 1, \ldots, n)$, and he called ε_i, "positive true rationals". These sums depended on ε_i and δ_i. Riemann defined $\int_a^b f(x)\,dx$ as the fixed limit A towards which these sums converge when δ_i converges towards zero and δ_i and ε_i are chosen arbitrarily.

Here, Riemann diverged from Cauchy in that, while Cauchy chose only continuous functions, he introduced integrable functions. He stated that if the above sums failed to converge towards a unique limit, the expression $\int_a^b f(x)\,dx$ was meaningless. For this reason, Riemann limited himself to bounded functions, sparking the inquiry into which functions are integrable—i.e., for which bounded functions do the sums uniquely converge.

Riemann answered this question in terms that may be broadly summarised as follows: Those functions are integrable for which the points at which the function's oscillation is greater than a given $\varepsilon > 0$ can be enclosed in intervals whose total length can be made as small as desired. This statement provided a sufficient condition of integrability, particularly for functions with poles.

In his theory, integrable functions constituted a category where a function's points of discontinuity were dense in every interval, irrespective of its size—though current understanding suggests these functions are not entirely discontinuous. Riemann offered an example of an integrable function defined by a convergent series, having discontinuities for specific rational x values.

$$1 + \frac{(x)}{1^2} + \frac{(2x)}{2^2} + \frac{(3x)}{3^2} + \cdots$$

where (nx) denotes the positive or negative difference between nx and the nearest integral, unless nx falls halfway between two consecutive integers. However, when nx falls midway between two consecutive integers, (nx) is set to 0. The series sum exhibits discontinuity for every rational x of the form $\frac{p}{2n}$ where p is an odd integer coprime to n.

Examples like this, and other integrable functions exhibiting an infinite number of maxima and minima, non-representable by a Fourier series, were thought-provoking. These inquiries deeply impacted Hermann Hankel, leading to his significant memoir "über die unendlich oft oszillirenden und unstetigen Functionen" (On Infinitely Oscillating and Discontinuous Functions). Hankel introduced the principle of "Condensation of Singularities," a contribution crucial to the independent theory of functions of a real variable. This distinction equally credits Riemann, historically one of the initiators of this theory, through his work on integrating discontinuous functions in papers related to "the representability of a function by Fourier series."

Riemann's example remains notable for mathematically formulating a discontinuous function that cannot be represented graphically.

1.3.1 Hankel's Work: Oscillation of f, Measure Theoretic and Topological Size of Sets

Hankel's Concept of Oscillation and Discontinuities

Hermann Hankel, in his pioneering work on functions and their properties, explored the oscillation of functions and the relationship between measure–theoretic and topological sizes of sets. Although both Hankel and Riemann originally defined the oscillation of a function as the difference between its maximum and minimum values within an interval, we adopt Darboux's later refinement of this concept. We define the oscillation of a function f in an interval I as:

$$\omega_f := \sup\{f(x)|x \in I\} - \inf\{f(x)|x \in I\},$$

Hankel also introduced the concept of a jump set of a function, defined as:

$$S_\sigma(f) := \{x \in \mathbb{R}|J_f(x) > \sigma\}$$

where $J_f(x)$ represents the jump of f at x, which is greater than some $\sigma > 0$.

Relationship Between Measure-Theoretic and Topological Sizes

Hankel aimed to establish a direct relationship between the measure-theoretic size and the topological size of a set. He posited that "topologically small sets", which are nowhere dense, should correspond to sets with "measure zero". Conversely, sets with measure zero should be considered topologically small. Hankel provided the following definitions in 1870:

1. **Dense Set:** An ensemble ("schaar") "fills up an interval", "if in the distance there is no interval however small not containing at least one point of the ensemble".

2. **Nowhere Dense Set:** A set lies "not filled up" but "scattered" on the interval, "if between any two points, however close, of the distance there is always an interval containing no point of that ensemble".

Hankel used the above definitions to characterise the sets $S_\sigma(f), \ \sigma > 0$. He arrived at the following statements, one of which was later proven incorrect:

1. If $S_\sigma(f)$ is nowhere dense in an interval (a, b), then the total length of the partial intervals I with $\omega_f(I) > 2\sigma$ can be assumed to be arbitrarily small.
2. Conversely, if the total length of the intervals I with $\omega_f(I) > \sigma$ can be assumed to be arbitrarily small, then $S_\sigma(f)$ is nowhere dense in (a, b).

The first statement is, in fact, false. If it were true, then by Riemann's criteria, any function f with its jump set $S_\sigma(f)$ being nowhere dense (as per the first statement) would be integrable. However, in 1875 H.J.S. Smith, an English mathematician, found functions with nowhere dense sets of discontinuities that were not integrable, thereby providing a counterexample. Darboux, in his paper,[6] commented on the work of Hankel. He stated:

> ... One of them, Mr. Hankel, published in 1870 the results of his studies on Riemann's Memoir. Unfortunately, the conclusions of his work are not beyond reproach, and a distinguished geometer, Mr. Gilbert, raised objections to Hankel's demonstrations that could also be directed towards the results. Hankel's principle of the condensation of singularities is stated in too absolute a manner, and, regrettably, that death did not allow this excellent geometer the time to reconsider the propositions he had given and to limit them, making them immune to any objection. I have taken it upon myself to revisit some of these propositions, and I believe I have rendered them immune to all criticism in the form I have presented them ....

This discrepancy led to the development of the theory of content because the "topological smallness" of sets was insufficient to determine the integrability, or lack thereof, of functions.

1.3.2 Darboux Upper and Lower Sums

In 1873–1874, Darboux revisited Riemann's work concerning the existence of continuous functions that did not possess derivatives. Prior to Riemann, it was universally accepted, without question, that every continuous function possesses a derivative. However, Riemann's work on trigonometric series demonstrated the existence of continuous functions without derivatives. In his paper,[7] Darboux showed that there exist infinitely many continuous functions without a derivative. Within this paper, Darboux reintroduced and extensively developed the definition of the definite integral in the Riemannian sense.

---

[6] *See* [4].
[7] See [4].

Darboux introduced the notions of the upper bound M, the lower bound m, and the oscillation Δ of a function f over an interval I. Here, the oscillation Δ is defined as $\Delta := M - m$. He partitioned the interval $[a, b]$ into sub-intervals determined by points:

$$a = x_0 < x_1 < x_2 < \cdots < x_n = b,$$

and formulated the sums:

$$\begin{aligned} M &:= M_1\delta_1 + \cdots + M_n\delta_n \\ m &:= m_1\delta_1 + \cdots + m_n\delta_n \\ \Delta &:= \Delta_1\delta_1 + \cdots + \Delta_n\delta_n \end{aligned}$$

His significant achievement was in demonstrating that M, m, and Δ converge toward definite limits as n approaches infinity. These limits depend solely on a, b, and f. These limits are expressed as:

$$\begin{aligned} \underline{\int_a^b} f(x)\,dx &:= \lim_{n\to\infty} \sum_{i=1}^{n} m_i\delta_i \\ \overline{\int_a^b} f(x)\,dx &:= \lim_{n\to\infty} \sum_{i=1}^{n} M_i\delta_i \end{aligned}$$

Darboux required that for a function f, the oscillation tends towards zero:

$$\lim_{n\to\infty} \Delta(n) = \lim_{n\to\infty} (M_i - m_i)\delta_i = 0$$

implying:

$$\begin{aligned} \overline{\int_a^b} f(x)\,dx - \underline{\int_a^b} f(x)\,dx &= 0 \\ \overline{\int_a^b} f(x)\,dx &= \underline{\int_a^b} f(x)\,dx \end{aligned}$$

and thus defining the integral as

$$\overline{\int_a^b} f(x)\,dx = \underline{\int_a^b} f(x)\,dx = \int_a^b f(x)\,dx.$$

According to this definition, every continuous function is integrable. If a given function $F(x)$ on an interval $[a, b]$ is differentiable, with a bounded and integrable derivative function f (i.e., $F'(x) = f(x)$ on $[a, b]$), then f satisfies the fundamental theorem of calculus:

$$F(x) - F(a) = \int_a^x f(y)\,dy \quad \forall x \in [a, b].$$

Thus, Darboux relaxed the conditions on f. He assumed f to be integrable, while Cauchy had to impose more stringent conditions on f, such as f being continuous, for the Fundamental Theorem of Integral Calculus to hold.

Consequences of Darboux's Definition

Various consequences can be deduced from the preceding definition of the integral.[8]

1. The integral of the function $f(x)$ will continue to exist and will not change in value if the values of $f(x)$ are altered for a limited number of x-values.
2. The integral
$$\int_a^x f(x)\,dx$$
is always a continuous function of x.
3. Let
$$F(x) = \int_a^x f(y)\,dy,$$
and suppose that $f(x)$ is continuous at $x = x_0$. Then, the function $f(x)$ will be the derivative of $F(x)$ for all values of x for which $f(x)$ is continuous.
4. Fundamental Theorem of Calculus, as shown above.
5. Let $\varphi(u)$ be a function of u such that
$$\frac{\varphi(u) - \varphi(v)}{u - v}$$
lies between the two finite limits α and β when u and v take all the values between A and B.
In this case, the oscillations of the function $f(x)$ always lie between A and B, and those of the function $\varphi(f(x))$ will clearly be of the same order in every interval. Consequently, the two functions $f(x)$ and $\varphi[f(x)]$ belong to the same class. For example, *if a function $f(x)$ is integrable, then the functions*
$$[f(x)]^2, \quad [f(x)]^n \quad \sqrt{1 + f^2(x)}, \quad \ldots$$
will also be integrable.

Darboux on Uniform Convergence and Discontinuities

We will discuss an important result concerning the uniformly convergent series and its application to the construction of functions with countable discontinuities in any interval.[9] Such

[8] See [4], part IV.
[9] See [4], part V.

highly discontinuous functions were constructed to demonstrate that even these functions are integrable according to Darboux's definition of integration.

He started with the discontinuous functions f, such that $f(x+h)$ and $f(x-h)$ have limits, when h approaches zero through positive values, that are different from $f(x)$. For instance, the function $E(x)$, which represents the greatest integer less than or equal to x, has the property that for any integer n, we have

$$E(n-0) = n-1, \quad E(n) = n, \quad E(n+0) = n.$$

Similarly, (x) is the function that represents the difference between x and the nearest integer. This function is indeterminate for $x = \text{integer} + \frac{1}{2}$; thus, he defined

$$(x) = 0 \quad \text{for} \quad x = \text{integer} + \frac{1}{2}.$$

This function is always continuous except at values of x of the form $n + \frac{1}{2}$, where n is an integer. For these values, we have

$$\left(n + \frac{1}{2}\right) = 0, \quad \left(n + \frac{1}{2} - 0\right) = \frac{1}{2}, \quad \left(n + \frac{1}{2} + 0\right) = -\frac{1}{2}.$$

It is to functions of this kind that he applied the following theorem:

Theorem 1.1 *If the terms of a uniformly convergent series*

$$f(x) = \varphi_1(x) + \cdots + \varphi_n(x) + \cdots$$

are functions $\varphi_n(x)$ for which $\varphi_n(x+0)$ or $\varphi_n(x-0)$ exist, then the series possesses the same property, and we have

$$f(x+0) = f(x) = \varphi_1(x+0) + \cdots + \varphi_n(x+0) + \cdots$$

$$f(x-0) = f(x) = \varphi_1(x-0) + \cdots + \varphi_n(x-0) + \cdots$$

It shows that a uniformly convergent series in a given interval can, in some sense, be treated like sums composed of a finite number of terms.

The preceding theorem allows us to define discontinuous functions within any interval. Consider the series

$$f(x) = \sum_{n=1}^{\infty} a_n \frac{E(nx)}{nx}$$

If the series of constants

$$A = a_1 + a_2 + \cdots$$

is absolutely convergent, the series $f(x)$ will have its terms bounded by those of the series A, and its remainder, for any x, will be less than that of A. Hence, it will be uniformly convergent within any interval. We, then, have

$$f(x \pm 0) = \sum a_n \frac{E(nx \pm 0)}{nx}$$

This function is continuous at any x-irrational and, discontinuous at any x-rational. Further, Darboux applied the result that if the terms of a uniformly convergent series are integrable, then the sum function is also integrable.

1.3.3 Ulisse Dini and Others: Counterexample

Ulisse Dini introduced the concept that, based on Riemann's definition, there may exist functions whose derivatives are not integrable. Specifically, it was proposed that if a non-constant function F defined on an interval $[a, b]$ has a derivative f (where $F' = f$) that vanishes on a dense set in $[a, b]$, then f cannot be integrable. According to Riemann's definition, if f was integrable, it would satisfy the Fundamental Theorem of Calculus:

$$F(x) - F(a) = \int_a^x f(y)\, dy;$$

However, since f vanishes on a dense set in $[a, b]$, Riemann's sums would yield:

$$\int_a^x f(x)\, dx = \lim_{n\to\infty} \sum_{i=1}^{n} f(\xi_i)\delta_i = 0.$$

In 1881, Vito Volterra provided an example of such a pathological non-constant function whose derivative vanished on a dense set. Similarly, in 1896, the Swedish mathematician Trosten Brodén published an example based on the principle of condensation of singularities, further illustrating this phenomenon.

1.3.4 Criticism of Riemann's Definition of Integration

1. Non-integrable derivative functions: As previously discussed, examples of functions whose derivatives were non-integrable were found (see Sect. 1.1). This was the initial shortcoming encountered in Riemann's definition of integrable functions. In the introduction of his thesis [5] Lebesgue remarks:

> Mr. Jordan, in the second edition of his *Cours D'Analyse* [2], conducted an in-depth study of these quantities. However, it seemed useful for me to revisit this study, and here is why. It is known that there exist non-integrable derivative functions when one adopts, as M. Jordan

does, the definition of integral given by Riemann. Thus, integration, as defined by Riemann, does not always solve the fundamental problem of integral calculus:

Find a function given its derivative.

This led M. Lebesgue to propose a definition of integration that would allow the integration of any bounded derivative. The integration of a bounded derivative would yield its primitive function, considered as a function of the upper limit of integration.

2. Term-by-term integrability of series: Another issue arising from Riemann's definition of integration was the inability to interchange the limit process and integration. Many mathematicians of the time overlooked this problem, considering the interchangeability of integration and the limit process as inherent. Thus, the term-by-term integration of a series,

$$\sum_{n=1}^{\infty} \int_a^b u_n(x)\,dx = \int_a^b \sum_{n=1}^{\infty} u_n(x)\,dx$$

was accepted without much scrutiny.

Weierstrass was the first to suggest that uniform convergence of series (rather than just point-wise convergence) of integrable functions, uniformly bounded between the convergence limits, was necessary to justify the interchange of the limit process and integration. Uniform convergence alone ensures the transfer of properties like continuity and integrability from the term functions u_n to the "sum" function u of the series. Essentially, uniform convergence became a crucial requirement for the interchange of the limit process and integration. Without assuming uniform convergence, conditions on the sum function u of the series were required.

Arzelà demonstrated that if a series of uniformly bounded and integrable functions (in Riemann's sense) converged to a sum function u that was also integrable in Riemann's sense (assumption of certain properties about the sum function), the interchange of the limit process and integration was permissible.

In Osgood's works, he required the sum function u of a series of continuous and uniformly bounded functions u_n to be continuous in order to justify the interchange of the limit process and integration. This condition was imposed on the sum function of the series. However, in 1899, Baire demonstrated that if uniform convergence is not assumed, and only the term functions u_n are uniformly bounded, then a series of integrable functions can converge to a function that is not integrable in the Riemann sense.

3. Multi-dimensional integral: The next problem stemming from Riemann's definition of integration pertained to integration in dimensions two or higher. One might assume that integration over multiple dimensions could be defined in a manner similar to one-dimension, using Riemann or Darboux sums. Just as an interval in one-dimension is partitioned into partial intervals for integrating a given function, one could partition the domain in two dimensions or more. However, a problem emerged:

Let $f : \mathbb{R}^2 \to \mathbb{R}$ be a function to be integrated. The task was to define $\int_A f(x, y)\,da$, where A is a domain in the plane $\mathbb{R}^2$. If one attempted to define it as

$$\int_A f(x, y)\,da = \sum f(x_i, y_i)a(R_{ij}) \quad \text{where} \quad (x_i, y_i) \in R_{ij},$$

dividing A into rectangles R_{ij}, a dilemma arose: which rectangles should be considered in the sum? Should it include only those that do not contain any points from the complement of A or all those that contain at least one point of A?

Ideally, one would expect that as the rectangles become finer (i.e., smaller), the total area of the rectangles touching the boundary should approach zero. However, Peano discovered curves that traversed every point of the domain. These findings led to the necessity of imposing appropriate conditions on A so that the integral of $f(x, y)$ could be properly defined over A.

1.4 Work of Camille Jordan

Jordan introduced the concept of a Jordan 'measurable' set in 1892. He took motivation from Cantor's work in set theory and provided a more general theory than Riemann's integration. Jordan developed the idea of inner and outer extent as a generalisation of the concept of upper and lower integrals of Riemann.

As discussed above, the theory of Riemann integration was not wholly satisfactory in higher dimensions. The main problem arose in the fact that the sum of the areas of rectangles, in a partition, that met the boundary of an arbitrary curve (or set) could not be made as small as desired. Even in the one-dimensional case, it was shown that there were nowhere dense sets that could not be enclosed in intervals of arbitrarily small lengths (see Example 1 of Sect. 1.5.1). It was further shown with the help of Peano's space-filling curves.

Camille Jordan explored the relationship between the 'measure' of sets and the integrability of functions over those sets. He focused on domains where functions of multiple variables could be integrated. For this analysis, Jordan partitioned the domain A and considered two types of rectangles: those that did not contain any points from the complement of the domain A^C and those that contained at least one point of the domain.

Using this approach, Jordan defined the inner and outer content of the domain. He proposed that a domain is measurable if the inner and outer contents converge to the same value as the size of the rectangles approaches zero through finer and finer partitions. Having established the concept of domain 'measure', Jordan utilised Riemann-Darboux sums to define the integral of a function, similar to the one-dimensional case. The only difference is that, in the multi-dimensional case, the measure of the domain replaced the length of intervals used in the one-dimensional case. With this background, it is appropriate to give the:-

1.4.1 Notion of Extent of a Set

Jordan considered the sets of dimensions 1, 2, and 3 and called their *extents* the *length, area and volume*.

For definiteness, he considered the sets of two dimensions, E, and took its points with coordinates (u, v). He represented the points geometrically on a plane. Then, he decomposed the plane into a grid of squares of side length r.

He considered two types of domains

1. S which is formed of the squares that were completely inside the set E and
2. $S + S'$ which is formed of squares completely inside the set E or meeting the boundary of E.

He denoted the area of the two types of domains respectively by S and $S + S'$ itself. Then, he varied the grid of squares by varying the edge length of the squares, by letting $r \to 0$, and showed that the respective areas S and $S + S'$ tended towards fixed limits.

First, he considered the domain S and kept $r \leq \rho$ (a fixed number), demonstrating that the area $S \leq A$. Similarly, for the domains $S + S'$, he showed that when $r \leq \rho$, the area $S + S' \geq a$. Thus, he concluded that as $r \to 0$, the area S increases to a definite limit A, and the area $S + S'$ decreases to a definite limit a. He referred to A *as the interior area of* E and a *as the exterior area of* E. He called a set to be *quarrable* when the area of the boundary of E—is zero, in which case $A = a$.

He also established the relationship between interior and exterior areas of sets and subsets. If $E' \subseteq E$, then the interior area of E exceeds the exterior area of E'. He considered E to be formed of a union of several partial sets $E_1, E_2, \ldots$. He established the following relationship between the interior and exterior areas of the sets $E, E_1, E_2, \ldots$

$$S \geq S_1 + S_2 + \cdots; \quad S + S' \leq S_1 + S_1' + S_2 + S_2' + \cdots$$

and by passing to the limit

$$A \geq A_1 + A_2 + \cdots; \quad a \leq a_1 + a_2 + \cdots$$

where $A, A_1, A_2, \ldots; a, a_1, a_2, \ldots$, respectively, represent the interior and exterior areas of $E, E_1, E_2, \ldots$. When all these sets are *quarrable*, then

$$A = A_1 + A_2 + \cdots; \quad a = a_1 + a_2 + \cdots$$

After these analytic considerations, Jordan considered the decomposition of planes into quarrable regions on exactly the same lines. These considerations were geometric in nature and led him to define the *interior and exterior extent* of the sets of different dimensions. He called a set to be *measurable* if the above two extents coincide.

Having defined the measure of a set, Jordan cleared the field for the integration of functions. For further analysis of integration, he defined:-

1.4.2 Integrable and Bounded Functions

Jordan started by defining the variable quantities $x, y, \ldots$ as those quantities that are not linked in any way and can take any value that can be taken when values of other variables are fixed. A function f is defined as a dependent variable on the values of $x, y, \ldots$. To each value of $x, y, \ldots$ in a set E, there corresponds a value of f. Having defined the functions in this way, putting some restrictive conditions on them was pertinent. Therefore, the *bounded functions* were defined.

Bounded functions: A function $f(x, y, \ldots)$ is considered bounded in a set E if the set of values it takes for all points $(x, y, \ldots)$ in E is bounded.

He stated that the *sum, the difference, and the product* of bounded functions are bounded, and if a function is bounded from below, then its reciprocal is also bounded. Jordan considered the bounded functions for the purpose of integration and began this analysis by defining the:-

Darboux Type Sums

He took a bounded function $f(x, y, \ldots)$ on a measurable domain E, whose measure is also denoted by E. He decomposed E into measurable elementary domains $e_1, e_2, \ldots$, and considered the maximum value M and the minimum value m of f on E as well as on these measurable elementary domains (respectively M_k and m_k). Darboux-type sums were formed:-

$$S = \sum M_k e_k; \quad s = \sum m_k e_k.$$

Using *Darboux's theorem*, Jordan showed that the sums S and s tended towards the same definite limits, irrespective of the method in which the domain E was partitioned when the diameter of the elementary domains tended towards zero.

Jordan showed that the sum S is bounded from below. The lower bound of these sums S was called *integral by excess*. Similarly, the sums s were shown to be bounded from above, and their upper bound was called *integral by defect*. Jordan was able to calculate the number of measurable elementary domains in which the domain E was to be divided, depending on the diameter of the domain and how close the sums S (s) should be to their lower bound (upper bound), i.e., *integral by excess* (*integral by defect*).

Jordan further noted that any measurable domain E can be decomposed into several measurable domains $E_1, E_2, \ldots$ which can be further divided into infinitely small elements $e_1, e_2, \ldots$, allowing for the formation of individual sums $\sum M_k e_k$ for each element. The corresponding sum for the entire domain E is obtained simply by adding these partial sums. Taking the limit of these sums, it was found that the integral by excess of a function f over the domain E equals the sum of the similar integrals for the individual constituent domains $E_1, E_2, \ldots$. This result holds analogously for the integral by defect.

Furthermore, he constructed an increasing sequence of measurable domains $E_1, \ldots, E_n, \ldots$ such that each domain is interior to both the following domains in the sequence and the original domain E. Additionally, the extents of these domains converge towards E itself. The integral of f over domain E (either by excess or by defect) is then equal to the limit of the integrals of f over the domains E_n as n approaches infinity. This is because the difference between two integrals is equivalent to the integral taken over the difference of their domains, $E - E_n$. The absolute value of this difference is bounded by $L(E - E_n)$, (where $L = \text{Max.}\{|M|, |m|\}$) which approaches zero as n approaches infinity.

Integrable Functions

Finally, an integrable function $f(x, y, \ldots)$ in the domain E was defined as one for which the two integrals by excess and by defect

$$T = \lim \sum M_k e_k; \quad t = \lim \sum m_k e_k,$$

coincide, i.e., for which $T = t$. This limit was called the *integral of function* f in the domain E. It was represented by the notation

$$I = \int_E f(x, y, \ldots)\, de;$$

From the above analysis, it was clear that the value of the integral was independent of the variables $x, y, \ldots$ but depended only on the nature of f and on the domain E.

In the above analysis, Jordan's approach was to enclose a set with finitely many rectangles. However, these finitely many enclosing rectangles were not sufficient for further analysis.

Borel took a different approach towards defining a 'measure' from Jordan's. He stated some properties (see Sect. 1.5.2) that a measure should have, which are discussed in the following section.

1.5 Work of Borel

Borel designed a remarkably simple construction for measurable sets, ensuring complete compatibility with the length of intervals. He crafted the definition to align with set-theoretic operations like addition and subtraction with their corresponding algebraic counterparts in measuring sets. Similarly, relationships like *greater than or equal to* or *less than or equal to* between measures are reflected in corresponding set-theoretic operations like $A \subseteq B$ or $B \subseteq A$, where A and B are Borel measurable sets. This elegant construction laid the foundation for further investigations into the intricate interplay between set theory and measure theory.

1.5.1 Synthesis of Borel Sets: Two Examples

Borel constructed a new type of set by removing countably many intervals from (0, 1). Specifically, he chose to remove intervals centered at rational points within the interval (0, 1). His primary interest laid in the study of *uncountable perfect sets*, which he found simpler to study compared to non-perfect sets. This focus was rooted in the approximation of algebraic and irrational numbers, naturally leading Borel to the consideration of uncountable perfect sets.

In the third chapter of his book [6], Borel defined derived sets, closed sets, perfect sets, and dense sets from a topological perspective. He provided insights into the size of sets (whether they are large or small, countable or uncountable) through this viewpoint. A brief outline of Borel's procedure for synthesizing such sets is given as follows:

Example 1. A Perfect, Uncountable, and Non-Dense Set

1. **Construction of set A.** Borel considered the rational numbers between 0 and 1, and associated with each of them $\frac{p}{q}$ the interval

$$\left(\frac{p}{q} - \frac{1}{q^3}, \frac{p}{q} + \frac{1}{q^3}\right) \tag{1}$$

In this way, he obtained a countable infinity of intervals situated on the line. Then, he considered the set A of those points of (0, 1) which did not belong to any of these intervals. He noted that each endpoint of any interval, having a rational coordinate, is the midpoint of another interval; hence, it was not necessary to specify whether the endpoints of an interval are considered part of it or not. In this way, he constructed the set A.

2. **Set A is Non-Empty.** Borel found a close connection between the consideration of set A and the question of the approximation of irrationals. To him it was clear that A was composed of the irrational numbers ξ having the property that, for any $\frac{p}{q}$,

$$\left|\xi - \frac{p}{q}\right| > \frac{1}{q^3}.$$

The existence of ξ was assured by the theory of continued fractions; such as, for example, the number $\frac{\sqrt{2}}{2}$. But this theory highlighted virtually nothing about the set A. Moreover, it relied on specific properties of rational numbers and was not easily extended to the approximation of irrationals by algebraic numbers of a determined class.

3. **A is Perfect and Non-Dense.** Borel first demonstrated that the set A was perfect. He argued that if a point a' in A' did not belong to A, then a' would have to be interior to one of the intervals (1) (without coinciding with its endpoints). This was absurd since these intervals did not contain any point of A. Thus, every point of A' belonged to A.

Borel further asserted that A was not dense in any interval. If A were dense in an interval, it would include all points of that interval. This was also absurd, since the points $\frac{p}{q}$ did not belong to A and there were some in every interval.

Thus, Borel provided an example of a perfect set A, that was not dense in any interval. Borel felt it important to delve deeper into the nature of such a set and, first of all, to demonstrate that it was uncountable without resorting to the theory of continued fractions.

4. **A is Uncountable.** Borel considered the interval $(0, 1)$ with length one. Then, he considered the interval (a_1, b_1) within the interval $(0, 1)$ with length α_1. It was clear that the set A of points in $(0, 1)$ that did not belong to (a_1, b_1) was composed of all the points in certain intervals whose total length was $1 - \alpha_1$. Assuming, for the sake of clarity, that $0 < a_1 < b_1 < 1$, these intervals were $(0, a_1)$ and $(b_1, 1)$.

   Next, he considered another interval, (a_2, b_2) with length α_2, within the interval $(0, 1)$, which had no common points with the interval (a_1, b_1). It was clear that the set A of points in the interval $(0, 1)$ that did not belong to either (a_1, b_1) or (a_2, b_2) was composed of all the points in a finite number of intervals, with a total length was $1 - \alpha_1 - \alpha_2$.

   More generally, if finite number of disjoint intervals $(a_1, b_1), (a_2, b_2), \ldots, (a_n, b_n)$, with respective lengths $\alpha_1, \alpha_2, \ldots, \alpha_n$ were removed from the interval $(0, 1)$, the set A of remaining points would consist of all the points in a finite number of intervals, with a total length of

$$1 - (\alpha_1 + \alpha_2 + \cdots + \alpha_n).$$

   Even if the intervals $(a_1, b_1), (a_2, b_2), \ldots, (a_n, b_n)$ were not disjoint, the conclusion remained the same, except that the sum of the lengths of the intervals whose points form A would be greater than or equal to $1 - (\alpha_1 + \alpha_2 + \cdots + \alpha_n)$. In any case, if it was assumed

$$\alpha_1 + \alpha_2 + \cdots + \alpha_n < 1,$$

   this sum was certainly not zero, and consequently, the set A had the power of the continuum.

   Next, Borel dealt with the case where countably many intervals (a_n, b_n) were removed from $(0, 1)$. For each integer n, he associated an interval (a_n, b_n) of length α_n. This meant he removed the points of a countable infinity of intervals from the interval $(0, 1)$. Additionally, he assumed the series of positive terms

$$s = \alpha_1 + \alpha_2 + \cdots + \alpha_n + \cdots$$

   was convergent and its sum s satisfied the inequality

$$s < 1.$$

   Then, he pondered what could be said about the set A consisting of points in the interval $(0, 1)$ that did not belong to any of the intervals (a_n, b_n)? He noted that this set might not be dense in any sub-interval of $(0, 1)$. It was also evident that it contained all points in the sub-intervals where it was dense. However, a preliminary question arose: did such a set A exist at all? Could it be deduced from the inequality

$$s < 1$$

that there were points in (0, 1) not belonging to any of the intervals (a_n, b_n)? Although this seemed almost obvious, Borel thought it useful to prove it rigorously, as doing so provided several important insights.

The first of these remarks was the following: it was to be demonstrated that there exist points not interior to certain intervals. He considered as belonging to an interval only the interior points, excluding the endpoints. He enlarged, at each endpoint, each interval (a_n, b_n) by a fraction ε of its length; i.e., he took $b_n b'_n = a_n a'_n = \varepsilon a_n b_n$, the points a'_n and b'_n being outside the interval (a_n, b_n).

The interval (a'_n, b'_n) had a length of $(1 + 2\varepsilon)a_n b_n$; therefore, the sum of all the intervals $(a'_1, b'_1), (a'_2, b'_2), \ldots$ is $s' = (1 + 2\varepsilon)s$; but, if

$$s < 1,$$

a positive number ε could be chosen such that

$$(1 + 2\varepsilon)s < 1.$$

Now, the hypothesis that every point of the interval (0, 1) was included in one of the intervals (a_n, b_n) (without excluding the endpoints) would have led to the result that every point was included in one of the intervals (a'_n, b'_n) (excluding the endpoints). Borel showed that this hypothesis was incompatible with the inequality

$$s' < 1.$$

He showed this by using the following result, in which it was expressly understood that the words "interior to an interval" excluded the end-points.

Theorem 1.2 *If on a bounded line segment, there is a countable infinity of partial intervals such that every point on the line is interior to at least one of the intervals, then there exists a finite number of intervals chosen from the given intervals that have the same property [every point on the line is interior to at least one of them].*

From this theorem, he concluded that if a finite number of intervals were such that all the points of a segment were interior to them, the sum of the lengths of these intervals was greater than the length of the segment. This was not possible if the finite intervals were chosen from an infinite number of intervals whose total sum was less than the length of the segment.

It was thus affirmed that the hypothesis

$$s' < 1;$$

or, equivalently, the hypothesis

$$s < 1$$

resulted in the definite existence of the set A.

Further, the set A was uncountable, otherwise, by denoting its points as $\alpha_1, \ldots$ and enclosing the point α_n in an interval of length $\frac{\varepsilon}{2^n}$, and adding these intervals to the given intervals; the sum s would have become

$$s_1 = s + \frac{\varepsilon}{2} + \cdots + \frac{\varepsilon}{2^n} + \cdots = s + \varepsilon,$$

and clearly, if

$$s < 1,$$

then, an ε could have been chosen so that

$$s_1 < 1.$$

Consequently, there would have existed points not belonging to the intervals (a_n, b_n) and certainly not coinciding with any of the points $\alpha_1, \ldots, \alpha_n, \ldots$. It was, therefore, proved that the set A was uncountable.

To go back to the case of considered intervals (1), one needed to substitute the intervals $(\frac{p}{q} - \frac{1}{q^3}, \frac{p}{q} + \frac{1}{q^3})$ in place of the intervals (a_n, b_n) and the proof followed.

Example 2. Construction of a Perfect, Non-Dense, Measure Zero, Borel Set Using Intervals

Borel considered the intervals

$$\left(\frac{p}{q} - \frac{\varepsilon}{q^3}, \frac{p}{q} + \frac{\varepsilon}{q^3}\right) \quad \text{for} \quad \begin{cases} q = 1, 2, \ldots, \infty \\ p = 1, 2, \ldots, q - 1 \end{cases}.$$

Their total extent, since there are $q - 1$ such intervals corresponding to the same denominator q, is given by:

$$2\varepsilon \sum_{q=1}^{\infty} \frac{q-1}{q^3} = 2\varepsilon \sum_{q=1}^{\infty} \left(\frac{1}{q^2} - \frac{1}{q^3}\right) = \varepsilon M$$

with M being an easy-to-calculate number and less than one. Moreover, if their common parts are not counted multiple times, we still have a countable infinity of intervals, whose total extent is less than εM.

Borel took $\varepsilon = \frac{1}{n}$, where n is an integer. He denoted by E_n the set of all points inside the corresponding intervals, and considered the sequence of sets:

$$E_1, E_2, \ldots, E_n, \ldots \tag{E}$$

It was clear that each set E_n contained all the points included in the subsequent sets. The set E_n was composed of all points inside intervals whose total length was less than $\frac{M}{n}$.

Next, Borel focused on the set E formed by the points that belong to all the E_n. This set E certainly included all the points $\frac{p}{q}$, but it was not obvious a priori that it contained other points. This set E also had the remarkable property that its points could be enclosed in a series of intervals whose total length is as small as desired. For all points of E were points of E_n and, therefore, were included in intervals of a total length less than $\frac{M}{n}$. Borel showed that E had the cardinality of the continuum.

It sufficed to show that the set of transcendental numbers ξ[10] belonged to the set E; i.e., to show that each of these numbers ξ belonged to one of the sets E_n. Each of these numbers ξ was such that there were infinitely many values of p and q satisfying the inequality

$$|\xi - \frac{p}{q}| < \frac{1}{q^4}.$$

Since there were infinitely many such values of q, it was assumed $q > n$ and, thus

$$|\xi - \frac{p}{q}| < \frac{1}{nq^3} = \frac{\varepsilon}{q^3},$$

which proved that ξ belonged to the set E_n.

1.5.2 Definition of Measurable Set

Borel's analysis motivated him to give a new definition of measurable sets and the measure of sets. Borel, in his book [6], Borel states in Chap. 3, p. 47

> Here are some new definitions: If a set E has measure s, and contains all the points of a set E' whose measure is s', then the set $E - E'$, formed by the points of E that do not belong to E', will be said to have measure $s - s'$. Furthermore, if a set is the sum of a countable infinity of sets with no common part, its measure will be the sum of the measures of its parts. Finally, if the sets E and E' have measures s and s', respectively, and E contains all the points of E', then the set $E - E'$ will have measure $s - s'$. In other words, the first definition states that the difference of two sets with measures s and s' is the set of points that belong to the first set but not to the second set, and has measure $s - s'$. The second definition states that the measure of a countable sum of sets with no common part is the sum of the measures of the individual sets. The third definition states that if two sets have measures s and s', and the first set contains all the points of the second set, then the difference of the two sets has measure $s - s'$. These definitions are used in the field of measure theory, which is a branch of mathematics that deals with the size of sets.

[10] See the book [6], p. 28.

Borel always restricted himself to the interval (0, 1) and worked with the points of this interval. He constructed *special* sets by meticulously adding and subtracting intervals within this interval, culminating in a set, say A. This intricate process of set-theoretic manipulation continued, adding and subtracting intervals to A, and even considering the addition and subtraction of (countable) infinitely many such sets $A_1, A_2, \ldots$. These resulting sets, that emerged from this meticulous process, were called *Borel sets*.

Borel only defined measure on these particular sets, calling them *measurable*. He admitted that defining measure on other sets, those not born from his established process, was possible, but he deemed it unworthy of his pursuit. Such sets, he argued, lacked the fundamental property he deemed essential: a clear correspondence between set-theoretic operations like addition, subtraction, and subset or super-set relations and their corresponding algebraic counterparts (addition, subtraction, and less than or greater than relations).

This meticulous approach had significant consequences. Important sets, such as all *bounded perfect sets*, were demonstrably measurable. Even a general method for constructing such sets was unveiled, involving the subtraction of intervals from the domain of interest, which is the interval (0, 1).

In essence, Borel meticulously crafted a measurable world within the confines of his chosen domain, laying the foundation for further explorations in the realm of measure theory. His dedication to rigour and precision established a cornerstone upon which future mathematicians could build.

1.6 Work of Lebesgue

The need to attribute numerical values to sets emerged from studies in function theory. This endeavour was equivalent to assigning a *measure* to these sets. However, the measure could not be defined for every set; Lebesgue's approach to defining measure encompassed sets that possessed characteristics akin to *segment lengths or polygonal areas*. In collaboration with Borel, Lebesgue devised a definition of measure of a set based on its fundamental properties. This definition of measure extended to spaces of multiple dimensions.

From the idea of measuring sets composed of points on a plane, he derived the notion of area for planar domains. Extending this concept to points in ordinary space led to the development of the idea of volume. Consequently, the integral of a continuous function was conceived as the area of a planar domain. This approach expanded to define the integral of discontinuous bounded functions as the measure of a specific set of points. Thus, a geometric interpretation of the integral emerged. Simultaneously, an analytical definition of the integral was introduced, wherein it represented the limit of a sequence of sums akin to those in Riemann's definition. Lebesgue referred to the functions to which the geometric definition applied as *summable*.

The class of Lebesgue's summable functions was very large. In his thesis [5], Lebesgue states that:

I am unaware of any function that is not summable; I do not know if such functions exist. All functions that can be defined using arithmetic operations and the limit transitions are summable. All functions integrable in the sense of Riemann are summable, and both definitions of the integral yield the same value. Any bounded derivative function is summable.

1.6.1 Problem of Measure

Lebesgue's initial definition described a bounded set as one where the distance between any two points remained bounded. In these bounded sets, he established operations like set addition and subtraction. However, he proposed that these set-theoretic operations should be performed only on a finite or countably infinite number of sets. With each bounded set, he associated a non-negative number termed its *measure*, subject to the following conditions:

1. There exist sets with non-zero measure.
2. Measure of two equal sets is equal.
3. The measure of the sum of a finite or countably infinite number of pairwise disjoint sets is the sum of the measure of these sets.

The problem of assigning a measure to a set, as defined above, became known as the *problem of measure.* Lebesgue aimed to address this problem exclusively for measurable sets, which shared properties akin to segments or areas. This problem of sets had different solutions, depending upon the type of point set under consideration—whether on a straight line or in a plane, for instance. Hence, the measures were called *linear measure* or *surface measure*, accordingly. Lebesgue attributed a unit measure (where measure $= 1$) to any set of non-zero measure. His rationale was based on the idea that if one measure system transforms into another by multiplication with the same number, these systems are essentially identical. Now, we shall briefly see how Lebesgue constructed the measurable sets, which had properties akin to the segment lengths and polygon area, etc.

Lebesgue began by examining a set comprising only a single point. He deduced that such a set possesses a measure zero. This conclusion arose from the notion that a bounded set, encompassing an infinite number of points, should have a finite measure. Furthermore, he asserted that the set of points on a segment AB should yield an invariant measure, irrespective of whether A and B belong to the set. It was crucial to establish that AB could not have a measure zero; otherwise, every bounded set would have a measure zero. Consequently, the finite non-zero measure of segment AB remained constant, irrespective of whether points A and B were part of the segment. Thus, it was demonstrated that the measure of a set comprising a single point must indeed be zero, aligning with the inherent 'zero-length' characteristic of a point.

Then, Lebesgue assigned a length 1 to segment AB and chose it to be the unit of length. Having done this, he assigned a number (its length) to each segment CD. He called this number the measure of the set of points of CD. If the length of CD is $l = \frac{r}{s}$ $(r, s \in \mathbb{N})$, then the length of CD can be thought of as having r number of AB joined end to end, and the total resulting length is cut into s pieces. The length of one such piece is the length of CD, and the measure of the set of points of CD is measure of the set of points of one such piece. Similarly, if l is irrational, then to every number $k < l$ corresponds a segment of length k contained in CD, and to every number $k > l$ corresponds a segment of length k containing CD.

To ensure the fulfilment of the 3rd condition of the measure problem, it is necessary that the total length of a segment, composed of either a finite or infinite series of disjoint segments, is equal to the sum of the individual lengths of those segments. This case is obvious when dealing with a finite number of segments, and it remains applicable even when the segments are countably infinite in number (refer to M. Borel's book [6]). With the definition of the length of segments and an illustrative geometric method for the construction of measurable sets, Lebesgue moved from the geometric approach to an analytical approach for determining the measure of a set, which is done by defining:-

1.6.2 Exterior and Interior Measure

Lebesgue considered a set E and enclosed its points in a finite or countably infinite number of intervals. There would exist infinitely many such intervals, which enclose the points of E. The sum of these 'enclosing intervals' is called E_1. The set E_1 contains the set E and therefore, it is reasonable to assume that $m(E) \leq m(E_1)$, where $m(E)$ and $m(E_1)$ are respectively measures of E and E_1. In other words, $m(E)$ is at most equal to the sum of the lengths of the intervals considered. The lower limit of this sum was called *exterior measure* or $m_e(E)$ by Lebesgue.

To define the *interior measure* of set E, denoted $m_i(E)$, Lebesgue assumed that all the points of set E belong to the segment AB. The complement of E, with respect to AB, denoted $C_{AB}(E)$, is the set $AB - E$. The measure of the set $C_{AB}(E)$ is at most $m_e(C_{AB}(E))$. Hence, the measure of E is at least $m(AB) - m_e(C_{AB}(E))$. Now, this number is independent of the choice of segment AB containing E, and is called the interior measure of E, defined as $m_i(E) = m(AB) - m_e(C_{AB}(E))$. Lebesgue arrived at the following inequality:

$$m_e(E) + m_e[C_{AB}(E)] \geq m(AB)$$

the exterior measure is never less than the interior measure. The definition of exterior and interior measure paved the way for defining the:-

1.6.3 Measurable Sets

Analytically, Lebesgue characterised sets as *measurable* when their exterior and interior measures coincided. In this case, the common value of these measures denoted the *measure* of the set, provided the problem of measure could be solved. If E was a measurable set, the defined number $m(E)$ would inherently satisfy the conditions of the problem of measure.

Another equivalent definition of measurable sets was proposed: *A set E was considered measurable if it was possible to enclose its points in intervals α while enclosing the points of its complement $C(E)$ in intervals β in such a manner that the total lengths of the common segments (intervals) between α and β are as small as desired.*

Further, it was proved that if $E_1, E_2, \ldots$ is a finite or countably infinite sequence of measurable sets, then the sum set E is measurable, and such a statement could be made only for measurable sets.

Indeed, Lebesgue never claimed that the problem of measure was not possible for the sets whose exterior and interior measures are unequal. However, the procedure he used to define a set consisted of two steps:

1. Forming the sum of a finite or countably infinite number of previously defined sets.
2. Considering the set of points common to a given finite or infinite number of sets.

and, these two steps applied to measurable sets, yield measurable sets.

1.6.4 Borel Measurable Sets: A Procedure

Now, we know that an interval is a measurable set. Finitely many additions and subtractions of intervals give measurable sets. Borel sets are obtained by applying the previous two procedures finitely many times along with the process of taking complements of the sets. The sets resulting from this procedure (the Borel sets) were called measurable by Borel and B measurable by Lebesgue. Furthermore, Lebesgue estimated their cardinality to be that of *continuum*. Examples of Borel sets include all sets formed by sums of intervals and closed sets whose complements are sums of intervals.

Another example of a Borel measurable set is the following *perfect set/ Cantor's set*:

The set E, formed of points:

$$x = \frac{a_1}{3} + \frac{a_2}{3^2} + \frac{a_3}{3^3} + \cdots .$$

Where a_i are equal to 0 or 2. This set is obtained by removing the middle third (interval) from the interval (0, 1) and from each subsequent resulting intervals. It is a perfect set, and hence it is B measurable. Its complement is formed of interval $\left(\frac{1}{3}, \frac{2}{3}\right)$ of length $\frac{1}{3}$, of two intervals $\left(\frac{1}{9}, \frac{2}{9}\right)$, $\left(\frac{2}{3} + \frac{1}{9}, \frac{2}{3} + \frac{2}{9}\right)$ of length $\frac{1}{3^2}$, of four intervals of length $\frac{1}{3^3}$, etc, therefore its (the complement's) measure is

$$\frac{1}{3} + 2\frac{1}{3^2} + 2^2\frac{1}{3^3} + \cdots = 1.$$

Consequently, E is of measure zero and has the cardinality of the continuum. Therefore, we form an infinite number of sets with points of E. All of them have an exterior measure zero and thus are measurable. The cardinality of the set of these sets is that of the set of sets of points.

Therefore, we conclude that there exist measurable sets (in Lebesgue's sense) that are not B measurable, and the cardinality of the set of measurable sets is that of the set of sets of points. Therefore, the class of Lebesgue measurable sets is wider than that of Borel.

1.6.5 Relation Between Jordan and Lebesgue Measurable Sets

The segment AB, which carries E, is divided into partial intervals. Let l be the sum of the lengths of those intervals where all points are interior to E, and L the sum of the lengths of those intervals containing points from E or its boundary. It is shown that when the division of AB is varied in any way so that the maximum length of the partial intervals tends towards zero, both numbers l and L tend towards definite limits, the interior and exterior extents of E.

From this definition, it follows that the exterior extent is at least equal to the exterior measure and that the interior extent is at most equal to the interior measure. Mathematically:

$$\text{interior extent}\,(E) \leq m_i(E) \leq m_e(E) \leq \text{exterior extent}\,(E).$$

Jordan called the sets whose exterior and interior extents are equal—measurable; these sets, which Lebesgue called J measurable, are therefore measurable in Lebesgue's sense, and the two definitions (Jordan's and Lebesgue's) of measure agree when both are applicable.

1.6.6 Relation Between Borel Measurable Set and Jordan Measurable Set

It can be said that the interior extent of E is the measure of the set of its interior points. The set of interior points is open; that is, it contains no points of its boundary. Therefore, it has as its complement, a closed set and hence is a B measurable set. The exterior extent of E is the measure of the set formed by E and its boundary, which is closed and thus B measurable. Therefore, for a set to be J measurable, it is necessary and sufficient for its *boundary to have zero measure.*

A closed set, with its exterior extent being its measure, can be affirmed as J measurable if it has zero measure. In particular, the perfect set defined in Sect. 1.6.1 is J measurable; the same goes for all those that can be formed with its points. Hence, the set of J measurable sets has the same cardinality as the set of point sets, and there exist J measurable sets that are not B measurable.

All this will imply that the cardinality of J measurable sets is greater than that of Borel measurable sets. Lebesgue compares the J and B measurable sets in one of his letters to M. E. Borel. He states thus:

> ... You mentioned that Jordan's definitions of measure are more general than yours. Indeed, Jordan associates two numbers with every set, but these two numbers seem to me to have no other property than their existence, except in the case where they coincide, and then they match yours. Please tell me if I am wrong and whether, in your opinion, do Jordan's definitions have any interest beyond providing a simpler definition of measure, in the case of a somewhat unusual set?
>
> Please let me know, for the time when I will be writing, if you are aware of any other definitions of measure and if the one you have adopted has already been used. I firmly believe that this is the correct one, as I have applied it myself [6]; and it seems impossible, as you mentioned, to give up the ability to add measures as we do with sets. Unfortunately, it seems difficult to demonstrate that these are the only ones we can measure or even that these others exist [7]. ...[11]

1.6.7 Comparison Between Borel and Lebesgue Measurable Sets

In another interesting letter, Lebesgue writes to Borel to compare sets measurable in his and Borel's senses.

> We are absolutely in agreement, I believe. I have slightly modified the language, that's all. If we consider a set E measurable (in my sense) [9], and we add to it one of those sets (analogous to the one on page 67 [10]) that is non-measurable in your sense but measurable in mine (let it be e_1), we have a set $E + e_1$ measurable in your sense. e_1 has measure zero (in my sense). $E + e_1$ has the same measure (in my sense) as E. Therefore, with the remarks on pages 48 and 49, [11], E has a measure at most equal to that of $E + e_1$. Besides, e_1 is part of a set measurable in your sense and of measure zero, let it be e_2. So, by removing from the set $E + e_2$ the set e_1, we obtain a set E' measurable in your sense contained in E and having the same measure as $E + e_1$, so that E having a measure at least equal to that of E' has exactly the measure of $E + e_2$.
>
> Is this not what you wanted to convey to me? I may have expressed myself very poorly, but this was the small demonstration I intended to include in my note. I do not have the text of my note, but I hoped I had been clearer than I fear I was. It seemed useful to me to clearly specify that the sets I called measurable were not exactly the same as those you referred to as such. I also believe there is no harm in calling measurable (as I do) even those sets that the remarks on pages 48 and 49 allow us to measure. However, this changes some statements, for example, the one on page 110 [12]: the set of measurable sets has the power of the continuum [13]. With the extended sense: the set of measurable sets has the same power as the set of all point sets, since there are sets of null measure that have the power of the continuum.
>
> Perhaps we could, and I believe this is certain, limit ourselves to the sets you call measurable, provided we only consider the functions of the Baire set [14]; this is a question I had not

[11] *See, Letter 1*, [8].

considered: it is evidently resolved affirmatively. In sum, I do not know anything about it [15].[12]

1.7 Integration in Lebesgue's Sense

1.7.1 Definite Integral

Lebesgue stated the problem of integration from the geometric point of view as follows:

Let a curve C be given by its equation $y = f(x)$ (where f is a continuous positive function, and the axes are rectangular). We have to find the area of the domain bounded by an arc of C, a segment of Ox, and two parallels to the y-axes at the given abscissae a and b, $(a < b)$.

This area is called the definite integral of f taken between the limits a and b; it is represented by $\int_a^b f(x)\,dx$.

The conventional approach for the general case involves determining a domain's interior and exterior extents by dividing the plane into rectangles with sides parallel to the Oy and Ox axes. To draw these rectangles, start by drawing parallels to Oy, then cut the bands formed by lines parallel to Ox using ordinates that extend from one band to the next. When considering one of these rectangles, say R_1, to calculate one of the extents (exterior or interior), all rectangles R situated within the same band and enclosed between R_1 and Ox must be taken into account for calculations of extent. Therefore, these extents (exterior and interior) are calculated by taking limits of the sum of areas of rectangles with their bases along Ox.

Let $\delta_1, \delta_2, \ldots$ be the lengths of these bases; $m_1, m_2, \ldots, M_1, M_2, \ldots$ be the lower and upper bounds of f in the corresponding intervals. If we assume these δ are given, i.e., the parallels to Oy have been drawn, and if we choose the segments parallel to Ox in such a way as to draw rectangles of very small size (to obtain the most approximate values possible for the extents), we will obtain for these approximate values of extents:

$$s = \sum \delta_i m_i \quad S = \sum \delta_i M_i.$$

Thus, we know how to calculate the two extents of the domain. We show that they are equal. The problem we have posed, therefore, has a meaning, and we know how to solve it.

If f is a bounded function, Darboux showed[13] that the two sums s and S tend towards the perfectly determined limits. He called them *integral by defect and integral by excess.* When these two integrals are equal, which occurs for functions other than continuous functions, the function is called *integrable,* and the common limit of s and S is called since Riemann[14] *the definite integral of f taken between a and b.*

[12] *See, Letter 3*, [8].

[13] Darboux [4].

[14] Riemann [16].

Summable Function

Geometric viewpoint: To understand a summable function geometrically, Lebesgue associated the set E of points whose coordinates satisfy the following two inequalities:

$$a \leq x \leq b \quad 0 \leq y \leq f(y),$$

to every positive function f defined on (a, b).

The two sums s and S are obviously the approximate values of the interior and exterior extents of E and consequently, s and S indeed have definite limits, which are these extents. Thus, from the geometric viewpoint, the existence of integrals by defect and by excess is a consequence of the existence of a bounded set's interior and exterior extents. For the function f to be integrable it is necessary and sufficient that E is J measurable. The measure of E is the integral.

When the function f takes both positive and negative values, we associate, a set E of points whose coordinates satisfy the following three inequalities, with it

$$a \leq x \leq b \quad xf(x) \geq 0 \quad 0 \leq y^2 \leq f(x)^2.$$

In this case, set E is the sum of two sets, E_1 and E_2, formed from points with positive ordinates for E_1 and negative ordinates for E_2.[15] The integral by defect is the interior extent of E_1 minus the exterior extent of E_2. The integral by excess is the exterior extent of E_1 minus the interior extent of E_2. If E is J measurable (in which case E_1 and E_2 are also J measurable), the function is integrable, the integral being $m(E_1) - m(E_2)$.

These results immediately lead to the following generalisation: *If the set E is measurable (in Lebesgue sense), (in which case E_1 and E_2 are also measurable (in Lebesgue sense)) the definite integral of f, taken between a and b, is the quantity:*

$$m(E_1) - m(E_2).$$

Definition: *The corresponding function f is called summable.* For non-summable functions, if they exist, Lebesgue defined the lower and upper integral as equal to

$$m_i(E_1) - m_e(E_2), \quad m_e(E_1) - m_i(E_2).$$

These two numbers are contained between the integrals by defect and excess.

Analytic viewpoint: Since E is measurable, it is contained in a set E' and contains a set E'' where E' and E'' are B measurable and of measure $m(E)$, §7. Further, the reasoning which gave us this result proves that we can assume E' and E'' formed of segments parallel to Oy and have their base on Ox, that is, they correspond to two functions f_1 and f_2 such that $f_1 \geq f_2$.

---

[15] It does not matter, whether the points $x \in (a, b)$ for which $f(x) = 0$ are counted in E_1 or E_2.

Lebesgue showed that *the set of values of* x *for which* $f(x)$ *is greater than* $m > 0$ *is measurable.* Similarly, *the set of values of* x *for which* $f(x)$ *is less than* $m < 0$ *is measurable.*

Hence, it follows that the set of values of x for which $f(x)$ is less than or equal to $m > 0$ (greater than or equal to $m < 0$) is measurable. Therefore, the set of points for which, either $a \geq f(x) > b > 0$, or $0 > c > f(x) \geq d$, or $e \geq f(x) \geq g$, $g \cdot e < 0$ is measurable, and by letting b tend towards a, or d towards c or e and g towards 0 it was seen that the set of points for which y has a given value ($y = k$) is measurable. In short, irrespective of the signs of a and b, *if* f *is summable, then the set of values of* x *for which*

$$a > f(x) > b$$

is measurable.

Conversely, *if for any* a *and* b *the set of values of* x *for which* $a > f(x) > b$ *is measurable, and if the function* f *is bounded, then it is summable.*

Geometric Definition of Integral of Summable Functions

Lebesgue's work distinguishes him from his predecessors, whose work will be discussed here. His revolutionary idea of looking at the integration process by partitioning the co-domain of a function is as follows.

Lebesgue divided the intervals of variation of $f(x)$. He considered $a_0, \ldots, a_n$ as the points of division, and e_i $(i = 0, 1, \ldots, n)$ as the set of values of x for which $f(x) = a_i$ (*See* [5]).

Let e'_i $(i = 0, 1, \ldots, n-1)$ be the set of values of x for which $a_i < f(x) < a_{i+1}$.

The points in the set E attached to $f(x)$, corresponding to the values of x belonging to e_i, form a measurable set in the plane with a measure of $|a_i| \cdot m_l(e_i)$ (where $m_l(e_i)$ denotes a linear measure).

Therefore, E contains a set of measure

$$\sum_0^n |a_i| m_l(e_i) + \sum_1^n |a_{i-1}| m_l(e'_i)$$

and is contained in a set of measure

$$\sum_0^n |a_i| m_l(e_i) + \sum_1^n |a_i| m_l(e'_i).$$

These two measures differ by less than $(a_n - a_0)\alpha$, where α is the maximum of $a_i - a_{i-1}$; therefore, it can be made as close as desired, and E is measurable; thus, f is summable.

Additionally, it is known how to calculate the measure of E; therefore, *if* f *is positive, the integral is the common limit of two sums*

$$\sigma = \sum_{0}^{n} a_i m_l(e_i) + \sum_{1}^{n} a_{i-1} m_l(e_i'),$$

$$\Sigma = \sum_{0}^{n} a_i m_l(e_i) + \sum_{1}^{n} a_i m_l(e_i')$$

when $a_{i-1} - a_i$ tends towards zero. Now if f is not always positive, the limit of the sum of those terms of σ or Σ which are positive gives a measure of the set that is denoted by E_1 (see Sect. 1.7), and the limit of the sum of the negative terms gives $-m(E_2)$, therefore *in all cases σ and Σ define the integral.*

Analytic Definition of Integral of Summable Functions

It will be useful to show that analytic reasoning could have led to the consideration of summable functions and to what are now called their integrals.

Lebesgue considered a monotonically increasing continuous function $f(x)$ defined between α and β ($\alpha < \beta$) and varying between a and b. For x the following values are taken arbitrarily

$$\alpha = x_0 < x_1 < x_2 < \cdots < x_n = \beta$$

to which correspond the values of $f(x)$

$$a = a_0 < a_1 < a_2 \cdots < a_n = b$$

In the ordinary sense of the word, the definite integral is the common limit of the two sums

$$\sum_{1}^{n} (x_i - x_{i-1}) a_{i-1} \quad \text{and} \quad \sum_{1}^{n} (x_i - x_{i-1}) a_i$$

when the maximum of $x_i - x_{i-1}$ tends towards zero.

However, x_i is determined by a_i, and $x_i - x_{i-1}$ tends towards zero if $a_i - a_{i-1}$ tends towards zero. Therefore, *for defining the integral of a monotonically increasing function $f(x)$ the values a_i can be specified. This means that instead of dividing the interval of variation of x, we can divide the interval of variation of $f(x)$.*

Proceeding in the same way, first in the simple case of continuous functions varying in any interval, and having only a finite number of maxima and minima, one is easily led to this property in the case of any continuous function. Let a continuous function $f(x)$ be defined in (α, β) and varying between a and b, ($a < b$). The following are chosen arbitrarily

$$a_0 = a < a_1 < a_2 \cdots < a_n = b;$$

$f(x) = a_i$ for the points of a closed set e_i, ($i = 0, 1, \ldots n$); $a_i < f(x) < a_{i+1}$ for the points of a set, sum of intervals, e_i', ($i = 0, 1, \ldots, n-1$); the sets e_i, e_i' are measurable.

The two quantities

$$\sigma = \sum_{0}^{n} a_i m(e_i) + \sum_{1}^{n} a_i m(e_i'),$$

$$\Sigma = \sum_{0}^{n} a_i m(e_i) + \sum_{1}^{n} a_{i+1} m(e_i')$$

tend towards $\int_a^b f(x)\,dx$ when the number of a_i increase in such a way that the maximum of $a_i - a_{i-1}$ tends towards zero.

This property being obtained, it can be used to define the integral of $f(x)$. However, the two quantities σ and Σ have a meaning for functions other than continuous, namely summable functions. Lebesgue showed that for these functions, σ and Σ have the same limit, which is independent of the choices of a_i; this limit would be, by definition, the integral of $f(x)$ taken between α and β.

When, between the a_i, new points of division are introduced, σ does not decrease, Σ does not increase, therefore σ and Σ have limits. They are equal because $\Sigma - \sigma$ is at most equal to $(\beta - \alpha)$ multiplied by the maximum of $(a_i - a_{i-1})$.

Next, the interval of variation of $f(x)$ was partitioned in another way using points b_i. In this mode σ' and Σ' are the corresponding values of σ and Σ. Again, σ'' and Σ'' are the corresponding values from the division mode in which a_i and b_i are used simultaneously. The pair of inequalities

$$\sigma \le \sigma'' \le \Sigma'' \le \Sigma'$$
$$\sigma' \le \sigma'' \le \Sigma'' \le \Sigma$$

proves that the six sums σ, σ', σ'', Σ, Σ', Σ'' have the same limit.

Hence, the existence of the integral is demonstrated. When adopting this slightly different approach, it is apparent that Riemann's definition of integral consistently agrees with the previous one. This agreement is supported by the fact that the points of discontinuity in integrable functions constitute a set of measure zero.[16] Let $f(x)$ be an integrable function and let E be the set of points for which we have

$$a \le f(x) \le b$$

where a and b are any two numbers. The limit points of E that do not belong to E are the points of discontinuity; therefore, they form a set e of measure zero. $E + e$ being closed is measurable, e is measurable, and so is E. This is sufficient to conclude that f is summable.

If in an interval of length l, the maximum of f is M and the minimum m, the integral is contained between lM and lm. Hence, it follows that the integral of a summable function

[16] Riemann stated this property in the following way: For a function to be integrable, it is necessary that "the total sum of the intervals, for which the oscillations are greater than σ, for any σ, can be made infinitesimally small".

is included between the integrals by defect and excess, and in particular, the two integral definitions coincide when they are both applicable.

Cardinality of Set of Summable Functions

Lebesgue showed that the elementary arithmetic operations applied to summable functions give back summable functions. Specifically, the sum of any finite number of summable functions is a summable function and the integral of the sum function is the sum of the integrals of individual term functions. Similarly, the product of two summable functions is a summable function and the reciprocal of a summable function, if it is bounded away from 0 is a summable function. The kth root of a summable function, if it exists, is summable. If f and φ are two summable functions, and if $f(\varphi)$ exists, then it is summable.

Another important result was deduced, that if a bounded function f is the limit of a sequence of summable functions f_i, then f is summable.

These results led to the definition of an important class of summable functions—*the polynomials*. Since $y = h$ and $y = x$ are summable functions, then px^q is summable; hence, all polynomials are summable.

Ever since Weierstrass, it has been known that every continuous function is summable. However, apart from continuous functions, there exist other functions that are limits of polynomials, as studied by M. Baire, referred to as class one functions [17]. Consequently, these first-class functions also fall under the category of summable functions. The limits of first or second-class functions are also summable, etc. All functions from the set denoted by E by M. Baire (p. 70 Loc. Cit.) are summable.

These results offer numerous examples of summable functions that are both discontinuous and non-integrable (in Riemann's sense). We can generate such examples using the following method: The reasoning from Sect. 1.7.1 showed that every integrable function is summable, proving that if, after ignoring a set of measure zero, a set remains where a function is continuous at each point, that function is summable. For instance, if f and φ are two continuous functions and F is defined as equal to f except at the points of a set E of measure zero, where we have $F = f + \varphi$. However, if φ is never zero and E is dense in all intervals, in that case all points become points of discontinuity for F, thereby rendering it non-integrable (in the sense of Riemann).

This approach enables the construction of summable functions that form a set equal in cardinality to the set of functions.

Term-by-Term Integration of Series

The calculation of the integral of a given function presents the same difficulties as the calculation of the measure of a given set. Most of the discontinuous functions that we have considered so far in Analysis were defined by series, therefore it may be worthwhile to note the following result:

If a sequence of summable functions, with integrals $f_1, f_2, f_3, \cdots$, has a limit f, and if $|f - f_n|$ remains, for any n, less than a fixed number M, then f has an integral which is the limit of the integrals of the functions f_n.[17]

Indeed, we have

$$f = f_n + (f - f_n).$$

Both functions on the right-hand side have integrals, and so does f, and the integral of f is the sum of the integrals of f_n and $f - f_n$. Now, an upper limit of the second integral is to be found.

A positive number Σ is arbitrarily chosen. Let e_n be the set of values of x for which one does not have,

$$|f - f_{n+p}| < \varepsilon$$

for every positive or null value of p, e_n is measurable.

Let E be the measurable set in which the integrals are taken:

$$\left|\int (f - f_n)\, dx\right| \leq M \cdot m(e_n) + \varepsilon\, [m(E) - m(e_n)]\, .$$

However, each set e_n contains all sets with greater indices, and there exists no point common to all e_n. Therefore, $m(e_n)$ tends towards zero with $\frac{1}{n}$ and consequently the same goes for

$$\left|\int (f - f_n)\, dx\right|$$

When f is bounded the proposition can be stated thus: When a sequence $f_1, f_2, \ldots$ of summable functions, bounded from above in absolute value in their set, has a limit f the integral of f is the limit of integrals of functions f_n.

Here is another form of the statement relative to the general case.

When the set of the remainder of a convergent series of functions having integrals is bounded from above in absolute value, the series is term-by-term integrable.

As a very particular case, the theorem on the integration of uniformly convergent series follows from the above.

1.7.2 Indefinite Integral

The Indefinite Integral and Its Continuity

We call the indefinite integral of a function $f(x)$, having a definite integral in an interval (α, β), a function $F(x)$ defined on (α, β) and such that, for any a and b contained between

[17] The most interesting particular case of this theorem, the one where f and the f_i are continuous functions, has already been obtained, using very different considerations by M. Osgood in his Memoire on non-uniform convergence (American Journal 1894).

α and β, we have:

$$\int_a^b f(x)\,dx = F(b) - F(a).$$

From this equality, we deduce:

$$F(x) = \int_a^x f(x)\,dx + F(a).$$

Therefore, every function having an indefinite integral admits an infinite number of indefinite integrals which differ only by a constant $F(a)$.

The indefinite integral is a continuous function.[18] This is obvious if the function $f(x)$ is bounded. The following inequality must be shown to prove it in a general case, which is stated without proof.

$$|F(a+h) - F(a)| = \left|\int_a^{a+h} f(x)\,dx\right| < \varepsilon,$$

for any a, and $h > 0$ small enough.

Indefinite Integrals and Primitive Functions

If M and m are the maximum and minimum of $f(x)$ in $(a, a+h)$ we have:

$$mh < \int_a^{a+h} f(x)\,dx < Mh$$

hence,

$$m < \frac{F(a+h) - F(a)}{h} < M,$$

therefore for $x = a$, if $f(x)$ is continuous at this point, $F(x)$ has a derivative equal to $f(a)$.

If $f(x)$ is continuous for every value of x, $F(x)$ is any one of the functions that has $f(x)$ as its derivative, i.e., one of the primitive functions of $f(x)$.

Thus, in the case of continuous functions, there is an identity between the search for primitive functions and the search for indefinite integrals of a given function. This well-known result still holds when dealing with a function whose derivative is integrable in the sense of Riemann.[19] However, there exist derivative functions that are not integrable in Riemann's sense[20]; given one of these functions, its primitive function cannot be calculated using integration in the sense of Riemann.

It was shown that any bounded derivative function has an indefinite integral that is one of its primitive functions. Therefore, the primitive function of a given bounded function can be calculated if it exists.

[18] We can also add that it is of bounded variation; this variation being at most equal to $\int |f(x)|\,dx$. The proof is the same as given by M. Jordan, *Cours D'Analyse* §81 [18].

[19] See Darboux [4].

[20] M. Volterra gave the first example of such functions. *Giornale de Battaglini, t. XIX*, 1881.

In the case of unbounded derivative functions, it was demonstrated that if they have integrals, there is an identity between their primitive functions and their indefinite integrals.

Primitive of Bounded Derivative Functions

The derivative of a function $f(x)$ is the limit of the expression:

$$\frac{f(x+h)-f(x)}{h}=\varphi(x)$$

when h tends towards zero, which, when h is fixed, represents a continuous function; therefore, the derivative is the limit of continuous functions, and so it is summable.

Let us assume that the derivative f' is always less than M, in absolute value. By virtue of the theorem of finite increment $\varphi(x)=f'(x+\theta h)$, therefore the functions $f(x)$ are bounded in their set, and as a result, we have (see Sect. 1.7.1)

$$\int_a^b f'\,dx=\lim\int_a^b \varphi(x)\,dx=\left[\lim_{h\to 0}\frac{1}{h}\int_x^{x+h} f(x)\,dx\right]_a^b$$

therefore,

$$\int_a^b f'\,dx=f(b)-f(a).$$

Every bounded derivative function admits its primitive functions as its indefinite integrals; this result remains true even if it is a right or left bounded derivative or a limit towards which $\varphi(x)$ tends for certain values of h tending towards zero.

The Necessary and Sufficient Condition for Existence of Integral of a Derivative Function

The primitive functions discussed above are of bounded variation.[21] Lebesgue showed that *the necessary and sufficient condition for the integral of the derivative (whether bounded or not) of a differentiable function to exist is that the function must be of bounded variation. If this condition is met, the function is one of the indefinite integrals of its derivative.*

Further Results and An Example

Some more results can be derived from the necessary and sufficient condition for the integral of $|f'|$ to exist, which may be obtained by resuming the above reasoning.

The total positive variation of $f(x)$ between a and x is equal to the integral of $f'(x)$ extended to the set of points for which f' is positive, which means to the integral

$$p(x)=\frac{1}{2}\int_a^x (f'+|f'|)\,dx.$$

Similarly, for the negative variation, we have:

[21] Lebesgue used some of the properties of these functions here (See Jordan, *Comptes Rendus de l'Academie des Sciences* 1881 and *Cours d'Analyse* [18].)

$$-n(x) = \frac{1}{2}\int_a^x (f' - |f'|)\,dx.$$

And as we have:

$$f(x) - f(a) = p(x) - n(x)$$
$$f(x) = f(a) + \int_a^x f'(x)\,dx$$

Thus, when a function $f(x)$ is given, we know how to find out if it is derivative of a function of bounded variation, and, if it is so, we know to find out its primitive functions. If $f(x)$ is bounded, its primitive functions, if they exist, are of bounded variation, and we can find them.

An Example of a Summable Unbounded Function That Does Not Have an Integral

The function $f(x)$ given by

$$f(x) = x^2 \sin\frac{1}{x^2} \quad \text{for} \quad x \neq 0$$
$$f(0) = 0,$$

is continuous and of unbounded variation. The sum of the variations of $f(x)$ is

$$\sum \frac{1}{k\pi + \frac{\pi}{2}}$$

which is a divergent series. This function admits a derivative

$$f'(x) = 2x\sin\frac{1}{x^2} - \frac{2}{x}\cos\frac{1}{x^2}, \quad \text{for} \quad x \neq 0$$
$$f'(0) = 0.$$

$f'(x)$ provides an example of an unbounded function that is summable but does not have an integral. The integral $f(x)$ of the summable function $f'(x)$ cannot be found by using previous methods.

It is interesting to note that the classical definition of the integral of a function that becomes infinite in the vicinity of a point allows us to find $f(x)$ when $f'(x)$ is known. In cases where the function to be integrated is unbounded, the definition Lebesgue has adopted is not a generalisation of the classical definition; it differs from this definition but agrees with it when both apply. It would also be very easy to generalize the notion of a definite integral so that the classical definition and the one given by Lebesgue become special cases of a more general definition.

We have seen that every indefinite integral is continuous. If now we consider this property as one of the parts of the definition of indefinite integrals, we are led to the following result:

A function $f(x)$ defined on (α, β) has an indefinite integral $F(x)$ in this interval, if there exists one and only one continuous function $F(x)$ up to an additive constant, such that

$$F(b) - F(a) = \int_a^b f(x)\,dx$$

for all a and b, between α and β, in such a way that the right-hand side has a meaning.[22]

Remarks of M. E. Borel on Lebesgues's Work

M. E. Borel in his Memoire[23] speaks about the work of H. Lebesgue. He says thus:

> It has been particularly painful for me to find myself compelled to discuss questions of priority with Mr. Lebesgue; for twenty years, in fact, I have been several times directly or indirectly aware of discussions of priority between Mr. Lebesgue and various mathematicians, and while I often thought that Mr. Lebesgue's opponents were right, I never believed that Mr. Lebesgue was wrong. This stance, which has often led his opponents to accuse me of unfairness, is based on my opinion of his good faith and the nature of his mind. I have no reason to change this opinion just because I find myself personally involved: patere legem quam fecisti. I am therefore led, after explaining why I believed I was right, to indicate why I am convinced that Mr. Lebesgue is not wrong in his judgment of his own work.
>
> To do this, I must try to explain what, in my eyes, characterizes Mr. Lebesgue's talent; I will not try to confine it in narrow formulae nor to specify in a few sentences his considerable contribution; even if I believed myself capable of doing so, I would not, because reading the passages of his Memoir, where he gives some details about his work and the stages of his thought, proves to all those who might have been unaware that, if he were to write a detailed Notice on his work, he would render the greatest service to Science. But I believe I can say, without fear of contradiction, that Mr. Lebesgue's talent consists mainly of a great capacity for assimilation, rare originality of thought, and exceptional creative power; all his works are original and belong entirely to him, are entirely personal to him, because he never takes an idea from someone else as it is, to develop it and use it; if a foreign idea is useful to him, it is in the way a powerful tree creates leaves and fruits with all the nutrients it finds in the soil.
>
> My teacher Jules Tannery often quoted a sentence from Liouville; after comparing long demonstrations to short ones, he concluded: "In short, long demonstrations have a great advantage, which is to be long, and short demonstrations have a great advantage, which is to be short." And Liouville recalled, it seems, when he was in a good mood, the joke attributed to Lagrange: "Mathematics is like the pig, everything about it is good." Mr. Lebesgue's demonstrations have a very great advantage, in that they are complicated; this complexity brings an extraordinary richness, a richness of thought such that Mr. Lebesgue hardly manages to fit all his ideas into a book or a Memoir. There are many ideas that he does not express; there are others that are only hinted at or incidentally mentioned. Therefore, Mr. Lebesgue alone knows all the richness of his thought and work; it is not surprising, therefore, if he perhaps more frequently than others have to make claims of priority. They are certainly always justified in his eyes, and that is why I have always assured those who complained about him that they had no right to hold it against

[22] Compare this definition with the one given by M.Jordan of the definite integral of an unbounded function. *Cours d'Analyse, pp.* 46–94 [2].

[23] See, pp. 71−92 [19].

him; I address myself today to this exhortation, in order to soften the pain caused by certain passages of his Memoir.

1.8 Work of Stieltjes

Stieltjes in his paper of 1894, [20] *Recherches sur les fractions continues,* did the path-breaking work by creating an analytic theory of continued fractions. In this paper, he set the connections between the function theory and integral theory which were essential for a better understanding of an important class of continued fractions. Stieltjes had been working on the problem of summation of divergent power series for quite a while. In his thesis of 1886, [21] Stieltjes did fundamental work on remainders in several asymptotic series. Then, in the period 1889–1890 he published quite a few papers containing examples of continued fraction expansions for asymptotic series. Such series arose as formal power series expansions of definite integrals of the form:

$$\int_0^{\infty} \frac{f(u)\, du}{z+u},$$

where $f(u) > 0$, and the continued fractions are of the form:

$$\cfrac{1}{z+\cfrac{a_1}{1+\cfrac{a_2}{z+\cfrac{a_3}{\ddots}}}},$$

where $a_p > 0$. The continued fractions can be transformed into

$$\cfrac{1}{z+b_1-\cfrac{p_1}{z+b_2-\cfrac{p_2}{z+b_3-\cfrac{p_3}{\ddots}}}},$$

where $b_k > 0$ and $p_k > 0$ are functions of a_k.

In the paper [20], a general theory of continued fractions was developed, which pondered the questions dealing with convergence and its links with the definite integrals and divergent power series. It was here that need was felt to generalise the already-known concepts of integration (Lebesgue or Riemann). This work further led to the development of a general *convergence continuation theorem* for sequences of analytic functions.

In his research on continued fractions of type (Stieltjes continued fractions),

$$\cfrac{1}{k_1 z + \cfrac{1}{k_2 + \cfrac{1}{k_3 z + \cfrac{1}{\ddots}}}},$$

in which k_p are real and positive and z is a complex variable, Stieltjes found that in some cases the value of continued fraction has the form

$$\int_0^\infty \frac{f(u)\,du}{z+u},$$

where $f(u)$ is a positive function of u, while in some other cases, the value of the continued fraction has the form

$$\sum_{p=1}^{\infty} \frac{L_p}{z+x_p},$$

where the $L_p > 0$ and $0 < x_1 < x_2 < x_3 < \cdots$. Further, in other cases, the value of the continued fraction may be a sum of an integral and an infinite series of the above form. This situation led Stieltjes to define an integral of the form[24]

$$\int_0^\infty \frac{d\varphi(u)}{z+u},$$

encompassing all three types of functions. This integral is indeed the Stieltjes integral, which we see in the modern form.

1.9 Work of Denjoy

Riemann integral holds significance when the integrated function is continuous or when its points of discontinuity form a set of measure zero. On the other hand, the Lebesgue integral applies to any function f that is measurable and bounded or summable. Both integrals, calculated between a and x, yield a continuous function of x, say $F(x)$ with a fixed, and their derivative is f, excluding a set of points of measure zero. However, there exist derivative functions that are integrable neither in the Riemann sense nor in the Lebesgue sense. Denjoy developed a calculation method that specifically worked for any derivative function, resulting in a continuous function with f as its derivative.

Denjoy extended the Lebesgue integral, establishing a clear link between totalisation and integration in a clear, coherent, and expressive language. Integration primarily focuses on the definite integral, while totalisation emphasises on the indefinite total. Totalisation aligns more closely with the integration of differential equations than with the computation of

[24] Chapter VI of Stieltjes [20].

quadrature. Notably, the definite total aligns with Duhamel's definite integral discussed in Chap. 7.

Denjoy initially defined a sequence of finite or transfinite operations called Totalisation, performed on a function $f(x)$ assumed to be finite throughout the interval (a, b). If these operations are feasible $f(x)$ is considered totalisable. Totalisation associates $f(x)$ with a continuous function $F(x)$ over the entire (a, b), representing the indefinite total of $f(x)$. $F(x)$ is determined up to an additive constant. For any interval (l, m) in (a, b), the difference $F(m) - F(l)$ of $F(x)$ in (l, m) represents the definite total of $f(x)$ in (l, m). Hence, totalisation serves the dual purpose of obtaining a function (indefinite total) or a value (definite total).

Operations of Totalisation

Arnaud Denjoy in his Memoire [22], considered the totalisation of f in an interval (a, b), $a < b$ (which is for calculating a definite total and for calculating an indefinite total it may be (a, x)) and established some rules, concepts, definitions, and conditions. He applied these rules to f (function to be totalised) indefinitely many times. The following concepts may be defined.

1. **Non-integrability of f at a point.** f is said to be non-integrable at a point if f is non-integrable over any interval containing this point. If a function is integrable at a point, then there exists an interval containing this point in which f is integrable.
   If P is a given perfect set and φ is a function equal to f on this set and zero outside P, then f is integrable (non-integrable) either (a) over P, or (b) within an interval containing points from P or, (c) at a point of P, if φ is integrable (non-integrable) over that interval or at that point. The points of P where f is non-integrable form a closed set.
2. **Absolute convergence of series in an interval.** Denjoy associated a number A_n with each of the contiguous intervals u_n of perfect set P. Denjoy called the series A_n to be *absolutely convergent in an interval* i if the series of numbers A_n corresponding to the intervals u_n interior to i is absolutely convergent. Therefore, series A_n is absolutely convergent in any interval interior to i.
   Absolute convergence of series at a point. He called the series A_n to be absolutely convergent at a point M of P, if M is interior to an interval i where the series A_n is absolutely convergent.
   The points of non-absolute convergence form a closed set.

Denjoy established a set of following definitions:

1. The totalisation of f in an interval where it is summable is the Lebesgue integral of f on this interval.
2. If the totalisation of f (say V) is defined over a sequence of intervals (a_1, a_2), $(a_2, a_3), \ldots, (a_{n-1}, a_n)$, then totalisation over the interval (a_1, a_n) say $(V(a_1, a_n))$ is the sum $V(a_1, a_2) + V(a_2, a_3) + \cdots + V(a_{n-1}, a_n)$.

3. If f is summable over a perfect set P in an interval (α, β) if V is calculated in every interval u'_n not having any point of P in its interior (i.e contained in contiguous intervals of P, i.e., u_n), and if $W(u_n)$ is the supremum of $|V(u'_n)|$ when u'_n varies over the entire u_n, the series $W(u_n)$ is convergent. Then

$$V(\alpha, \beta) = \sum V(u_n) + \int_P f,$$

where $V(u_n)$ is the totalisation of f between the origin and end point of u_n.

Denjoy called f totalisable in the interval (a, b) if it satisfied the following conditions.

1. For any perfect set P in (a, b) the set of points of P where f is non-summable over P, is non-dense on P.
2. **Continuity of** V:- If $V(c', d')$ has been calculated for any $c' < d'$ both c', d' interior to (c, d), $V(c', d')$ tends to a limit when c' tends towards c and d' tends towards d.

$$V(c, d) = \lim V(c', d').$$

3. For any perfect set P, if V has been calculated in every interval u_n contiguous to P, the set of points of P where the series W_n is not convergent is non-dense on P.

With these concepts, definitions and conditions, we find the indefinite and definite total of f in the following.

Indefinite Totalisation

Denjoy assumed the indefinite totals $V_k(x)$ were known in the intervals (a_k, b_k) such that

$$a < \cdots < a_2 < a_1 < b_1 < b_2 < \cdots < b$$

a_k tend towards a and b_k tend towards b. Then, he formed the continuous function $V(x)$ equal to $V_1(x)$ in (a_1, b_1), equal to

$$[V_k(x) - V_k(a_{k-1})] + [V_{k-1}(a_{k-1}) - V_{k-1}(a_{k-2})] + \cdots + V_1(a_1)$$

in (a_k, a_{k-1}) and equal to

$$[V_k(x) - V_k(b_{k-1})] + [V_{k-1}(b_{k-1}) - V_{k-1}(b_{k-2})] + \cdots + V_1(b_1)$$

in (b_{k-1}, b_k), using definition 2, indefinite total $V(x)$ in (a, b) can be found.

Definite Totalisation

Denjoy took a perfect set P contained in an interval (α, β). He assumed that the totals of $f(x)$ were known in various intervals (l, m) contiguous to P, with respect to (α, β), and he supposed that the series $\sum[V(m) - V(l)]$ provided by these totals is convergent and that $f(x)$ is summable over P. Then, using Definition 3

$$\int_P f\,dx + \sum[V(l) - V(m)]$$

is called the definite total of $f(x)$ in (α, β).

1.10 Summary

Prelude to Lebesgue's Integration: A Tapestry of Historical Threads. Before Henri Lebesgue revolutionised integration theory, mathematicians had explored various avenues to express and quantify the concept of the area under a curve. This chapter lays the foundation by examining key ideas that paved the way for Lebesgue's seminal work.

1. Beyond Riemann's Reach: Expanding the Class of Integrable Functions. Riemann's integral, though powerful, seemed to reach its limit at functions with countably infinite discontinuities distributed in a specific manner. However, the Fourier problem of representing arbitrary functions using trigonometric series sparked further investigation. Dirichlet and Riemann tackled the question of term-by-term integration of such series, but it was Lebesgue who ultimately solved it, broadening the realm of functions amenable to analysis.

2. Unlocking the Fundamental Theorem: Measurable Derivatives Take Centre Stage. The fundamental theorem of calculus, the problem of finding the primitive of derivative function, faced hurdles with certain functions possessing bounded derivatives, yet being non-Riemann integrable. Lebesgue's theory provided the key, enabling the integration of functions with summable (integrable) derivatives. Volterra's example of a differentiable but non-Riemann integrable function exemplifies the need for this refined approach.

3. From Jordan's Extent to Lebesgue's Measure: Quantifying Length and Area. Before Lebesgue, Camille Jordan delved into the concept of "extent" to quantify the length of curves and area of domains. Around the same time, Borel worked on constructing Borel's set and Borel measure, and subsequently, Lebesgue did fundamental work on measure theory. Lebesgue's definition of measure provided a powerful tool for quantifying not only length and area but also more general geometric notions.

Development of the Function Concept and Riemann Integration: a cornerstone of mathematical analysis

4. From Explicit Formulae to Abstract Notions: Breaking Free from Algebraic Chains. Euler's 1748 definition of a function as an analytical formula composed of algebraic operations laid the foundation. However, the quest for representing arbitrary functions, particularly in the context of partial differential equations, propelled a paradigm shift. Mathematicians like Euler, D'Alembert, Bernoulli, and Lagrange contributed significantly to this evolving understanding. The wave equation and its solution sparked a pivotal debate, ultimately settled by Fourier: a function need not have an explicit expression. Discontinuities, in the modern sense, were admitted into the realm of admissible functions determining the initial shape of the string. Dirichlet, in his 1829 memoir, further expanded this class by including functions that were bounded, piecewise monotonous, piecewise continuous, and even possessing infinite discontinuities within finite intervals. His iconic "Dirichlet function," with distinct values for rational and irrational x, exemplified functions defying representation by series.

5. Riemann's Integration: Taming Discontinuities with Sums. Section 1.3 focuses on Riemann's ground-breaking work on integration, born from his doctoral studies under Dirichlet. Initially fascinated by representing functions using the Fourier series, he proved the convergence of Fourier coefficients for bounded and integrable functions. Subsequently, Riemann diverged into defining integration through Cauchy-Darboux-type sums. This approach enabled integrating functions with densely packed discontinuities within the integration interval. Here, the distribution of singularities held paramount importance in Riemann's analysis. Functions with reducible sets of discontinuities and zero mean oscillation were deemed integrable. The integration process involved partitioning the domain of the function (unlike Lebesgue's later focus on the co-domain), followed by summation and limit-taking procedures.

6. Critiques and Refinements: Pushing the Boundaries of Integration. The subsequent subsections analyse critiques of Riemann's definition. One critical limitation was its inability to explain the non-solvability of the fundamental theorem of calculus, as illustrated by Dini's counterexample. Term-by-term integration of series for Riemann integrable functions also faced restrictions, demanding additional assumptions on the series' sum function. Furthermore, Riemann's approach proved inadequate for resolving the multidimensional integral problem without ambiguity.

7. Jordan's Extent and Borel's Measurable Sets: Stepping Stones to Lebesgue. Section 1.4 dives into Camille Jordan's concept of "extent," a precursor to Lebesgue's measure theory. Jordan assigned two numbers to sets—exterior and interior extents—enabling integration definitions based on this framework. This led to the concepts of integration by excess and defect, further paving the way for measure-theoretic advancements.

Section 1.5 explores the contributions of Émile Borel, who focused on constructing and characterising measurable sets. He defined a method for constructing "Borel sets" through operations like addition, subtraction, and intersection with intervals. He then provided a measure specifically for Borel sets, capturing size in both topological and measure-theoretic senses (illustrated by examples in Sect. 1.5.1).

Crucially, Borel acknowledged that his definition of measure might not encompass all possible sets. However, he emphasised that Borel sets retain key properties similar to intervals, making them a valuable stepping stone towards a more comprehensive measure theory.

8. Problem of Measure: As Visualised by Lebesgue. Section 1.6 unveils Henri Lebesgue's revolutionary approach to measure and integration. He began by formulating the problem of measure based on a few key axioms. Unlike Borel, Lebesgue defined exterior and interior measures of sets differently by enclosing the set and its complement in progressively smaller intervals and taking limits. Sets with equal inner and outer measures were deemed Lebesgue measurable.

Significantly, Lebesgue designed his measure theory to inherit the properties of line segment measures (without common points) and allow countable additivity. He even estimated the cardinality of sets measurable under each framework (Jordan, Borel, Lebesgue), demonstrating that while Jordan and Lebesgue have the same cardinality, some Lebesgue-measurable sets are not Borel-measurable.

9. Lebesgue's Revolutionary Ideas of Integration. Section 1.7 dives into Lebesgue's original and fundamental integration theory. Here, he partitions the co-domain of the function, leveraging the concepts of measurable functions, summable functions, and domain measure. This novel approach encompasses Riemann integration as a special case, overcoming limitations like the non-solvability of the fundamental theorem and problematic term-by-term integration of series. Lebesgue further explores indefinite integration, highlighting its connection to differentiation and establishing it as the inverse of the differentiation process.

10. Stieltjes Integration: From Theory of Continued Fractions to Integration. Section 1.8 explores Thomas Joannes Stieltjes's generalisation of Lebesgue and Riemann integrals. He integrated functions against functions of bounded variation (the integrator), using continued fractions as a springboard for this novel method. This work had profound ramifications for developing the theory of distributions, where the distributional derivative of functions of bounded variation became known as the "Radon measure."

11. Denjoy's Work: Concept of Totalisation. Finally, Sect. 1.9 discusses Arnaud Denjoy's work, another generalisation of Lebesgue and Riemann integrals. Denjoy's "totalisation" method focuses on integrating functions that are not measurable in the Lebesgue sense, primarily addressing indefinite totalisation.

References

1. Lebesgue, Henri Leon. 1928. *Leçons sur l'intégration et la recherche des fonctions primitives: professées au Collège de France*, vol. 6. Gauthier-Villars et cie.
2. Jordan, Camille. 1894. *Cours d'analyse de l'École polytechnique*, vol. 2. Gauthier-Villars.
3. Lejeune Dirichlet, G. 1829. Sur la convergence des séries trigonométriques qui servent à représenter une fonction arbitraire entre des limites données.
4. Darboux, Gaston. 1875. Mémoire sur les fonctions discontinues. *Annales scientifiques de l'École normale supérieure* 4: 57–112.
5. Henri, L. 1902. *Lebesgue, "Intégrale, Longueur, Aire" .fre*. PhD thesis, PhD Thesis. Université de Paris
6. Borel, Émile. 1928. Leçons sur la théorie des fonctions. Gauthier-Villars et fils.
7. Jordan, Camille. 1892. Remarques sur les intégrales définies. *Journal de mathématiques pures et appliquées* 8: 69–99.
8. Lebesgue, Henri. 1991. Lettres d'henri lebesgue à émile borel. *Cahiers du séminaire d'histoire des mathématiques* 12: 1–506.

9. Banach, Stefan. 1923. Sur le problème de la mesure. *Fundamenta Mathematicae* 4 (7): 33.
10. Banach, Stefan, and Kazimierz Kuratowski. 1929. Sur une généralisation du problème de la mesure. *Fundamenta Mathematicae* 14 (1): 127–131.
11. Lebesgue, H. 1902. Sur les transformations de contact des surfaces minima. Bulletin of Mathematical Sciences, II. Sér. 26: 106–112. Bulletin of Mathematical Sciences, II. Sér. 26: 106–112.
12. Lebesgue, Henri. 1903. Sur le problème des aires. *Bulletin de la Société Mathématique de France* 31: 197–203.
13. Lebesgue. H. 1902. Un théorème sur les séries trigonométriques. *Comptes rendus de l'Académie des Sciences* 134: 585–587.
14. Lebesgue, Henri. 1903. Sur les séries trigonométriques. *Annales scientifiques de l'École normale supérieure* 20: 453–485.
15. Montel, Paul. 1907. Sur les suites infinies de fonctions. *Annales scientifiques de l'École Normale Supérieure* 24: 233–334.
16. Riemann, Bernhard. 1873. Sur la possibilité de représenter une fonction par une série trigonométrique. *Bulletin des sciences mathématiques et astronomiques* 5: 20–48.
17. Baire, René. 1899. Sur les fonctions de variables réelles. *Annali di Matematica Pura ed Applicata* 1898–1922 (3): 1–123.
18. Jordan, Camille. 1983. Cours d'analyse de l'École polytechnique, vol. 1. Gauthier-Villars et fils.
19. Borel, Émile. 1919. L'intégration des fonctions non bornées. *Annales scientifiques de l'École Normale Supérieure* 36: 71–92.
20. Stieltjes, T.-J. 1894. Recherches sur les fractions continues. *Annales de la Faculté des sciences de Toulouse: Mathématiques* 8: J1–J122.
21. Stieltjes, T.-J. 1886. Recherches sur quelques séries semi-convergentes. *Annales scientifiques de l'École Normale Supérieure* 3: 201–258.
22. Denjoy, Arnaud. 1912. Une extension de l'intégrale de m. lebesgue. Comptes rendus de l'Académie des Sciences Paris 154: 859–862.

The Integration Before Riemann

2

2.1 The Integration of Continuous Functions

The integration was first defined as the inverse operation of differentiation; it is the operation that permits to solve the problem of primitive functions:

Find the functions $F(x)$ which admit as its derivative a given function $f(x)$.

We know that, if the solution to this problem is possible, it is possible in infinitely many ways, and that all the primitive functions $F(x)$ of the same function $f(x)$ differ only by an additive constant. We propose to find any one of such functions $F(x)$.

At the time when the problem of primitive functions was posed under the form that I indicate, that is, the times of Newton and of Leibnitz, the word *function* was quite poorly defined. We thus call, most often, a quantity y linked to the variable x by an equation where a certain number of symbols of operations appear which we usually see. The foremost of these operations are: arithmetic operations (addition, subtraction, multiplication, division, extraction of the roots), trigonometric operations (with the signs sin, cos, tan, arc sin, arc cos, arc tan), logarithmic operations, and exponentials (with signs log, a^x).

For a large number of functions expressed in this manner, we could have expressed the primitive functions in the same manner, so that it appeared to some that any function admits a primitive function. Moreover, we could respond to anyone who doubted this proposition.

Let (Fig. 2.1) the curve Γ, $y = f(x)$, represent the given function $f(x)$; the axes being rectangular. Let us suppose, for simplicity, $f(x)$ positive; let aA, bB be two parallel lines to the y-axis, of abscissa a and b. These two parallel lines, the arc AB of Γ, the segment ab of Ox, bound a domain of area $S(x)$. By evaluating the increment $bBCc$ of this area, we see that $f(x)$ is the derivative of $S(x)$.[1]

[1] For the proof as well as for the case where $f(x)$ is not always positive, *see* the classical treatise of differential and integral calculus.

R. Jain, *Lebesgue's Theory of Integration*,
https://doi.org/10.1007/978-981-96-1169-0_2

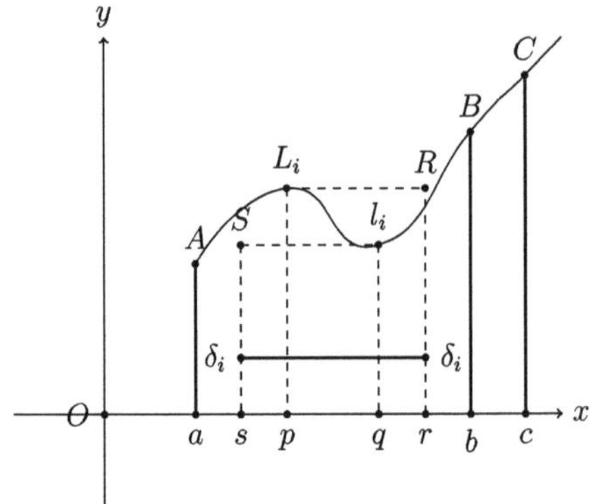

Fig. 2.1 Curve Γ

Let us note that in the previous considerations the word *function* has already received a considerable extension. The relation between $S(x)$ and x is in fact a geometric relation and not an algebraic, trigonometric, or logarithmic one anymore. Such relations were still considered as defining the functions. We only carefully distinguished between the geometric figures defined with the help of laws expressible by geometric equalities and the figures which were not defined in this way. The curves $y = f(x)$ of the first kind or the geometric curves defined the functions $f(x)$. The curves of the second kind or arbitrary curves did not define the true functions. When we used the word function for these two types of correspondence between y and x, we distinguished the first from the second by calling them *continuous functions*.[2]

There was also an intermediate category of functions, the ones that were represented with the help of several arcs of the geometric curves. We would more readily consider them as forming parts of functions.

The *continuous* functions were the true functions. Thus, we gave quite a restricted sense to the word *function* because we believed that any continuous function, whether defined geometrically or not, could be represented by an analytic expression of the nature of those previously discussed, and we believed that it was impossible for the functions which were not continuous.

But Fourier showed that the trigonometric series, which represented the continuous functions, in the extended case, could also be used to represent discontinuous functions, formed of parts of functions. In particular, a function 0 from 0 to π, equal to 1 from π to 2π, admits a convergent trigonometric expansion. Thus, the only criterion enabling to distinguish the true functions from false ones disappeared. It was necessary, either to extend the sense of the word *function*, or to restrict the category of algebraic, trigonometric, exponential expressions that could be used to define functions.

Cauchy remarked that the difficulties that resulted from the research of Fourier arose even when we used very simple expressions, that is, the functions appeared as continuous or

[2] This continuity is known as Eulerian continuity.

discontinuous, according to the method used to define them. Cauchy cited, as an example, a function equal to $+x$ for x positive, to $-x$ for x negative. This function is not continuous, it is formed of parts of two continuous functions $+x$ and $-x$. It appears on the contrary as continuous when it is defined $+\sqrt{x^2}$.

To preserve the words *continuous function* in their original sense, it would therefore have been necessary to consider only very particular analytic expressions.[3] Cauchy preferred instead to modify the definitions considerably.

For Cauchy *y is a function of x when, to each state of magnitude of x, there corresponds a perfectly well-determined state of quantity of y.*

This definition appears to be the same as the one later given by Riemann. However, in reality, the correspondences that Cauchy considered are still those that can be established using analytic expressions. After defining functions, Cauchy adds: a functions is said to be explicit if the equation which links x to y is solved in y, and implicit if it is not. The fact that the correspondences are established using analytic expressions never appeared in reasonings of Cauchy, so that the properties and proofs obtained by Cauchy apply immediately to the functions satisfying the Riemann's definition.[4]

For Cauchy, *a function $f(x)$ is continuous at value x_0 if, for any positive number ε, we can find a number $\eta(\varepsilon)$ such that the inequality $|h| \leq \eta(\varepsilon)$ implies*

$$|f(x_0 + h) - f(x_0)| \leq \varepsilon;$$

the function $f(x)$ is continuous in (a, b) if the correspondence between ε and $\eta(\varepsilon)$ could be chosen independent of any number x_0 in, (a, b).

We recognize here the definitions that are now classic.

To prove the existence of primitive functions of continuous functions, it is sufficient to consider the geometric proof indicated previously. In this proof, we have appealed to the concept of area. This concept, already quite delicate when it comes to the domains bounded by simple geometric curves such as circles or ellipses, becomes even more delicate when it comes to domains involved in the proofs which concerns us.

The curves Γ which bound these domains are not necessarily geometric curves anymore, they could be formed of the parts of the geometric curves ($y = +\sqrt{x^2}$). Therefore, we know that these curves could be complicated without knowing where this complication stops. Also,

---

[3] This is what Méray did when he assigned to the word *function* a meaning very close to the one given to *continuous function* in the past. Méray defined the function through its Taylor series and its analytic continuation. By adopting of Méray definition, the existence of the primitive functions follows immediately from the properties of the entire series.

However, if we apply Méray definition to complex variable functions, we find ourselves necessarily led to, as M. Borel pointed out to me, consider the discontinuous functions of a real variable. For example, when a Taylor series is convergent on its circle of convergence, its values, on this circle, could define two real variable discontinuous functions.

[4] I do not mean by this, that the definition of Cauchy is less general than Riemann's but only that, if there existed functions satisfying Riemann's definition without satisfying Cauchy's, they would not be excluded from the reasoning.

Cauchy thought it necessary to clarify what should be understood by the number $S(x)$ in the previous proof. All he had to do for this was to repeat the operations that were ordinarily used to calculate the approximate value of $S(x)$ considered as area and to demonstrate that these calculations led to a limit.[5] We thus have the, now classical, proof of the existence of primitive functions.

Let (a, X) be the considered interval. Let us partition (a, X) into partial intervals using increasing numbers

$$a_0 = a, a_1, a_2, \dots, a_{n-1}, a_n = X;$$

and form the sum

$$S = (a_1 - a_0) f(x_1) + (a_2 - a_1) f(x_2) + \cdots + (a_n - a_{n-1}) f(x_n),$$

where x_i is any number included between a_{i-1} and a_i. We prove that S tends to a definite number $S(x)$ when the maximum of $a_i - a_{i-1}$ tends to zero in any manner.

The number $S(x)$ thus obtained is called *the definite integral of* the function $f(x)$ in the interval (a, X). Since Fourier, we represent it by the notation $\int_a^X f(x)\,dx$. This symbol up to now made sense only in positive intervals (a, X), $(X \geq a)$; by definition, we set

$$\int_a^X f(x)\,dx + \int_X^a f(x)\,dx = 0.$$

It is obvious that we have, for any a, b, c,

$$\int_a^b + \int_b^c + \int_c^a = 0.$$

Let us again remark that if L and l are upper and lower limits of $f(x)$ in (a, b), $\int_a^b f(x)\,dx$ is bounded between $L(b-a)$ and $l(b-a)$. The continuous function $f(x)$ takes all the values between l and L, it includes the values l and L, we can write

$$\int_a^b f(x)\,dx = (b-a) f(\xi),$$

[5] Very often, in mathematics, we take, as Cauchy did here, the procedure of calculation of a number as its very definition. So much so that some mathematicians do not accept other definition of a number than the one that permits its calculation.

The value of such a straightforward approach, as demonstrated by Cauchy, lies in its ability to reduce the number of prerequisites needed to start our reasoning. It is said that Descarte brought down geometry to algebra; however, would not have been true, if Cauchy, by his definition of the integral, had not given a logical construction of the concepts hitherto deduced from geometric intuition—such as area, the volumes, etc.

Then, there was a progress of immense philosophical importance; but, as the work of Cauchy did not bring any enrichment to the concept of integral, its mathematical interest is little. For this reason, Cauchy presented it primarily as a pedagogical tool.

ξ being included between a and b[6]; this is the mean value theorem.

The number $S(x)$ now being defined in a precise manner, we prove the existence of the primitive function of $f(x)$ without difficulty. In fact, we have

$$\frac{S(x_0+h)-S(x_0)}{h}=\frac{1}{h}\int_{x_0}^{x_0+h} f(x)\,dx = f(x_0+\theta h)$$

equality which proves that the function $S(x)$ is continuous and has $f(x)$ as its derivative.

The function $S(x)$ which figures in the previous proof or more precisely the function

$$S(X)+K=K+\int_a^X f(x)\,dx = K_1+\int_\alpha^X f(x)\,dx,$$

in which K and K_1 are the arbitrary constants and α a value of x taken in the interval of definition of $f(x)$, is called *the indefinite integral* of the function $f(x)$ and is denoted by $\int f(x)\,dx$. We see that the indefinite integral of the function $f(x)$ is the most general function $F(x)$ such that we have, for any α and β in the interval of definition of $f(x)$,

$$F(\beta)-F(\alpha)=\int_\alpha^\beta f(x)\,dx. \tag{2.1}$$

We also see that, for continuous functions, there is an identity between the indefinite integrals and the primitive functions.[7]

2.2 Integration of the Discontinuous Functions

In what preceded, the definite integral appeared as an element that enabled us to calculate the primitive function. On the contrary, in practise, the primitive functions serve to calculate the definite integrals. These definite integrals are limits of sums. Their number of terms increases indefinitely while the absolute values of these terms tend to zero. They are encountered in a large number of questions of analysis, geometry and mechanics.[8] For the calculation

---

[6] This proof does not exclude the equality $\xi=a, \xi=b$. In some cases it is good to prove that we could choose ξ different from a and b; the proof is immediate.

The considered theorem is the mean value theorem for the function

$$S(x)=\text{const.}+\int_a^x f(x)\,dx;$$

it provides, in fact, an expression of increment $S(b)-S(a)$ undergone by $S(x)$ when we pass from a to b.

[7] This would not hold true if we had not introduced the constant K in the definition of the indefinite integral.

[8] The simplest application of the concept of integrals is in the quadrature of plane domains. Because of this application, we often track the concept of definite integral back to Archimedes and to the

of some of these limits of sums, for example, for the definition and the calculation of the area bounded between a curve and its asymptotes, the integration of the continuous functions alone is not sufficient anymore. We are thus led to focus on the integration of functions that are infinite at some points or in the neighborhood of some points. On the other hand, in certain applications of definite integrals—such as calculating the coefficients of the trigonometric series representing a given function—it seems advantageous to define the integral of a function which, while remaining finite, is discontinuous at some points. Ever since the introduction of the notion of the definite integral, this concept has been extended to certain discontinuous functions.

We were led to the definition that will be given later by assuming as a principle the identity, observed in the case of continuous functions, between the indefinite integral and the primitive function. Let us consider the function $f(x)$ which, for $x \neq 0$, is equal to $\frac{1}{\sqrt[3]{x}}$. The only *continuous* functions which admits, except for $x = 0$, a derivative equal to $f(x)$ are given by the formula $K + \frac{3}{2}\sqrt[3]{x^2}$. We have said that $F(x) = \frac{3}{2}\sqrt[3]{x^2}$ was the indefinite integral of $f(x)$, and agreed that the formula (2.1) would give the definite integral of $f(x)$ in an arbitrary interval (α, β).

Let again the function $f(x)$ (considered by Fourier) be equal to -1 for x negative, to $+1$ for x positive.[9] The only *continuous* functions that admit $f(x)$ as its derivative, except for the singular value $x = 0$, are the functions (considered by Cauchy) $K + \sqrt{x^2}$. If we consider these functions as indefinite integrals, we deduce, by formula (2.1), the value of the definite integral of $f(x)$ in any interval.[10]

Cauchy stated in a very precise manner the definition, of which, two applications have just been seen. For him, *if a function $f(x)$ is continuous in an interval (a, b), except at a point c, in the neighborhood of which $f(x)$ is bounded or not,*[11] *we can define the integral of $f(x)$ in (a, b) if the two integrals*

$$\int_a^{c-h} f(x)\,dx \quad \text{and} \quad \int_{c+h}^{b} f(x)\,dx$$

tend to a definite limit when h (positive) tends to zero, then we have by definition

---

quadrature of the parabola. It is true that a lot of quadrature had been done before the introduction of integral calculus, but the mathematicians did not attach any particular importance to special domains whose areas must be calculated to obtain definite integrals. The importance of these domains became quite evident only after the introduction of the concept of the derivative.

[9] This function, not defined for $x = 0$, admits, as we know, a trigonometric expansion. We could also denote it by $\frac{\pm\sqrt{x^2}}{x}$.

[10] It is good to add that the definite integrals, that we could thus attach to the two kinds of discontinuous functions we consider, permit us to express the coefficients of trigonometric expansion of the functions using the formulae of Euler and Fourier which are applied in case of continuous functions.

[11] Cauchy did not deal with the values of the function for point $x = c$. Moreover, for him, if $f(x)$ tends to a definite value when x tends to c, this limit value is $f(c)$. If $f(x)$ does not tend to a unique limit, $f(c)$ is any one of the values included between the smallest and the largest of the limits of $f(x)$. In some Memoirs, P. Du Bois Reymond has taken these conventions.

$$\int_a^b f(x)\,dx = \lim_{h\to 0}\left[\int_a^{c-h} f(x)\,dx + \int_{c+h}^b f(x)\,dx\right].$$

If[12] in (a, b) there exist several points of discontinuity, we partition (a, b) into a sufficient number of partial intervals so that, in each of them, there exist not more than one singular point, if that is possible. Then we take the sum of the numbers thus obtained.

These definitions are linked to the known criteria regarding the existence of integrals of functions that are infinite around a point.

For the research related to the theory of functions and in particular for the study of trigonometric series, Lejeune–Dirichlet extended the concept of trigonometric series. The research of Lejeune–Dirichlet, which he stated himself, has never been published. But, according to Lipschitz, it can be summed up as follows.

Let a function $f(x)$ be defined on a finite interval (a, b), in which it is to be integrated. Let e be the set of points of discontinuity of $f(x)$. If e contains only a finite number of points, we would apply the definitions of Cauchy.

According to Lipschitz, the case that Dirichlet studied is the one where the derived set e' of e contains only a finite number of points,[13] as it is seen in the example, for the function $\frac{1}{sin\frac{1}{x}}$, where e' contains only $x = 0$.

The points of e' then divide (a, b) into a finite number of partial intervals. Let (α, β) be one of them. In $(\alpha + h, \beta - k)$, there are only a finite number of points of e. If, in this interval, the definitions of Cauchy are not applicable, we say that the function does not have an integral in (a, b). If, on the contrary, they are applicable, we consider the integral $\int_{\alpha+h}^{\beta-k} f(x)\,dx$ and we simultaneously let h and k tend to zero according to an arbitrary. If we do not obtain a definite limit, $f(x)$ does not have an integral in (a, b). If, on the contrary, we have a definite limit, we set

$$\int_\alpha^\beta f(x)\,dx = \lim_{\substack{h\to 0\\ k\to 0}} \int_{\alpha+h}^{\beta-k} f(x)\,dx.$$

The integral in (a, b) is, by definition, the sum of the integrals in the intervals (α, β).

We see that the definition of Dirichlet is based on the same principles as that of Cauchy. The general definition which evolves from these principles can be stated thus:

A function $f(x)$ has an integral in a finite interval (a, b) if there exists in (a, b) one and only one continuous function $F(x)$, up to an additive constant, such that we have

---

[12] Cauchy also dealt with the case where the right-hand side of this equality would have a meaning, without the two integrals which appear in it having the limits. In this case, he called this right-hand side the *principal value of the integral* $\int_a^b f(x)\,dx$.

[13] For the definition of derived sets and the properties of the reducible sets, *see* the Note at the end of the volume. In the recently given translation of Memoir of Lipschitz (*Acta mat.*,t.36), M. Montel observes that Lipschitz explicitly admitted only that the derived set is not dense (see below). However, Lipschitz believed he could conclude that it contained only a finite number of points. This error further obscured the intended meaning of Lipschitz's text.

$$\int_{\alpha}^{\beta} f(x)\,dx = F(\beta) - F(\alpha), \tag{2.2}$$

in any interval where $f(x)$ is continuous. $F(x)$ is the indefinite integral of $f(x)$ and we set

$$\int_{a}^{b} f(x)\,dx = F(b) - F(a).$$

For this definition to be applicable, it is first necessary that there exists a continuous function $F(x)$ satisfying the formula (2.2). This amounts, in both cases treated by Cauchy and Dirichlet, to assuming the existence of the limits used in the definition. We assume this condition satisfied and we would seek how should the singular points of $f(x)$ be distributed for this function to have an integral. From the point of view which concerns us, the singular points of $f(x)$ are those which are not interior to any interval in which $f(x)$ is continuous. Therefore, these are the points of e and of e'. These points form a set which we denote by E. Any limit point of points of E, by its very definition, is also a point of E. Therefore, E contains all its limit points. These are the sets that Jordan called *perfect* and M. Borel called *relatively perfect*. We would call such sets as *closed sets*, in conformity with the usage which is now universal.

For the formula (2.2) to completely define $F(x)$, it is necessary that, in any interval, there exists another interval where $f(x)$ is continuous. Therefore, the set E must be such that, in any interval, we could find an interval that does not contain points of E. This is what we mean by saying that E must be non-dense in every interval.[14]

This property of E is by no means sufficient. For stating the necessary and sufficient property that E must satisfy, it is necessary to resort to the properties of the derived sets.

The closed set E has successive derived sets E', E'', ..., E^{ω}, ... We know that, if one of the derived sets is empty, E is called reducible-it is a countable set. Otherwise, one of the derived sets is perfect, E and any of its derived sets have the cardinality of the continuum.[15]

These are the properties which will be useful to us. Let us suppose that there exists a function $F(x)$ satisfying the equality (2.1) in all intervals where $f(x)$ is continuous and let us investigate if $F(x)$ is well-determined; when this is the case, the equality (2.1) would be used in the definition of the integral.

We will rely on this obvious remark: if the integral $\int_{\alpha}^{\beta} f(x)\,dx$, which appears in the left-hand side of (2.1), has a meaning in all the intervals which do not contain any of the

[14] P. Du Bois Reymond, to whom we owe the distinction between two remarkable classes of sets—those we call *sets dense* in every interval on one hand, and *sets non-dense* in every interval on the other—referred to the former as *systèmes pantachiques or pantachies* and the latter as *systèmes apantachiques or apantachies*. It was also Du Bois Reymond who gave the general method of formation of closed sets and the apantachies. This procedure consists of removing a suitably chosen finite or countable number of intervals from an interval. On the subject of closed sets and of non-dense sets *see* BOREL, *Leçons sur la theorie des fonctions, Chap. III*.

[15] *See* the Note at the end of the Volume.

finite points $x_1, x_2, \dots, x_n$, the different continuous functions $F(x)$ satisfying the equality (2.2) could differ only by a constant.

If E contains, only a finite number of points, $F(x)$ is therefore well-determined, and hence the definition of Cauchy.

The left-hand side of (2.2) now has a meaning in any interval not containing points of E'; therefore, if E' has only a finite number of points, $F(x)$ is well-determined, and hence the definition of Dirichlet–Lipschitz.

We pass from there to the case where either E'', or E''', ..., or E^n contains only a finite number of points.

In any interval where E^ω does not have points, $F(x)$ is well-determined.[16] As a result, the left-hand side of (2.2) has a meaning in such an interval. From there we conclude that $F(x)$ is well-determined when E^ω has only a finite number of points. As a result, we pass to the case where either $E^{\omega+1}$, or $E^{\omega+2}, \dots$ have only a finite number of points and then to the case where $E^{2\omega}$ possesses this property, and so on.

Thus, we see that, if E is reducible, $F(x)$ is well-determined, so that our definition is applicable. Then, there exists an integral obtained by repeated application of the method of Cauchy–Dirichlet.

To have examples of such functions on which this method is applicable, it is sufficient to take a reducible set E, arrange its points in a simply infinite sequence, $x_1, x_2, \dots$, and form the series

$$f(x) = \sin\frac{1}{x - x_1} + \frac{1}{2}\sin\frac{1}{x - x_2} + \cdots + \frac{1}{2^p}\sin\frac{1}{x - x_{p+1}} + \cdots .$$

Now[17] suppose that the set E of singular points of $f(x)$ is not reducible. We would see that if there exists a function $F(x)$ satisfying the condition (2.2) in any interval where $f(x)$ is continuous, there exist an infinite such $F(x)$.

Let E^α be that derived set of E which is perfect. E^α is obtained by removing from the considered interval (a, b), the points *interior* to the intervals $\delta_1, \delta_2, \dots$, which form a countable sequence if E is non-dense in every interval. This is the only case we are interested in.

Let us define a function $\varphi(x)$ by the condition of being zero for $x = a$, equal to 1 for $x = b$. At all the points of δ_1, $\varphi(x) = \frac{1}{2}$. At all the points of δ_2, $\varphi(x) = \frac{1}{4}$, if δ_2 is between a and δ_1; and $\varphi(x) = \frac{3}{4}$, if δ_2 is between δ_1 and b. In a general way, having assigned to $\varphi(x)$, in $\delta_1, \delta_2, \dots, \delta_{n-1}$, the values $\Delta_1, \Delta_2, \dots, \Delta_{n-1}$, we assign to $\varphi(x)$, in δ_n, the value $\frac{\Delta_i + \Delta_j}{2}$, i and j being the indices of those of two intervals which contain δ_n.

Any point of E^α is the limit point of some intervals δ_n. It is easy to see that if the points of $\delta_{\alpha_1}, \delta_{\alpha_2}, \dots$ tend to x, $\Delta_{\alpha_1}, \Delta_{\alpha_2}, \dots$ tend to a definite limit. We take this limit as the value

[16] Because, in such an interval, one of the sets E^n has only a finite number of points.

[17] According to the properties of uniformly convergent series, $f(x)$ has all points of E as its points of discontinuity. We easily see that the previous series is term-by-term integrable.

For examples of reducible sets, *see* the Note at the end of the volume.

of $\varphi(x)$. Thus, $\varphi(x)$ is determined everywhere. It is a non-constant continuous function in (a, b). However, it is constant in any interval not containing points of E. So, there exists a function $F(x)$, satisfying equality (2.2) in any interval where there are no points of E. $F(x) + \varphi(x)$ also satisfies this condition.

Now, if we note that the sets which (Sect. 2.2) are denoted by E and e are reducible simultaneously,[18] we see that *for the adopted definition to be applicable, it is necessary and sufficient that the set of points of discontinuity of the function to be integrated* $f(x)$ *be reducible and that there exists a continuous function* $F(x)$ *satisfying* (2.2) *in the intervals where* $f(x)$ *is continuous.*

[18] It is quite necessary to note that e can be countable without E being countable; in such a case, e is a countable yet non-reducible set. Such is the case of the set of rational numbers.

The Definition of Integral Given by Riemann

3

3.1 Properties Relating to Functions

The functions on which the previous definitions apply can have an infinite number of points of discontinuity; but these points are still exceptional, in the sense that they form a non-dense set. Dirichlet incidentally encountered the function

$$\chi(x) = \lim_{m\to\infty} [\lim_{n\to\infty} (\cos \mathrm{m}!\pi x)^{2n}],$$

which is discontinuous at every point, since it is zero for x irrational, equal to 1 for x rational. The considerations of Cauchy and Dirichlet therefore do not apply to all the functions in the sense of Cauchy. Riemann[1] showed, by an example, how the use of the series would permit to construct the functions whose points of discontinuity form an everywhere dense set. Therefore on such functions, the previous definitions cannot apply.

Let (x) be the difference between x and its nearest integer; if x is equal to an integer plus $\frac{1}{2}$, we take $(x) = 0$. The function thus defined is called *excess of* x; it is a function in the sense of Cauchy because it admits a Fourier expansion, proceeding along the trigonometric lines of multiples of $2\pi x$, which is everywhere convergent. Let us consider the function, in the sense of Cauchy,

$$f(x) = \frac{(x)}{1^2} + \frac{(2x)}{2^2} + \frac{(3x)}{3^2} + \cdots;$$

we immediately see that if x is not of the form $\frac{2p+1}{2n}$ (n and $2p+1$ being co-prime) $f(x)$ is continuous.[2] On the contrary, if x is of the indicated form, when x increases to $\frac{2p+1}{2n}$, $f(x)$ converges to a limit which we denote

[1] On the possibility of representing a function by a trigonometric series (*Bulletin des Sciences Mathematique,* 1873, *OEuvres des Riemann*).

[2] We would rely on the uniform convergence of the series representing $f(x)$.

R. Jain, *Lebesgue's Theory of Integration*,
https://doi.org/10.1007/978-981-96-1169-0_3

$$f\left(\frac{2p+1}{2n}-0\right)$$

and[3] which is

$$f\left(\frac{2p+1}{2n}-0\right)=f\left(\frac{2p+1}{2n}\right)+\frac{\pi^2}{16n^2};$$

when x decreases to $\frac{2p+1}{2n}$, $f(x)$ converges to

$$f\left(\frac{2p+1}{2n}+0\right)=f\left(\frac{2p+1}{2n}\right)-\frac{\pi^2}{16n^2}.$$

In any interval, $f(x)$ has a point of discontinuity. Therefore, the considerations of previous chapter are not applicable to $f(x)$.

By using a procedure similar to that of Riemann, it was possible to form a number of examples of functions that are very discontinuous. By using the now classic concept of uniformly convergent series, it is easy to give a general statement: a uniformly convergent series of discontinuous functions f_n defines a function f which admits for points of discontinuity all the points of discontinuity of functions f_n, provided that each of these points is a point of discontinuity for only one single function f_n. When it is not so, as in the example of Riemann, it is necessary to find if the different discontinuities, that we encounter for the considered values, do not compensate themselves in such a manner that f becomes continuous.

We often have the opportunity of applying a similar procedure when, knowing the functions f_n which exhibit a certain singularity at the isolated points A_n, we want to construct a function that exhibits this singularity in the whole interval. We try to obtain the desired result by taking a uniformly convergent series of functions f_n, such that the corresponding points A_n form an everywhere dense set. This method of construction is named *principle of condensation of singularities*.[4]

The example of Riemann shows that the functions, to which the definition methods examined in the previous chapter cannot be applied, do not constitute a very particular class in the set of functions in the sense of Cauchy. And since the restriction[5] imposed, according to Cauchy, on the functions $f(x)$, namely the relation between $f(x)$ and x is analytically expressible, never played a role in our reasoning. It neither simplified the statements, nor the solutions of the problems that we have proposed. Therefore there is no disadvantage in saying, according to Riemann: *y is a function of x if, to each value of x, there corresponds a well-determined value of y, regardless of the method which led to establish this correspondence.* It is this definition that we would now adopt. Only, instead of always supposing

[3] This notation is due to Dirichlet.

[4] This proof is due to Hankel. Hankel believed that he could make general reasoning about this method. However, what we must retain is limited to the immediate applications of the known properties of uniformly convergent series.

[5] I do not have to inquire here if this restriction is real or deceptive.

that x can be taken arbitrarily in an interval (a, b), we would, at times, suppose that x must be taken in a certain set E, at the points of which the function y is thus defined, without it being defined for all the points of interval. For example, the function $\left[\frac{1}{x}\right]!$ is defined for the set of inverses of the positive integers.

Before taking up the study of the integration of functions in the sense of Riemann, I would give those properties of the functions which would be useful to us in what follows. If we know that a function always remains confined between the two finite numbers A and B, we say that it is bounded.[6] We would mostly limit ourselves to the study of bounded functions[7]. When the function is bounded, it admits an *upper limit* L and a *lower limit* l. These numbers are defined, as we know, by the condition that (l, L) is the smallest interval containing all the values of $f(x)$. $\omega = L - l$ is said to be the *oscillation of* $f(x)$.

Let A be a limit point of the set E on which $f(x)$ is defined.[8] Let δ_1 be an interval containing A; in this interval there exist points of E; they form a set e_1. The function $f(x)$ defined on e_1 admits the upper and lower limits, L_1, l_1, an oscillation ω_1. Let δ_2 be an interval containing A and contained in δ_1, the numbers corresponding to δ_2 are L_2, l_2, ω_2; and we obviously have

$$l_1 \leq l_2 \leq L_2 \leq L_1, \quad L_2 - l_2 = \omega_2 \leq \omega_1.$$

If we consider the intervals $\delta_1, \delta_2, \delta_3, \ldots$ all containing A, one containing the others, and the lengths of which converges to zero, we have a sequence of upper limits and lower limits satisfying the inequality

$$l_1 \leq l_2 \leq l_3 \leq \cdots \leq L_3 \leq L_2 \leq L_1$$

The l_i on one hand, the L_i on the other, converge therefore to the two limits l and L $(l \leq L)$ and the ω_i converge to

$$\omega = L - l.$$

We would see that the numbers, L, l, ω, thus obtained, are also the limits of numbers L'_i, l'_i, ω'_i, corresponding to the intervals δ' containing A and whose two extremities converge to A in any manner as i goes to infinity. In other words, they are independent of the choice of the intervals δ_i and we can assume that these intervals are not necessarily contained in

---

[6] It is well understood that an unbounded function can however be always finite; such is the case with function

$$f(0) = 0, \quad f(x) = \frac{1}{x} \quad \text{for} \quad x \neq 0.$$

If we only know that a function is always smaller than a fixed number, we would say that it is bounded from above.

[7] We often find that the questions are very simple to treat when they are limited to the bounded functions, whereas they tend to become very complicated for most general functions. I have carefully indicated in the following if the theorems obtained are valid for all functions or only for the bounded functions. Ofter, we omit to indicate explicitly that the functions considered are bounded.

[8] A does not necessarily belong to E.

one another. In fact, i being chosen arbitrarily, if j is sufficiently large, δ'_j is contained in δ_i; if k is sufficiently large, δ_k is contained in δ'_j; therefore we have

$$l_i \leq l'_j \leq l_k \leq L_k \leq L'_j \leq L_i,$$

which is sufficient to prove the property.

The numbers L, l, ω are called the maximum or the upper limit, the minimum or the lower limit and the oscillation of the function at A. A is a point of continuity or of discontinuity, according to ω is zero or positive, that is, according to L and l are equal or unequal.[9]

If x_0 is the abscissa of A and if we agree to consider only the values of x greater than x_0, $(x > x_0)$, we obtain the maximum M_d, the minimum m_d and the oscillation ω_d *from the right side of A, beyond A*. If $\omega_d = 0$, that is, if $M_d = m_d$, $f(x_0 + 0)$, exists and is equal to M_d; *the function is said to be right continuous at point x_0 to the right of x_0*. If $M_d = m_d = f(x_0)$, the function $f(x)$ is said to be *right continuous at point x_0*. We similarly define the numbers M_g, m_g, ω_g.[10]

If ω_d and ω_g are zero, that is, if $f(x_0 + 0)$ and $f(x_0 - 0)$ exist, the discontinuity is said to be of *first kind* otherwise it is said to be of *second kind*.

All these definitions could be given for the unbounded functions; nothing would change, except that the numbers defined would not be necessarily finite anymore.

To the previous concept, we can attach the concept of *limit of indetermination* which would be often useful to us. This concept is due to P. Du Bois Reymond.

A method of calculation gives, in some conditions, a definite number φ. In the other conditions, on the contrary, it no longer gives a definite number, but depending on how it is applied, it gives different numbers forming a set A. We can then, either say that the method no longer gives a number, or say that the method gives one of the numbers of A for φ. The number φ is thus considered as indetermined. The smallest interval which contains all the points of A, either in its interior, or coinciding with its extremities, has its origin and its extremity as*the lower limit and the upper limit of indetermination* of numbers φ. These limits may be finite or infinite, they do not necessarily make part of A. For example, we give the expression

$$\varphi = \lim_{n \to \infty} x^n,$$

where n is integer. φ is zero for $|x| < 1$; for calculating φ in this case we can choose arbitrarily a sequence of increasing integers $n_1, n_2, \ldots$ and take the limit of the corresponding sequence x^{n_i}. If x is no longer included between -1 and $+1$, by so operating and suitably choosing the

[9] These definitions are related to very important concepts of lower semi-continuous functions and upper semi-continuous functions introduced by M. Baire. These are the functions f which are equal at each point, respectively, to the number l or L attached to f at this point.

[10] The previous definition is that of maximum, minimum, oscillation of $f(x)$ to the right of x_0, x_0 being excluded. We also often consider similar numbers, including x_0; it is necessary, then, to take the values of x equal to or greater than x_0, $(x \geq x_0)$. It is this method of operating that is linked to the concept of *right-continuity at point x_0*.

integers n_i, we could still have limit, but this limit would depend sometimes on the choice of the integers n_i. For $x = -1$, the set A contains only $+\infty$ and $-\infty$,[11] which are the two limits of indetermination.

For $x = 1$, φ is equal to 1. For $x > 1$, φ is equal to $+\infty$. The concept of the limits of indetermination can often be replaced by the simpler concept of *smallest* and *largest limit*, a concept that we attribute to Cauchy.

Let us suppose that the number φ is defined as the limit of a function $\psi(\lambda)$ for $\lambda = \lambda_0$; λ takes either all the possible values or only those from a certain set of which λ_0 is a limit point (the previous example reduces to this case if we take $\lambda = \frac{1}{n}$, where n is integer, and $\lambda_0 = 0$). The function $\psi(\lambda)$ is not defined for $\lambda = \lambda_0$, but we know that it has *a minimum or limit inferior l and a maximum or limit superior L*[12] *for* $\lambda = \lambda_0$; these numbers, finite or not, are respectively *the smallest and the largest of the limits* that we can obtain when, in $\psi(\lambda)$, *we let* λ converge to λ_0. l and L are the two limits of indetermination defined previously; but, in the case we deal with, these numbers are contained in the set A of limit values, while, in the general case, they either make part of A or of the derived set A' of A.

But it is also possible, and we shall soon see examples, that the function $\psi(\lambda)$ is no longer a well-determined function, but a *function of several determinations*.

We say that we have such a function if, to each value of λ, taken from a set where the function is defined, corresponds a set of numbers; each of these numbers is represented by notation $\psi(\lambda)$. What has been said in relation to the limit superior and the limit inferior for the single valued functions, is applicable without any change to the functions of several determinations. $\psi(\lambda)$ therefore has *a limit inferior l and a limit superior L for* $\lambda = \lambda_0$, which are respectively, *the smallest and the greatest of the limits* that one can attain by choosing a sequence of numbers λ_i converging to λ_0 and by suitably choosing the corresponding numbers $\psi(\lambda_i)$. These two numbers are *the limits of indetermination of the limit of $\psi(\lambda)$ when λ converges to λ_0.*[13]

Now, let us go back to the study of functions.

There is a very simple relation between the oscillations relative to the intervals contained in (a, b) and the oscillations at the different points of (a, b). We can express it thus:

If, at every point of (a, b), the oscillation is at most equal to ω, then in any interval of length λ, interior to (a, b), the oscillation is less than $\omega + \varepsilon$ when λ is sufficiently small; ε being any positive number.

---

[11] **Translator's note:** The set A contains only $+1$ and -1.

[12] These denominations, limit inferior and limit superior, are the ones adopted by M. J. Hadamard.

[13] Du Bois Reymond simply said $\ll$ the limits of indetermination of $\psi(\lambda)$ for $\lambda = \lambda_0 \gg$. This idea of defining the value of function at a point of discontinuity is due to Du Bois Reymond (note 1, Sect. 2.2). I believe that it is better to adopt the language of text, more in conformity with the modern ideas on the determination of functions.

The multiple-valued functions, or the multiple-form functions, have been very less studied up to now. The only work of any extent concerning them is the recently submitted before the Faculté des Science de Paris by M. F. Vasilesco.

If it was otherwise, we could find a couple of points a_p, b_p, such that $b_p - a_p$ converges to zero and we had

$$|f(b_p) - f(a_p)| > \omega + \varepsilon.$$

The set of numbers a_p, have at least, one limit point α. If we take a sequence of numbers a_p converging to α, the numbers b_p also converge to α, therefore at α the oscillation is at least $\omega + \varepsilon$. There is, then, a contradiction with the hypothesis.

The property is proved. In case where $\omega = 0$, it reduces to a well known fact: a function continuous at all the points of an interval is continuous in this interval.[14]

The converse of our property is not true. Let a function be equal to -1 for x negative, to $+1$ for x positive, zero for x zero. Its oscillation at $x = 0$ is 2 and, however, if we use the point of division $x = 0$, the function has an oscillation equal to 1 in each of the intervals obtained.

We would now define the mean oscillation of a bounded function $f(x)$ defined on a finite interval (a, b). Let us partition the interval (a, b) into partial intervals $\delta_1, \delta_2, \ldots, \delta_n$. Let ω_i be the oscillation of $f(x)$ in the interval δ_i, the extremities of the δ_i may or may not be considered as making part of the interval. And let us form the following quantity

$$A = \frac{\delta_1\omega_1 + \delta_2\omega_2 + \cdots + \delta_n\omega_n}{b - a}$$

If Ω is the oscillation of $f(x)$ in (a, b), $\omega_1, \omega_2, \ldots, \omega_n$ being at most equal to Ω, A is at most equal to Ω. Therefore, if we divide δ_i into partial intervals $\delta_i^1, \delta_i^2, \ldots, \delta_i^{p_i}$ at which the corresponding oscillations are $\omega_i^1, \omega_i^2, \ldots, \omega_i^{p_i}$, we have

$$\sum_{j=1}^{j=p_i} \delta_i^j \omega_i^j \leq \delta_i \omega_i$$

By subdividing the intervals δ_i we therefore replace A by a smaller or at most equal number.

Let us consider two series of divisions of (a, b) into partial intervals; to the divisions, D_i, of the first series correspond the numbers $A_1, A_2, \ldots$, to, Δ_i, of the second, the numbers $\alpha_1, \alpha_2, \ldots$. Let us suppose that, for each of the two series, the maximum of the length of the intervals used in the ith division converges to zero as $\frac{1}{i}$[15]; in these conditions we would see that A_i and α_i have the same limit.

Let us compare A_i and α_j. The intervals of the division Δ_j are of two types: some, the intervals d, containing the points of the division D_i in their interior; while the others, the

---

[14] It is this property that we state: the continuity is uniform. We express it by saying that the quantity $\eta(\varepsilon)$ can be chosen uniformly in the considered interval, that is, independent of the variable x; see Sect. 2.1.

The general theorem that we have proved here is due to M. Baire.

[15] The points of division used in the ith division are not necessarily used in the $(i + 1)$th; in other words, for passing from one division to the next, we do not sub-divide the intervals of this division; we mark the new intervals without worrying about those which are previously used.

intervals d', are entirely contained in the intervals of D_i. The contribution of the intervals d to the numerator of α_j is at most $n\lambda_j\Omega$, if n is the number of points of division of D_i and λ_j the maximum of the length among the intervals of Δ_j. The intervals d' are part of the division Δ'_j, obtained by combining the points of division of D_i and Δ_j, therefore their contribution to the numerator of α_j is at most equal to $(b-a)A'_j$, where A'_j is the number analogous to A and related to Δ'_j. But, since we know that A'_j is at most equal to A_i, we deduce

$$\alpha_j \leq A_i + n\lambda_j\Omega.$$

All the α_j, starting from a certain index, are smaller than $A_i + \varepsilon$, $(\varepsilon > 0)$; therefore their largest limit is at most $A_i + \varepsilon$ and, since i and ε are arbitrary, the largest limit of α_j is at most equal to the smallest of A_i. Nothing would stop us from interchanging A_i and α_j in the reasoning. Therefore, all the limits of A_i and of α_j are equal, A_i converges to a definite limit. This limit ω is *the mean oscillation of the function in* (a, b).

It is necessary to note what we have proved: A_i converges uniformly to ω; that is, when the length of all the intervals are smaller than a certain number λ, the number A differs from ω only by a quantity smaller than ε chosen in advance.

3.2 Conditions of Integrability

Given these definitions, let us arrive at the definition of the integral as given by Riemann.

Riemann brought his attention on the operative procedure which allows to calculate the integral with as close an approximation as we want, in case of continuous functions. Then, he asked, in what cases does this procedure give a definite number,[16] when applied to the discontinuous functions.

Let $f(x)$ be a bounded function defined on a finite interval (a, b). Let us divide the interval (a, b) into partial intervals $\delta_1, \delta_2, \ldots, \delta_n$ and let us choose arbitrarily, for any i, a point x_i in δ_i or coinciding with one of the extremities of δ_i. Let us consider the sum

$$S = \delta_1 f(x_1) + \delta_2 f(x_2) + \cdots + \delta_n f(x_n).$$

[16] Cauchy applied his method of defining integral only to the functions which were considered *a priori* as 'interesting': the continuous functions. Now, on the contrary, any function would be considered interesting to which the method of definition is applied.

Hence, on one hand, a new classification of functions arises, and on the other, an enrichment of the concept of integral follows. If we compare the results of Cauchy and that of Riemann (note 1, Sect. 2.1), it is necessary to note the mathematical character of the progress made due to the later.

The way in which this progress was attained: delimiting the domain of application of a definition when we do not introduce *a priori* any restriction on its use, has frequently been used since a century.

Let us continuously increase the number of intervals δ and choose them in such a manner that the maximum of their length converges to zero.[17] Then, if S converges to a definite limit, independent of the intervals and of the chosen points x_i, Riemann said that the function $f(x)$ is *integrable* and has for integral, in (a, b), the limit of S.

When $\delta_1, \delta_2, \ldots, \delta_n$ are chosen, the number S is not completely determined; its lower limit and upper limit are:

$$\underline{S} = \sum l_i \delta_i, \quad \overline{S} = \sum L_i \delta_i,$$

where l_i and L_i represent the lower limit and upper limit of $f(x)$ in δ_i. Let us set $L_i - l_i = \omega_i$, then

$$\overline{S} - S \leq \overline{S} - \underline{S} = \sum \delta_i \omega_i.$$

For S to converge to a definite limit, it is first necessary that $\overline{S} - \underline{S}$ converges to zero. However, $\sum \delta_i \omega_i$ converges to $(b - a)\omega$, where ω is the mean oscillation of $f(x)$. Therefore, *for $f(x)$ to be integrable, it is necessary that it be of mean oscillation zero.*

This condition is sufficient. To prove it, it is sufficient to prove that $\overline{S}$ has a well determined limit, since $\overline{S} - S$ converge to zero. Let us suppose, to conduct this study, that we argue not on the function f, but on $f + k$, k being such a constant that $f + k$ *is never negative*.

Let, as previously, (Sect. 3.1), the two sequences of divisions $D_1, D_2, \ldots; \Delta_1, \Delta_2, \ldots$ be such that the maximum of the length of partial intervals converges in each sequence to zero. This maximum is λ_j for Δ_j. Let $\overline{S_1}, \overline{S_2}, \ldots; \overline{\Sigma_1}, \overline{\Sigma_2}, \ldots$ be the numbers similar to $\overline{S}$ corresponding to these divisions.

Let us compare $\overline{S_i}$ and $\overline{\Sigma_j}$. Let us partition the intervals of Δ_j into two categories, as it was done in the study of the mean oscillations (Sect. 3.1). The intervals d contribute, in $\overline{\Sigma_j}$, at most equal to $n\lambda_j L$, where L is the maximum of $f(x)$ in (a, b). All the intervals d' are part of Δ'_j, corresponding to the number $\overline{\Sigma'_j}$. Therefore, the contribution of the intervals d' in $\overline{\Sigma_j}$ is at most equal to $\overline{\Sigma'_j}$. But, Δ'_j is obtained by breaking up the intervals of D_i. In these conditions it is obvious, that $\overline{\Sigma'_j}$ is at most equal to $\overline{S_i}$. From all of this we get

$$\overline{\Sigma_j} \leq \overline{S_i} + nL\lambda_j.$$

From this inequality we conclude, as previously, that $\overline{S_i}$ and $\overline{\Sigma_j}$ have the same limit and even that they converge uniformly to this limit. The property is proved for $f + k$, therefore it is true for f, because, in passing from f to $f + k$, we increase all the sums $\overline{S}$ by $k(b - a)$.

In the following, it is important to note that we have proved the existence of a limit for $\overline{S}$, without making any assumption on the bounded function $f(x)$. The condition that $f(x)$ has mean oscillation zero appears only when, from the existence of a limit for $\overline{S}$, we deduced the existence of a limit for S.

We can transform the obtained integrability condition: it is necessary and sufficient that the sum $\sum \delta_i \omega_i$ converges to zero. It is equivalent to saying that the intervals δ_i, in which ω_i

[17] It is understood, of course that, in passing from one division to the next, we are not bound to use the points of division already used.

is greater than an arbitrarily chosen positive number ε, for sufficiently large i, have a total length λ as small as we want, because we have:

$$\lambda\varepsilon \leq \sum \delta_i \omega_i \leq (b-a-\lambda)\varepsilon + \lambda\Omega,$$

Ω being the oscillation of $f(x)$ in (a, b). We thus have the statement given by Riemann:

For a bounded function to be integrable over (a, b), it is necessary and sufficient that we could divide (a, b) into partial intervals such that the sum of the lengths of those intervals in which the oscillation is greater than ε, can be made as small as we want, for any $\varepsilon > 0$.

If such a division is possible, there exist an infinite number of them in any sequence of divisions such that the maximum length of partial intervals converges to zero, since for any such sequence, $\sum \delta_i \omega_i$ always converges to the same number.

From this property of $\sum \delta_i \omega_i$, it also follows that, if to a sequence of divisions of the given nature, there correspond the numbers $\underline{S}$ and $\overline{S}$ having the same limit, we can derive the integrability of the considered function.

The form given by Riemann to the condition of integrability shows that although the function is integrable, it does not highlight the role of the points of discontinuity of the function. Paul Du Bois Reymond has highlighted this role by a transformation of the condition of integrability. The statement of Paul Du Bois Reymond assumes that the definition of *integrable groups* is known.

A set of points of a straight line constitutes an integrable group, if the points of the set could be enclosed in a *finite* number of segments, the sum of lengths of which can be made as small as we want.[18]

A finite number of points constitute an integrable group, but the converse is not true.

Let us consider the set Z of points whose abscissa are given by the formula

$$x = \frac{a_1}{3} + \frac{a_2}{3^2} + \frac{a_3}{3^3} + \cdots,$$

In which all the a's are either 0 or 2. This set is obtained by removing from the interval $(0, 1)$, first the points interior to the interval $(\frac{1}{3}, \frac{2}{3})$, then the points interior to the intervals $\left(\frac{1}{3^2}, \frac{2}{3^2}\right)$ and $\left(\frac{2}{3} + \frac{1}{3^2}, \frac{2}{3} + \frac{2}{3^2}\right)$, then the points interior to the intervals $\left(\frac{1}{3^3}, \frac{2}{3^3}\right)$ and $\left(\frac{2}{3^2} + \frac{1}{3^3}, \frac{2}{3^2} + \frac{2}{3^3}\right)$ and $\left(\frac{2}{3} + \frac{1}{3^3}, \frac{2}{3} + \frac{2}{3^3}\right)$ and $\left(\frac{2}{3} + \frac{2}{3^2} + \frac{1}{3^3}, \frac{2}{3} + \frac{2}{3^2} + \frac{2}{3^3}\right), \cdots$ etc. Therefore we always divide each remaining interval in three equal parts and we remove the middle part. After n such operations, there remains 2^n intervals. These 2^n intervals can be used to enclose[19] the points of Z. However, they have a total length $\frac{2^n}{3^n}$. Therefore, Z is an

[18] We can, whenever we want, consider that a point is enclosed in an interval, either if it is in the interior of this interval or coincides with its extremities; or if it is in the interior of the interval, excluding the extremities. The two corresponding definitions of the integrable groups are obviously identical. For passing from the first to the second it is sufficient to stretch the intervals, and their two extremities, as little as desired.

[19] To 'enclose' is taken here in the larger sense.

integrable group. This construction of Z also shows that Z is perfect, therefore it has the cardinality of continuum.

It is obvious that the set formed by the union of points of two integrable groups is an integrable group.

Here is the statement of Du Bois Reymond:

For a bounded function to be integrable, it is necessary and sufficient that, for any $\varepsilon > 0$, the points where the oscillation is greater than ε form an integrable group.

Suppose f is integrable, then we can divide (a, b) into partial intervals such that those intervals in which the oscillation is greater than ε have a total length smaller than η. A point where the oscillation is greater than ε cannot belong to an interval where the oscillation is not greater than ε. Therefore, such a point is necessarily one of the points, which are used in the division of (a, b), or else it is in the intervals of length η. The points of division being finite in number, the points where the oscillation is greater than ε could be enclosed in a finite number of intervals of total length 2η, and, as η is arbitrary, they form an integrable group.

Conversely, we suppose that the points of oscillation greater than ε form an integrable group. We can therefore enclose them in a finite number of intervals of total length η. Let us use these intervals I for the division of (a, b) and let I' be the other intervals. In each I', there are no more points of oscillation greater than ε, each of these intervals can therefore be divided into partial intervals I'' in each of which the oscillation is at most 2ε. The only intervals, with oscillations greater than 2ε, are therefore some of the intervals I; their total length is at most η and that is sufficient, according to the *criterion* of Riemann, for asserting that f is integrable.

In the previous statement, we can replace the set $G(\varepsilon)$ of points where the oscillation is greater than ε by the set $G_1(\varepsilon)$ of points where the oscillation is not smaller than ε, because $G(\frac{\varepsilon}{2})$ contains $G_1(\varepsilon)$ which itself contains $G(\varepsilon)$.

The set $G_1(\varepsilon)$ enjoys one of the properties which would allow us one last transformation of the condition of integrability: $G_1(\varepsilon)$ is closed. Indeed, if A is a limit point of $G_1(\varepsilon)$, any interval containing A contains points of $G_1(\varepsilon)$ and f has an oscillation at least equal to ε in this interval.

For the new statement of the condition of integrability, I will introduce a concept that will be encountered later: that of a set of measure zero. This is a set whose points can be enclosed in a finite number or *a countably infinite number* of intervals total length of which can be made as small as we want.

A point, and an integrable group are examples of sets of measure zero. The set E formed by the union of a finite number or a countably infinite number of sets E_n of measure zero is obviously also of measure zero[20]; any countable set is of measure zero. It is enough to

---

[20] Because we can enclose E_n in a countably infinite number of intervals α_n of total length $\frac{\varepsilon}{2^{n+1}}$ and the set E, the union of sets E_n, could be enclosed in countably infinite number of intervals $\alpha_1, \alpha_2, \ldots$ of total length

$$\sum \frac{\varepsilon}{2^{n+1}} = \varepsilon.$$

show the difference between a set of measure zero and an integrable group: the first can be everywhere dense, the second is always non-dense.

Let $f(x)$ be an integrable function, its points of discontinuity are those of the set obtained by the union of the integrable groups $G(1), G(\frac{1}{2}), G(\frac{1}{3}), \dots$; therefore they form a set of measure zero.

Now, let $f(x)$ be a bounded function whose points of discontinuity form a set of measure zero. $G_1(\varepsilon)$, being a subset of this set, is of measure zero, and it is closed. We will show later that this is sufficient to assert that $G_1(\varepsilon)$ is an integrable group.[21] It follows that f is integrable. Therefore:

For a bounded function to be integrable, it is necessary and sufficient that the set of its points of discontinuity be of measure zero.[22]

As an example of discontinuous integrable function, Riemann cites the function

$$f(x) = \frac{(x)}{1} + \frac{(2x)}{4} + \frac{(3x)}{9} + \cdots$$

Its integrability follows from the fact that the only points of discontinuity, being of the form, $x = \frac{2p+1}{2n}$, form a countable set, therefore of measure zero; or still, from the fact that, the oscillation being $\frac{\pi^2}{8n^2}$ for $x = \frac{2p+1}{2n}$, the points at which the oscillation is greater than ε are in finite number.

To have a function of an uncountably infinite number of discontinuities, let us consider the set Z which has been defined previously (Sect. 3.1). The function $f(x)$, the characteristic function of Z, which is zero between 0 and 1 for all the points, except for the points of Z where it is equal to 1, is integrable. Its points of discontinuity form in fact the integrable group Z. Z, being perfect, is of the cardinality of the continuum.[23]

Now, if we want that, in any interval, there is a set of uncountable points of discontinuity, it is enough to apply the principle of condensation of singularities. We could consider, for example, the function

$$\varphi(x) = \frac{f(x)}{1} + \frac{f(\frac{x}{2})}{2^2} + \frac{f(\frac{x}{3})}{3^2} + \cdots$$

The only points of discontinuity are, according to the property of uniformly convergent series, those or some of those of functions $f(x), f(\frac{x}{2}), \dots$; therefore they form a set of measure zero and φ is integrable.

---

[21] See Sect. 8.2.

[22] In my thesis, I used this property as a sufficient condition for integrability. In the first edition of this book I observed that it was also a necessary condition. Independently, M. Vitali, obtained the converse (*Rend. del R.* 1st. *Lomb.*, série II, XXXVII, 1904).

[23] The two functions which precede are not integrable by the method of Cauchy–Dirichlet, since the set of their points of discontinuity is not reducible.

3.3 Properties of Integral

The reasoning which precedes is general, it allows us to show that:

A uniformly convergent series of integrable functions is an integrable function.

In fact, the points of discontinuity of the sum function belong to the set E formed by the points of discontinuity of different terms. The singular points of different terms form a set of measure zero, therefore E is of measure zero and the series represents an integrable function.

In particular, a sum of two terms being a series whose only first two terms are not identically zero, *the sum of two integrable functions is an integrable function.* Similarly, *the product of two integrable functions is integrable,* because the points of discontinuity of product are points of discontinuity for at least one of the factors. Also similarly, *if f is integrable and $\frac{1}{f}$ is bounded, $\frac{1}{f}$ is integrable; if f is integrable, the* m*th arithmetic root of f, if it exists, is integrable; if f is positive and integrable and φ is integrable, f^{φ} is integrable;* etc.

The composition $f(\varphi)$, applied to the integrable functions, can on contrary give a non integrable functions.

Let us take for f a function equal to 1 everywhere, except for $x = 0$, where it is zero. f has only one point of discontinuity and is therefore integrable. φ would be zero for x irrational and φ would be $\frac{1}{q}$ for $x = \frac{p}{q}$ (p and q being co-primes). φ is integrable since its points of discontinuity, being those of rational abscissa, form a countable set.

The function $f(\varphi)$ is the non-integrable function $\chi(x)$ of Dirichlet (Sect. 3.1), since all its points are points of discontinuity.

We can sharpen the first two theorems which have just been obtained. Let f and φ be two integrable functions; let us partition the interval where they are given in parts $\delta_1, \delta_2, \ldots, \delta_n$ in which we choose the values $x_1, x_2, \ldots, x_n$. We have

$$\sum \delta_i [f(x_i) + \varphi(x_i)] = \sum \delta_i f(x_i) + \sum \delta_i \varphi(x_i);$$

Now the three sums which appear in this equality are the approximate values of the integrals of $f + \varphi$, f, φ; therefore the integral of $f + \varphi$ is the sum of the integrals of f and φ.[24] In other words:

The integral of a sum is the sum of the integrals. We suppose, of course, that it is for a true sum, that is, sum of a finite number of terms and not of a series.

For arriving at the case of a uniformly convergent series, it would be advantageous for us to use the *theorem of means.*

Let $f(x)$ be a function bounded between l and L in (a, b). The integral of f is, as we know, the limit of the sum $S = \sum \delta_i f(x_i)$, but we have

---

[24] It suffices to slightly modify the wording in order to simultaneously demonstrate the integrability of $f + \varphi$, which is assumed to have been previously established in the text's exposition.

$$(b-a)l = \sum \delta_i l \leq \sum \delta_i f(x_i) \leq \sum \delta_i L = (b-a)L$$

Therefore S, and as a result its limit, the integral, is bounded between $(b-a)l$ and $(b-a)L$; the integral is therefore of the form $(b-a)\mu$, where μ lies between l and L; this is the theorem of means.

What distinguishes it from the finite increment theorem, proved for continuous functions, is that it is impossible for us to assert that μ is one of the values that f takes in (a, b).

From this theorem it follows that, if the modulus of f is smaller than ε, the integral of f is smaller in modulus than $|b-a|\varepsilon$.

This being said, let a function f be the sum of a uniformly convergent series of integrable functions

$$f = u_1 + u_2 + \cdots + u_n + \cdots$$

Let s_n be the sum of the first n terms, r_n be that of remaining terms, F, U_n, S_n, R_n be the integrals of f, u_n, s_n, r_n. We have

$$S_n = U_1 + U_2 + \cdots + U_n,$$

according to the theorem on the integration of a sum. This same theorem shows that

$$F = S_n + R_n.$$

However, when n is greater than n_1, r_n is smaller than ε in modulus, and R_n is smaller than $|b-a|\varepsilon$ in modulus. When n is greater than n_1, $|F - S_n|$ is smaller than $|b-a|\varepsilon$. The series $\sum U_n$ is therefore convergent to the sum F.

A uniformly convergent series of integrable functions is term-by-term integrable.

The previous theorems are proved only in the case where the interval (a, b) is a positive interval $(b > a)$, since the integral has been defined only in this case. We complete the definition as previously.

The integral in (a, b) is always denoted $\int_a^b f(x)\,dx$. The complementary definition is expressed by the equality

$$\int_b^a f(x)\,dx + \int_a^b f(x)\,dx = 0.$$

It is obvious that the previous theorems proved for the positive intervals are also true for the negative intervals.

I add that we immediately verify the following:

$$\int_a^b f(x)\,dx + \int_b^c f(x)\,dx + \int_c^a f(x)\,dx = 0.$$

3.4 Integration by Lower Sum and Upper Sum

The definition we are concerned with, has been obtained by applying the methods of integral calculus for continuous functions to discontinuous functions. We know that among bounded functions there exist non-integrable functions for which this method does not lead to a definite number. Nevertheless, we can use this procedure to attach two well defined numbers with each bounded function.

We have seen (Sect. 3.2) that the sums $\overline{S} = \sum \delta_i L_i$ converges to a definite limit when the numbers δ_i converge to zero in an arbitrary manner. This limit is one of the two numbers, which we are dealing with. We call it *upper sum integral* and we represent it by the symbol $\overline{\int_a^b} f(x)\,dx$ which is stated: upper sum integral from a to b of $f(x)$.

Similarly, we can prove the existence of a limit for the sums $\underline{S} = \sum \delta_i l_i$. Moreover, while studying the mean oscillations, we saw that $\sum \delta_i \omega_i$ converges to a definite limit $(b-a)\omega$ and as we have

$$\overline{S} - \underline{S} = \sum \delta_i \omega_i,$$

the existence of the limit of $\underline{S}$ is proved.[25] *This lower sum integral* would be denoted by $\underline{\int_a^b} f(x)\,dx$.

These two numbers have been defined for the first time, in a precise way, by Darboux.[26] To complete their definitions, given only for $b > a$, we set

$$\overline{\int_a^b} + \overline{\int_b^a} = 0, \quad \underline{\int_a^b} + \underline{\int_b^a} = 0.$$

It is necessary to remark that, in a negative interval, the upper sum integral is smaller than the lower sum integral.

We always have

$$\overline{\int_a^b} + \overline{\int_b^c} + \overline{\int_c^a} = 0, \quad \underline{\int_a^b} + \underline{\int_b^c} + \underline{\int_c^a} = 0;$$

now, the interval of integration being positive, we have

$$\overline{\int (f+\varphi)} \leq \overline{\int f} + \overline{\int \varphi}, \quad \underline{\int (f+\varphi)} \geq \underline{\int f} + \underline{\int \varphi}.$$

As seen through reasoning similar to that of Sect. 3.3 but without using the same relationships where inequality signs are replaced by equality signs; the inequality signs are

---

[25] We could also deduce the existence of this limit from the existence of the upper sum integral for $-f$.

[26] Annales de l'École Normale Superieure, 1875.

indispensable. For example, let us take $f(x) = \chi(x)$ (Sect. 3.1), and $\varphi(x) = -\chi(x)$; we would have in (0, 1),

$$\overline{\int} f = 1, \quad \overline{\int} \varphi = \overline{\int} f + \varphi = 0, \quad \underline{\int} f = \underline{\int} f + \varphi = 0, \quad \underline{\int} \varphi = -1.$$

The[27] integral has been defined as the limit of the number

$$S = \sum \delta_i f(x_i)$$

When the maximum length λ of the intervals δ_i converges to zero. Let us set $S = \psi(\lambda)$, we would thus define a multiple valued function (Sect. 3.1). The limits of indetermination of $\psi(\lambda)$ are the two integrals by lower and upper sums, as λ converges to 0. This suggests that these two integrals often reveal us the limit superior and limit inferior of a number when we know that this number is given by the integral $\int f\,dx$ whenever f is integrable.

To better study the limit of indetermination of S, it is necessary to determine the set A of all the limit points of S.[28] For the case of integration, we have this property which I have the privilege of stating: Any number included between the upper sum and lower sum integrals is one of the limits of the sum S, when λ converges to zero.[29]

---

[27] We could use in a similar manner an arbitrary non-integrable function as an example in which the signs of inequality are indispensable.

[28] In certain cases, we have determined not only the set of limits of a function $\psi(\lambda)$, but also the *frequency* of each of these limits. This has been done, notably, for summation of certain divergent series (see BOREL, Leçons sur les séries divergentes, p. 5).

[29] See LEBESGUE, *Ann. de l'Ecole Norm. sup.*, 1910. As an exercise concerning the upper sum and lower sum integrals, we could prove that, $f(x)$ being a function of bounded mean oscillation ω in (a, b) and whose limit inferior, limit superior and oscillation at x are $l(x)$, $L(x)$ and $\omega(x)$, we have

$$(b-a)\omega = \overline{\int} f(x)\,dx - \underline{\int} f(x)\,dx = \overline{\int} L(x)\,dx - \underline{\int} l(x)\,dx = \overline{\int} \omega(x)\,dx.$$

The same relations are true if, in the definition of $L(x), l(x), \omega(x)$, we exclude the value x of the variable, or if, by these notations, we denote the limit superior, limit inferior and oscillation to the right or to the left of x, x being excluded or not. (*See* the note 2, Sect. 3.1).

4 Geometric Definition of the Integral

4.1 The Measure of the Sets

In the first chapter, the definition of the integral was linked to certain areas; we will now investigate if, through a similar geometric approach, we could arrive at the general definition of Riemann. We will see that it is possible, so that Riemann's integral appears to be the natural generalisation of Cauchy's integral, whether we place ourselves in the geometric or analytical point of view.[1] First, I will associate sets of numbers that will serve as analogues of lengths, areas, volumes corresponding to line segments, plane domains, or spatial domains. The first definition of these numbers is credited to Cantor; I would adopt Jordan's method of presentation which has simplified and completed the definition given by Cantor.[2] Let E

---

[1] In what follows, I assume that the Euclidean length of a segment and the Euclidean area of a polygon are defined.

To avoid any difficulties, it is easy to consider a point as a set of three numbers x, y, z; a displacement as a change of coordinates whose coefficients satisfy known conditions. Then by definition, the distance between two points (a, b, c) and (α, β, γ) is

$$+\sqrt{(a-\alpha)^2+(b-\beta)^2+(c-\gamma)^2}.$$

The function defined in this way is, up to a multiplicative constant, the only function of two points which remains invariable under displacements and satisfies the relation

$$f(P, Q) + f(Q, R) = f(P, R),$$

when Q is on the segment PR. This is where the importance of length number comes from.

The area of a polygon is defined by the theorems of elementary Geometry; the importance of this number is justified in the same way as that of length. (*See the Géométrie élémentaire* of M. Hadamard, note D, or the *Géométrie* of M. M. Gérard et Niewenglowski.)

[2] In the case of a set of points in space, the definition which Cantor used (*Acta mathematica, t. IV*) can be stated thus: from each point M in a set E as centre, draw a sphere of radius ρ. The set of points

R. Jain, *Lebesgue's Theory of Integration*,
https://doi.org/10.1007/978-981-96-1169-0_4

be a bounded set[3] of numbers or, if we want, of points on a straight line. Let (a, b) be one of the intervals containing E. Let us divide (a, b) into a *finite* number of partial intervals. Let λ be the maximum of the lengths of these intervals. I denote by A the sum of the lengths of the partial intervals which contain points of E and by B the sum of the lengths of those whose all points belong to E.[4] Jordan showed that A and B converge to two perfectly definite limits when λ converges to zero. For us, the existence of these limits is obvious, because A and B are the approximate values of the integrals by upper and lower sums of the function ψ equal to 1 for the points of E, zero for the other points.[5]

The limit of A is called *the exterior extent of* E, $e_e(E)$; that of B is *interior extent*, $e_i(E)$.

When these two extents are equal, we would say that the set is J measurable, that is, by the method of Jordan, and of extent[6]

$$e(E) = e_i(E) = e_e(E);$$

in this case, the function ψ attached to E is integrable in Riemann's sense and its integral is $e(E)$ in (a, b).

Let us interpret the condition of integrability of ψ. The points of discontinuity of ψ are points of E which are limits of points not belonging to E, and the limit points of E which do not belong to E. These points are called, by Jordan, the *frontier points* of E; their set is

---

interior to these spheres form one or several domains whose volume (in the ordinary sense of the word) is obtained by a triple integral. Let $f(\rho)$ be this volume; the limit of $f(\rho)$, when ρ converges to zero, is the volume of E.

This definition is equivalent to that of the exterior extent given by Jordan (t. 1 of the 2nd edition of his *Cours d'Analyse*).

Minkowski used the number $f(\rho)$. In the case where E is formed from points of a curve, Minkowski considered the ratio $\frac{f(\rho)}{\pi\rho^2}$; if it has a limit, Minkowski called it *the length of the curve.* The area of the surface is defined by the ratio $\frac{f(\rho)}{2\rho}$.

We see that the number $f(\rho)$ could be used in the theory of sets. The above discussion seems to suggest that it could be used in different manners depending on the number of dimensions of E. Moreover, Cantor indicated in his memoir that the concept of volume was useful to him in the defining the number of dimensions of continuous set. In many questions, such a definition seems quite useful; however, Cantor did not publish his work on this subject. The recent works concerning the number of dimensions of a set, raises strong doubts about whether Cantor could have arrived at these important results in this way.

Related to the number $f(\rho)$, we can consult a very interesting book: *Theory of sets of points,* published since the first edition of its Lessons by Mr. and Mrs. W. H. Young.

[3] That is, a set whose all numbers are included between two finite limits.

[4] We can give two meanings to the two expressions ≪ an interval containing points ≫ and ≪ all points of an interval ≫ as in the term ≪ enclosed ≫ (*see* note 19 Sect. 3.2). It does not matter whether we adopt one or the other.

[5] M. de la Vallée Poussin defined the exterior extent and the interior extent with the help of ψ.

[6] The word *extent* is deliberately used here; the word *measure*, that we often use as synonym of the extent, would be defined later.

the *frontier* set of E. Therefore, for a set to be J measurable, it is necessary and sufficient that its frontier set forms an integrable group.

This condition can be transformed if we note that, by definition, for an integrable group, A converges to zero. Therefore, an integrable group is a set of exterior extent zero or, if we want, a J measurable set and of extent zero. The previous method could be applied to set formed by points of space of several dimensions only if we had studied a priori, the multiple integrals by upper sum and lower sum. Such a study does not present difficulties, but it is simpler to use Jordan's method which is, in short, the proof of the existence of these integrals in the particular case of the function ψ.

Let us consider a bounded set of points E in the plane, that is, the set of coordinates of the points of E is bounded. Such a set is entirely contained in a suitably chosen square, of area R. Let us divide the plane into small squares with maximum diagonal length of λ. Let A be the sum of areas of the squares which contain the points of E and B be the sum of the areas of those squares whose all points belong to E. Both A and B are smaller than R. It is necessary to show that they converge to definite limits when λ converges to zero. For that, let us first consider a sequence of divisions $D_1, D_2, \ldots$, on which the corresponding numbers $A_1, B_1, A_2, B_2, \ldots$, and such that the corresponding λ values converge to zero. Also, consider a sequence of division Δ_j on which the corresponding numbers α_j and β_j, and such that the corresponding λ_j numbers converge to zero.

Let us compare A_i and α_j. The squares in Δ_j that contribute to α_j are of two types: The squares d which contain in their interior, the points of the edges of the squares of D_i contributing to A_i, and the others are the squares d'. The points of the squares d form a set which is contained in the set of points which are at a distances less than λ_j from at least one of the points on the edges of the squares of D_i.

If only a single square of D_i of perimeter $4c$ contributed to the sum A_i, this set could be decomposed into domains whose sum of areas, in the elementary sense of the word, would be $8c\lambda_j + (\pi - 4)\lambda_j^2$ for $c > 2\lambda_j$; more generally, if in D_i the sum of the perimeters of the squares contributing to the sum A_i is l, the corresponding set could be divided into domains whose sum of areas is at most $2l\lambda_j$. This value is also the maximum contribution of the squares d to the sum α_j.

As for the squares d', they obviously give a contribution at most equal to A_i, therefore we have

$$\alpha_j \leq A_i - 2l\lambda_j,$$

and that is sufficient[7] to show that α_j and A_i converge to the same limit $\mathfrak{A}$.

The number $\mathfrak{A}$, whose existence has just been shown, is the exterior extent of E, $e_e(E)$; but this is a surface extent. This distinction is important to note, because any set of points on a straight line has an exterior surface extent zero and could have any exterior linear extent.

[7] Let us compare with the reasoning of Sect. 3.1.

Similarly, we would show that B_i and β_j converge to the same limit $\mathfrak{B}$. We can also note that, if to the division Δ_j and to the set of points of the square of area R which does not belong to E, we associate two numbers α'_j and β'_j, similar to α_j and β_j, we have

$$\alpha'_j + \beta'_j = R$$

and the existence, which we have just proved, of the limit of α'_j shows the existence of the limit of β_j. This limit is the interior surface extent of E, $e_i(E)$.

As for the linear sets, we say that a set is J measurable and of extent $e(E) = e_e(E)$, if the exterior and interior extents are equal.

If we note that the squares which contribute to the sum A without contributing to the sum B are the ones that we must consider to obtain the exterior extent of the frontier of E, we see that the frontier of E has an exterior extent $e_e(E) - e_i(E)$; hence, the necessary and sufficient condition for a set to be J measurable is deduced.

I have already used the word *domain*, it needs to be specified here what one means by that.

A curve is a set of formulae

$$x = x(t), \quad y = y(t), \quad z = z(t);$$

where $x(t), y(t), z(t)$ are continuous functions defined on a finite interval (t_0, t_1). The points of the curve are those that are obtained by giving any definite value to t; the points which correspond only to one value of t are called *simple*; the others are called multiple. If the two points corresponding to t_0 and t_1 are identical, the curve is said to be *closed*; if the points t_0 and t_1 do not correspond to any other value of t, this point is not considered multiple.

If we replace t by a monotonically increasing or decreasing function of θ, we obtain a new curve which is not considered as different from the first; but two curves, to which correspond the same set of points, could be different; it is the case of two curves, defined on $(-\frac{\pi}{2}, +\frac{\pi}{2})$, by $x = \sin t$, $y = 0$, $z = 0$, and $x = \frac{2t}{\pi}\sin^2\frac{\pi^2}{4t}$, $y = 0$, $z = 0$.

In case of a closed curve, we could make the transformation $\theta = \frac{t-t_0}{t_1-t_0}$ and consider the function of θ obtained as periodic and of period 1. Then, to define the curve, it would be sufficient to prescribe it in any interval of extent 1 and not necessarily in (0, 1) only. Finally we could, in this interval, replace θ by a monotonically increasing or decreasing function of τ. All the curves thus obtained are considered identical.

Jordan first proved, in the second edition of his *Cours d'Analyse,* that a closed curve without multiple points separates the plane into two regions[8]; we will accept this result,

[8] *See* also the *Traite d'Analyse* of M. de la Vallée Poussin. In this second edition, I should mention many other references, as the recent works on this topic and related questions are numerous. I would content myself by referring to the article of M. Zoretti: *Researches récentes sur la théorie des fonctions (Encyclopédie des Sciences Mathématiques, II, I*emph),and to the first volumes of the *Fundamenta Mathematicæ.*

which appears intuitively so obvious that we initially have some difficulty to understand why it needs to be proven.

The points of the interior region constitute what we call the *the domain bounded by the curve*. Relative to the points of this curve, we could make two conventions: consider them as points of the domain or not, which is generally of little importance.

The frontier of a domain is formed by the closed curve used to define it.

When the exterior and interior extents of the domain are equal, the domain is said to be *quadrable* and its surface extent is called its area.[9]

For a domain to be *quadrable*, it is necessary that its frontier curves be of exterior extent zero; such a curve is called a *quadrable curve*. A square is obviously quadrable.

From the definition of quadrable domains, it follows that nothing would have changed if we had assumed that the division Δ_j (Sect. 4.1) was a division into quadrable domains of diameter less that λ_j.

Now, here are examples of the various circumstances which we have just envisaged.

The integrable groups provide us with the first example of the linear J measurable sets. In particular, the set Z (Sect. 3.2) is of exterior extent zero. The same holds true, a fortiori, for any set formed using the points of Z; all these sets are therefore J measurable and of extent zero. As Z has the cardinality of the continuum, it is possible to establish a bijective correspondence between the points of Z and those of an interval, so that any set of points in this interval corresponds to a set of points in Z. Therefore, the set of J measurable sets has a cardinality at least equal to that of the set of point sets and, since it obviously cannot have a higher cardinality, it has exactly that cardinality.[10]

Another example of a linear J measurable set is provided by a finite number of intervals. If we remove an integrable group from such a set, there remains a J measurable set, and the extent has not changed.

We easily see that the most general J measurable set differs from a J measurable set, formed by a countably infinite number of intervals, only by the addition of an integrable group G_1, and the subtraction of another integrable group G_2.[11]

It is also easy to mention the surface J measurable sets. Any bounded set Z_1, projecting itself on the x-axis, as the set Z, is a J measurable set of surface measure zero. The sets of exterior surface measure zero play the same role in the theory of double integrals, in the sense of Riemann, as integrable groups on the line; we can call them the *integrable groups of the plane.*

---

[9] Moreover, some authors always use, in place of the *linear extent* and *surface extent*, the words *length* and *area*.

[10] A very important theorem on the comparison of cardinality is used here which can be found in Note 1 of *Leçons sur la théorie des fonctions* of M. Borel; a proof due to M. F. Bernstein. Since this theorem is frequently used, we may state it as follows:

If a set E contains a set E_1 and is contained in a set E_2, where E_1 *and* E_2 *have the same cardinality,* then E, E_1 *and* E_2 *have the same cardinality.*

[11] If by the points of an interval we mean the *interior points* to this interval, the consideration of G_2 is futile.

A square is a surface J measurable set. Starting with squares and integrable groups in the plane, we can construct any J measurable set of plane, just as we did in case of line.

The integrable groups of the plane could be very different from the integrable groups of the line. Z_1 is, as Z, a discrete set, at least when each point of Z is the projection of only single point of Z_1. In other words, we cannot move continuously from one point to another in this set without passing through points that are not part of the set. However, an integrable group in the plane could be a *continuous* set, that is, a set such that any two of its points could be joined by a curve passing only through the points of the set. We know in fact that a segment, a polygonal line, a circumference, an ellipse have an exterior surface extent zero.

The curves which are integrable groups are the ones that we called quadrable.

To obtain a J non-measurable set, it is sufficient to take a set which is everywhere dense and contains no intervals if it is a set on the line or contains no domains if it is a set in the plane. Indeed, for such a set, the interior extent is zero, while the exterior extent is not. Therefore, set of points with rational coordinates (or the coordinate) is not J measurable.

P. Du Bois Reymond noticed that a set can be non-dense without being J measurable. Let us take a sequence of fractions $\alpha_1, \alpha_2, \ldots,$ such that the infinite product $P = \alpha_1 \times \alpha_2 \times \cdots$ is convergent and non-zero; we can take, for example, $\alpha_n = \frac{4n^2-1}{4n^2}$. Let us divide the interval (a, b) into three parts, with the middle one having a length $(b-a)(1-\alpha_1)$, and the two extreme ones being equal. Let us remove the middle interval and operate on the remaining two intervals as we did on (a, b), with α_1 being replaced by α_2, and so on. Let R be the set of points remaining after all the operations. If we use the successive divisions which has given R for calculating the exterior extent of R, we see that this extent is $P(b-a)$, which is different from zero. However, the interior extent is zero, since R is non-dense. Therefore R is not J measurable.[12]

An absolutely similar construction can be done in case of plane; we could, for example, divide a rectangle, by two series of three parallels to its edges, in nine rectangles and remove the interior points to that of the middle one, which we will choose in a manner that its area be $(1-\alpha_1)$ times that of the initial rectangle. Then we operate on each of the above rectangles by replacing α_1 by α_2; etc.

Among the J non-measurable sets of the plane, there are non-quadrable curves, that is, their exterior extent is not zero. However, not all non-quadrable curves are necessarily J non-measurable; in that case, they contain all the points of a square.

M. Peano constructed the first curve which passes through all the points of a square; M. Hilbert then indicated a simple geometric method which allows the construction of such curves. All these curves are non-quadrable.[13]

---

[12] If we had $\alpha_n = \frac{2}{3}$, we would have the set Z which is J measurable, because P would be zero.

[13] PEANO, *Sur une courb qui remplit toute une aire* (Math. Ann., Bd XXXVI).—HILBERT, *Ueber die stetige Abbildung einer Linie auf ein Flachenstuck* (Math. Ann., Bd XXXVIII). The curve of M. Hilbert is defined on page 23 of volume I of the second edition of *Traite d'Analyse de M.Picard.* The method of definition which will be indicated, which differs from that of MM. Peano and Hilbert,

To obtain a curve passing through all the points of a square $0 \leq x \leq 1, 0 \leq y \leq 1$, define a function of one parameter t varying from 0 to 1, I set

$$x = \frac{1}{2}\left(\frac{a_1}{2} + \frac{a_3}{2^2} + \frac{a_5}{2^3} + \cdots + \frac{a_{2n-1}}{2^n} + \cdots\right),$$
$$y = \frac{1}{2}\left(\frac{a_2}{2} + \frac{a_4}{2^2} + \frac{a_6}{2^3} + \cdots + \frac{a_{2n}}{2^n} + \cdots\right),$$

when

$$t = \frac{a_1}{3} + \frac{a_2}{3^2} + \frac{a_3}{3^3} + \cdots + \frac{a_n}{3^n} + \cdots$$

where the numbers a_i are equal to 0 or 2. Then t belongs to the set Z of Sect. 3.2.

Let some t be not contained in Z, then it belongs to one of the intervals which have been removed in the construction of Z; let (t_0, t_1) be this interval. At the points t_0 and t_1 of Z correspond the coordinates (x_0, y_0); (x_1, y_1); then we set, for any interval (t_0, t_1):

$$x = x_0 + \frac{x_1 - x_0}{t_1 - t_0}(t - t_0), \quad y = y_0 + \frac{y_1 - y_0}{t_1 - t_0}(t - t_0),$$

In (t_0, t_1), therefore, the curve is reduced to a segment.

Our curve is completely defined; but to refer to it as a curve, it is necessary to show that x and y are continuous functions of t in $(0, 1)$. For that, it is obviously sufficient to show it specifically for the functions x and y of t defined on Z. And this follows from the fact that, when t (belonging to Z) is sufficiently close to θ (also belonging to Z), the first $2n$ digits $a_1, a_2, \ldots, a_{2n}$ of t, written in system of base 3, are the same as those for θ. In other words, the first n digits of $x(t)$ and $x(\theta)$ on the one hand, and those of $y(t)$ and $y(\theta)$ on the other hand, are the same when we write these coordinates in the system of base 2.[14]

Our curve indeed fills the entire square, and it even passes several times through some points. We will easily show that it cannot be otherwise.[15]

What has just been done in the case of one and two dimensions can obviously be repeated in the case of any number of dimensions.

---

could be used for spaces of any number of finite dimensions and similarly for spaces of countably infinite number of dimensions (*see* LEBESGUE, *Journal de Mathematiques*, 1905).

[14] **Translator's Note:** The original text says 'system of base 2,' but this is likely a typo. Given the earlier reference to base-3 digits, 'system of base 3' seems intended.

[15] This follows from the earlier work of Lüroth *(Sitz. Phys. med. Soc. Erlangen, t.* 10*)*; for limiting the order of the multiplicity of singular points, *see* LEBESGUE, *Fundamenta Mathematicæ*, *II*, 1921.

We will find, in Chap. 7 (Sect. 5), an example of the application that can be made in some arguments involving Peano and similar curves.

The Peano curve is J measurable and of non-zero extent, but it cannot be used to bound a domain. There exist curves without multiple points which are non-quadrable; these curves are not J measurable, and they can be used to bound non-quadrable domains. See W.-F. OSGOOD, *A Jordan curve of positive area. (Trans. of Amer. Math. Soc.,* 1903*)* or H. LEBESGUE, *Sur le probléme des aires. (Bull. de la Soc. math. de France,* 1903*)*.

In particular, in the case of three dimensions, we will define the volume of a domain. That would require, a priori, the precise definition of a closed surface and, for defining a domain, studies similar to that of Jordan on closed curves.

4.2 Definition of the Integral

Let $f(x)$ be a continuous and positive function, defined on a positive interval (a, b), and let the domain $abBA$ be the one which we have attached to it (Fig. 2.1). Let us find if this domain is quadrable. For that, let us divide (a, b) into partial intervals $\delta_1, \delta_2, \ldots, \delta_p$. The largest rectangle, of base δ_i and all the points of which belong to the domain $abBA$, has a height equal to the lower limit l_i of f in δ_i. The smallest rectangle, of base δ_i and which contains all the points of domain which project onto δ_i, has a height equal to the upper limit L_i of f in δ_i.

From this follows that the two sums

$$\underline{S} = \sum \delta_i l_i, \quad \overline{S} = \sum \delta_i L_i$$

converge to definite limits which are the interior and exterior extents of domain, when the maximum of δ converges to zero. However, $\overline{S} - \underline{S}$ converges to zero, because the continuous functions are of mean oscillation zero; the domain $abBA$ is therefore quadrable.

If we use the initial method, if we call the *definite integral of f in (a, b)* the area of $abBA$, we arrive at Cauchy's integral. There is only a difference of form, between this definition and that of Cauchy.

In case where $f(x)$ is not always positive, the curve AB meets the x-axis finite or infinite number of times and we have two types of domains, the ones above the ox, the others below. Each of these domains is quadrable as explained earlier.

The sum of the areas of those which are above ox, minus the sum of the areas of those which are below, is by definition, the integral of $f(x)$.[16]

Let us now consider any function $f(x)$, defined on the positive interval (a, b). Let $E(f)$ be the set of points whose two coordinates are linked by only the condition that y cannot be exterior to the positive or negative interval $[0, f(x)]$. In other words, we have

$$yf(x) \geq 0 \quad \text{and} \quad 0 \leq y^2 \leq \overline{f(x)}^2$$

The x-axis partitions this set into two other sets: the points situated above ox form $E_1[f(x)]$, and those which are below ox form $E_2[f(x)]$. Regarding the points situated on ox, we put them indifferently either in E_1 or in E_2, that is not of much importance for what follows, because they form an integrable group of the plane.

[16] The two sums or the series which appear in this definition indeed exist, since the set of all domains can be enclosed in a circle of finite radius.

By analogy with the previous definition, it is natural to call *integral of* f the difference

$$I = e[E_1(f)] - e[E_2(f)],$$

when E_1 and E_2 are J measurable.

When a set is not J measurable, its extent can be considered as an indefinite number whose two limits of indetermination are the interior and exterior extents of the set; that leads to two limits of indetermination, for I

$$\underline{I} = e_i[E_1(f)] - e_e[E_2(f)], \quad \overline{I} = e_e[E_1(f)] - e_i[E_2(f)].$$

We will calculate these two limits of indetermination and for that we first assume that f is never negative, that is, E_2 does not contain any points. The calculation of interior and exterior extents of E (or E_1) is done as in the case where f is continuous, that is, these extents are the limits of numbers $\underline{S}$ and $\overline{S}$. The extents are therefore the integrals by upper sum and lower sum of f.

To study the general case let us set $f = f_1 - f_2$, where f_1 is equal to f when f is positive or zero, and equal to zero when f is negative. We then have, obviously,

$$e_i[E_1(f)] = \underline{\int f_1\,dx}, \quad e_e[E_1(f)] = \overline{\int f_1\,dx},$$

$$e_i[E_2(f)] = \underline{\int f_2\,dx}, \quad e_e[E_2(f)] = \overline{\int f_2\,dx},$$

therefore

$$\underline{I} = \underline{\int f_1\,dx} + \underline{\int -f_2\,dx}, \quad \overline{I} = \overline{\int f_1\,dx} + \overline{\int -f_2\,dx}.$$

It is, in general, impossible to replace sums of upper or lower integrals with the upper or lower integrals of the sum, because the maximum of a sum is, in general, less than the sum of the maxima of the terms of the sum, while the minimum is, generally, greater than the sum of the minima. But here, in any interval, the maximum (or the minimum) of $f = f_1 - f_2$ is indeed the sum of the maximums (or of minimums) of f_1 and $-f_2$. We can therefore write

$$\underline{I} = \underline{\int f\,dx}, \quad \overline{I} = \overline{\int f\,dx}.$$

We have thus found the Darboux integrals and we have their geometric significance.

Let us note that $E(f)$ is J measurable when both E_1 and E_2 are. And conversely, if $E(f)$ is J measurable, then so are E_1 and E_2. Thus, our geometric definition of the integral is applicable when E is J measurable. However, in this case, and only in this case, $\overline{I}$ and $\underline{I}$ are equal, that is, their integrals $\overline{\int f\,dx}$ and $\underline{\int f\,dx}$ are equal, and therefore:

For a bounded function f to be integrable in the sense of Riemann, it is necessary and sufficient that $E(f)$ be surface J measurable; in this case, we have

$$I = \int f\,dx.$$

The geometric definition of integral is completely equivalent to the analytic definition given by Riemann.

The Functions of Bounded Variation

5

5.1 The Functions of Bounded Variation

The concept of a linear measure is a generalisation of the notion of length of a segment; another generalisation leads to the definition of the length of an arc of a curve. By studying the questions related to the rectification of curves, we will have the opportunity to apply some of the results that we have obtained on the integration. We will see, at the same time, the importance of a class of functions defined by Jordan: the functions of bounded variation.

Let a bounded[1] function $f(x)$ be defined on a positive finite interval (a, b). Let us partition the interval (a, b) using the points

$$a = a_0 \le a_1 \le a_2 \le \cdots \le a_n = b;$$

the sum

$$\nu = |f(a_1) - f(a_0)| + |f(a_2) - f(a_1)| + \cdots + |f(a_n) - f(a_{n-1})|$$

is what we call *the variation of* $f(x)$ *for the system of points* $a_0, a_1, \cdots, a_n$. If, for any system of points of division, ν is bounded, the function is said to be of *of finite total variation* or, simply, *of bounded variation*. Finite or infinite total variation is, by definition, the greatest limit of ν, when the maximum λ of the length of partial intervals used, tends to zero. It is to be noted that if we add new points between the chosen points of division, we increase ν or, at least, we do not decrease it; thus by indefinitely adding new points, in a manner such that λ tends to zero, we have a sequence of numbers ν tending to a limit, whether finite or

[1] It is moreover obvious that an unbounded function cannot satisfy the following definitions.

R. Jain, *Lebesgue's Theory of Integration*,
https://doi.org/10.1007/978-981-96-1169-0_5

infinite, which is at least equal to a number ν from where we started. We can therefore say that the total variation of f is the upper limit of the set of numbers ν.[2]

We also see very simply that, in the previous definition, we can replace ν by

$$o = \omega_1 + \omega_2 + \cdots + \omega_n,$$

where ω_i is the oscillation of f in (a_{i-1}, a_i), including the extremities.

Because of this property, some authors call the functions with which we deal, the *functions of finite total oscillation;* the total oscillation being the upper limit of o.

A function of bounded variation is integrable; it is, indeed, of mean oscillation zero, since this oscillation is the quotient by $(b-a)$ of the limit of

$$\sum(a_i - a_{i-1})\omega_i \leq \sum \lambda\omega_i = \lambda \sum \omega_i = \lambda o \leq \lambda O,$$

when λ tends to zero. O being the total oscillation of $f(x)$.

Also, the integrability result of this proposition is obvious: *the points at which a function of bounded variation has an oscillation greater than $\alpha (\alpha > 0)$ are in finite number and, as a result, form an integrable group.*

Let us choose the numbers $\alpha_1, \alpha_2, \cdots$, which decrease towards zero. The points at which the oscillation is greater than α_n without being greater than α_{n-1} are in finite number; by varying n, we would see *that a function of bounded variation has at most a countably infinite points of discontinuity.*

The converse is not true; there even exist continuous functions of unbounded variation.

The oscillation of a sum $f_1 + f_2$ being, in an arbitrary interval, at most equal to the sum of the oscillations of f_1 and f_2 in this interval, therefore, the total oscillation of $f_1 + f_2$ is, at most, the sum of the total oscillations of f_1 and f_2. *Hence, the sum of two functions of bounded variation is a function of bounded variation.*

A similar reasoning would allow us to show that the operations done in Sect. 3.3 on integrable functions give the functions of bounded variation when they are done on the functions of bounded variation.

However, it is not true that a uniformly convergent series of functions of bounded variation necessarily yield a function of bounded variation. The property that replaces this is the following:

The limit to which a sequence of functions with total variations at most equal to M converges (uniformly or not) is a function whose total variation is also at most equal to M.

Indeed, let us take a division of interval; the corresponding variation for the terms of sequence converge to the variation related to the limit function and the division used; therefore, this variation is at most equal to M and the same applies to the total variation of the limit function.

The above allows us to cite the functions of bounded total variation. An increasing function $f(x)$ is, in fact, a function of finite total variation with total variation equal to

[2] And no more than the limit superior of the limit of indetermination of the numbers ν.

$f(b) - f(a)$; similarly, a decreasing function is of bounded variation. Consequently, the difference of two monotonically increasing functions is a function of bounded variation. We would now show the converse: *any function of bounded variation is the difference of two non-decreasing functions.*

Let us again consider the variation

$$\nu = |f(a_1) - f(a_0)| + |f(a_2) - f(a_1)| + \cdots + |f(a_n) - f(a_{n-1})|,$$

and let p be the sum of those increments $f(a_i) - f(a_{i-1})$ which are positive and $-n$ the sum of those which are negative. We obviously have

$$\nu = p + n, \quad f(b) - f(a) = p - n,$$

hence

$$\nu = 2p + f(a) - f(b), \quad \nu = 2n + f(b) - f(a),$$

p is the positive variation for the chosen division, n the negative variation. The last two equalities show that the limit superior V, P, N of v, p, n that we call *total variation, total positive variation, total negative variation,* are linked by the same relations as that of v, p, n.

This being said, let $V(x)$, $P(x)$, $N(x)$ be the three total variations in (a, x), $(x > a)$, we have

$$f(x) = f(a) + P(x) - N(x).$$

But $P(x)$ and $N(x)$ cannot decrease when x increases, therefore the statement of the theorem is proved.

We have, moreover,

$$V(x) = P(x) + N(x).$$

Hence it follows that, if one of the three variations is finite, so are the other two, since we have the two relations between the variation and the increments $f(b) - f(a)$

$$V = P + N, \quad f(b) - f(a) = P - N.$$

A function of bounded variation could be put in the form of a difference of two increasing functions in an infinite number of ways. If we add the same non-decreasing function $\lambda(x)$ to $P(x)$ and $N(x)$, we obtain two non-decreasing functions $P_1(x)$ and $N_1(x)$ such that we have

$$f(x) = f(a) + P_1(x) - N_1(x).$$

We easily see that the most general non-decreasing functions P_1 and N_1 satisfying this equality are the ones which we have just constructed; so *$P(x)$ and $N(x)$, among all the functions, $P_1(x)$ and $N_1(x)$, which are non-negative and non-decreasing, and satisfy the previous equality, are the smallest.*

To calculate the total variation of a discontinuous function as the limit of a sequence of variations ν, it is necessary to choose the division points in a very particular manner; for example, for a function which is zero everywhere, except at the origin, it is necessary that the origin be a point of division. For the continuous functions, on the contrary, we have this property: *the variation of a continuous function, relative to an arbitrary division, uniformly tends to the total variation of this function when the maximum λ of the lengths of the intervals used tends to zero.*

Indeed, let two sequences of divisions be $D_1, D_2. \cdots, \Delta_1, \Delta_2, \cdots$ for which λ tends to zero, and let λ_j be the value of λ for Δ_j. The maximum of the oscillation of $f(x)$ in an interval of extent λ_j is a number ε_j which tends to zero as λ_j does. Let us compare the variations ν_i, ν'_j, relative to D_i and Δ_j.

The intervals of Δ_j are always partitioned in two classes, let d' be those which do not contain any points of division of D_i. Let us consider all those d' which are between x_i and x_{i+1}, they cover an interval whose origin O_i is between x_i and $x_i + \lambda_j$ and whose extremity E_i is between $x_{i+1} - \lambda_j$ and x_{i+1}. The values of $f(x)$ for this origin and extremity differ from the numbers $f(x_i), f(x_{i+1})$ by ε_j at most. The contribution to ν'_j of the considered intervals is therefore at least

$$|f(E_i) - f(O_i)| \geq |f(x_{i+1}) - f(x_i)| - 2\varepsilon_j,$$

and the contribution of all the d' in ν'_j is at least equal to

$$\sum [|f(x_{i+1}) - f(x_i)| - 2\varepsilon_j] = \nu_i - 2n\varepsilon_j,$$

if the points of division of D_i are n in number. We have, a fortiori

$$\nu'_j \geq \nu_i - 2n\varepsilon_j,$$

and one of the limits of ν'_j is at least equal to one of the limits of ν_i. However, we can permute ν'_j and ν_i, therefore ν'_j and ν_i tend to the same definite limit.

The stated proposition is thus proved; and to make it clear, it is convenient to state it as follows: *the variation ν of a continuous function $f(x)$, for a division of the considered interval into partial intervals of length smaller than λ, differs from the total variation V of $f(x)$ by at most an infinitely small $\theta(\lambda)$ if V is finite, and if V is infinite, ν is greater than an indefinitely large $\Theta(\lambda)$.*

This expresses that ν tends to its limit, V, finite or not, with a sort of uniformity.

Under these given conditions, the two numbers p and $-n$ associated with the considered division, which are respectively the sum of positive increments and the sum of negative increments given by the partial intervals, tend to their limits P and $-N$ with the same type of uniformity.

Here is an immediate consequence of this property: *the three total variations of a continuous function of bounded variation are continuous functions.* It is enough to prove it for $V(x)$ since $P(x)$ and $N(x)$ are expressed immediately with the help of $f(x)$ and $V(x)$.

To calculate $V(x_0)$, I use a division $a_1, x_1, \cdots, x_n, x_0$; the variation ν corresponding to this division is equal to one, corresponding to $a, x_1, \cdots, x_n$ plus $|f(x_0) - f(x_n)|$, ν is therefore at most equal to

$$V(x_n) + |f(x_0) - f(x_n)| \leq V(x_0 - 0) + |f(x_0) - f(x_n)|,$$

because $V(x)$ is increasing; and, since $f(x_0) - f(x_n)$ tends to zero when the maximum of $x_{i+1} - x_i$ tends to zero, the value $V(x_0)$ is at most equal to $V(x_0 - 0)$. But $V(x)$ is an increasing function, therefore we have

$$V(x_0) = V(x_0 - 0),$$

the function is continuous at left.

Let us study the total variation of $f_1(x) = f(-x)$ between $-b$ and $-x$, $(x < b)$; this total variation is obviously equal to

$$V(b) - V(x).$$

Considered as a function of $-x$, it is continuous at left of $-x_0$; therefore, as a function of x, it is continuous at right of x_0. The function $V(x_0)$ is therefore continuous.

The second part of this proof essentially assumes that the function is of bounded variation. If $V(x)$ suddenly becomes infinite for $x > x_0$, and we will see that this is possible, the symbol $V(b) - V(x)$ would not have any sense for $x > x_0$.

Since, $P(x)$ and $N(x)$ are continuous functions, *any continuous function of bounded variation is the difference of two continuous non-decreasing functions.*

The variation ν, for the division D, has been defined only in case where D contains only a finite number of intervals; in the following, it is useful to study a case where D contains an infinite number of intervals. This is the case where the points of division of D form a closed reducible set E; then we call *variation* u, for this division, the sum of the series $\sum |f(x_i) - f(y_i)|$, extended over all the intervals (y_i, x_i) contiguous[3] to E.

We can compare the set of variations u which has just been defined with the set of variations ν defined previously.[4]

The set of u contains the set of ν, when we vary the reducible set E; therefore, the limit superior of the set u is at least equal to the limit superior of the set ν. It is sufficient to show that u is always smaller than the total variation for which it is to be proved that the limit superior of u is the total variation V.

Let (α, β) be an interval contiguous to E'. Let α_1 and β_1 be two points of E belonging to (α, β); the contribution of the subset of E in (α_1, β_1) for the calculation of u is at most

---

[3] An interval (y_i, x_i) is said *contiguous* to a set E if it does not contain points of E and if its extremities belong either to E or to E'. The proof of the contiguous intervals is due to M. R. Baire.

[4] Because we have not yet proved that the series defining u is convergent, it does not exclude the case that the variation u has a value $+\infty$.

equal to the contribution which it provides in V, since E contains only a finite number of points in (α_1, β_1). Let points α_1 and β_1 tend to α and β, the proposition remains true and we find that (α, β) furnishes in V a contribution at least equal to the one it gave in u. If, in (α, β), there were no points of E close to α, it would be necessary to take $\alpha_1 = \alpha$ and we would operate in a similar way if, in (α, β), there were no points of E close to β.

We could similarly prove that the proposition is true in an interval contiguous to E'' or $E''', \cdots$; but one of the derived sets of E being empty in (a, b), the proposition is true for (a, b).

Thus, the numbers u can replace the numbers ν.

When it comes to a continuous function the number u, as the number ν, tends uniformly towards the total variation, when the maximum λ of the length of intervals contiguous to E tends to zero.

Indeed, let D be a division corresponding to a closed and reducible set E and has a certain maximum λ. By suitably choosing a finite number of points of E, we obtain a division D_1 corresponding to a maximum of at most 2λ. Let u and ν_1 be the variations relative to D and to D_1. Let (a, b) be an interval appearing in D, we claim that its contribution to u is not smaller than its contribution to ν_1. This is obvious if (a, b) contains only a finite number of points of E; if it contains an infinite number but still only a finite number of points of E', we would argue as we just did, a moment ago. From there we would pass to the case where E'' has only a finite number of points in (a, b), etc.

Finally, we conclude: $u \geq \nu_1$. Therefore, $\nu_1 \leq u \leq V$, and, since ν_1 tends to V (finite or not) when λ tends to zero, our proposition is proved.

We could sharpen the statement of this proposition as we have done (Sect. 5.1) when we were only dealing with variations calculated with the help of a finite number of points of division.

The series u being convergent, the series $\sum[f(x_i) - f(y_i)]$, extended to all the intervals contiguous to E, is absolutely convergent and, by a reasoning similar to the previous one, we verify that its sum is equal to the increment $f(b) - f(a)$ of $f(x)$ in the considered interval (a, b).[5] We could thus speak of the sum of its positive terms and the sum of its negative terms, these two sums tend towards P and $-N$ when λ tends towards zero.

It is important to note that we cannot replace the reducible set E by an arbitrary non dense set without some of its previously mentioned properties ceasing to hold. Let, in fact, the function $\xi(x)$ be defined by

$$2\xi(x) = \frac{a_1}{2} + \frac{a_2}{2^2} + \frac{a_3}{2^3} + \cdots,$$

[5] The notation $\sum[f(x_i) - f(y_i)]$ assumes that the contiguous intervals have been numbered. This is possible in infinitely many ways, but none is preferable. So that the terms of our series are not actually arranged in a specific order; the series can, therefore, be convergent only if it is absolutely convergent.

We will encounter this fact in the sequel for all series of numbers associated with the intervals contiguous to a set.

when

$$x = \frac{a_1}{3} + \frac{a_2}{3^2} + \frac{a_3}{3^3} + \cdots,$$

where the numbers a are equal to 0 or 2. Then x belongs to the set Z. We immediately verify that, for the two extremities of an interval contiguous to Z, ξ takes the same value; we require ξ to remain constant in an interval. $\xi(x)$ is now defined everywhere; it is a non decreasing function and yet, we find zero for u, if the points of Z are among the points of division used.

We have seen that the finite or infinite total variation of a function is the limit superior of the numbers ν for this function, and therefore also of its corresponding numbers u. Let us show that, if the total variation of $f(x)$ is infinite, there is a reducible set E for which the corresponding numbers u are infinite.

Let us suppose that there is an interval (a, b) and let x increases from a to b. If, from a to x_0, $f(x)$ is of bounded variation, $f(x)$ is a fortiori of bounded variation from a to $x < x_0$; therefore, when x varies in (a, b), x attains a value ξ which is either the first for which the total variation from a to ξ is infinite, or the last for which this variation is finite; ξ could moreover be identified with a, or b.

In the first envisaged case $f(x)$ would have an unbounded total variation in any interval $(\xi - h, \xi)$; in the second case its total variation would be infinite in any interval $(\xi, \xi + h)$. We could choose in (ξ, b) the points $b > b_1 > b_2 > \cdots > b_p > \xi$, such that the variation ν in (ξ, b) for this system of points exceeds $o + 1$, o being the oscillation of $f(x)$ in (a, b). We therefore have

$$|f(b_p) - f(\xi)| + |f(b_{p-1}) - f(b_p)| + \cdots + |f(b) - f(b_1)| > o + 1,$$
$$|f(b_p) - f(\xi)| \leq o;$$

therefore

$$|f(b) - f(b_1)| + |f(b_1) - f(b_2)| + \cdots + |f(b_{p-1}) - f(b_p)| > 1.$$

Let us choose in (ξ, b_p) the points $b_p > b_{p+1} > \cdots > b_{p+q} > \xi$, such that the variation ν in (ξ, b_p) calculated with the help of these points exceed $o + 1$; since, in (ξ, b_{p+q}) the points

$$b_{p+q} > b_{p+q+1} > \cdots > b_{p+q+r} > \xi,$$

such that they give for variation ν a value greater than $o + 1$ in (ξ, b_{p+q}), and so on.

It is clear that the set E formed from point ξ and this infinite sequence $b_1, b_2, \cdots$ answers the question, since the series $\sum[f(b_i) - f(b_{i+1})]$ is divergent.

I would conclude by giving some examples of various specialities which have been mentioned.

The function $x \sin \frac{1}{x}$ is equal to $(-1)^{K+1} \frac{1}{K\pi - \frac{\pi}{2}}$ for $x = \frac{1}{K\pi - \frac{\pi}{2}}$, therefore, if we use these values of x for calculating u in the interval $(0, \frac{1}{\pi})$, we find

$$u = \frac{1}{K\pi - \frac{\pi}{2}} + \frac{2}{2K\pi - \frac{\pi}{2}} + \frac{2}{3K\pi - \frac{\pi}{2}} + \cdots,$$

and the function is of unbounded variation although it is continuous.

For a continuous function, zero for x negative, equal to $x \sin \frac{1}{x}$ for x positive, the total variation from -1 to x jumps briskly from 0 to ∞ when x is greater than zero.

The function $x \sin \frac{1}{x}$ has an infinite number of maxima and minima, but this condition is not sufficient for a function to be of unbounded variation. The function $x^2 \sin \frac{1}{x^{\frac{4}{3}}}$ admits one and only one maximum or minimum, in each interval $\left(\frac{1}{(K\pi)^{\frac{3}{4}}}, \frac{1}{[(K+1)\pi]^{\frac{3}{4}}}\right)$; if we note that the absolute value of this maximum or of this minimum is at most $\frac{1}{(K\pi)^{\frac{1}{2}}}$, we see that the function is of finite total variation which is at most equal to $2\sum \frac{1}{(K\pi)^{\frac{3}{2}}}$.

The previous two functions have an infinite number of maxima or minima only in the neighborhood of the origin; if we want this to be the case around any point, it is necessary to apply the principle of condensation of singularities. It is necessary to use this principle in a rather specific way because the limit to which the functions of bounded variation uniformly tend can be of unbounded variation, and because the maxima and the minima are not necessarily conserved under addition.

Let us consider the two functions, defined on $(-1, +1)$,

$$a(x) = x \sin \frac{\pi}{x}, \quad b(x) = x^2 \sin \frac{\pi}{x^{\frac{4}{3}}};$$

the one and the other vanishes at -1 and $+1$, the first is of infinite total variation V, the second is of bounded total variation V. $f_1(x)$ would denote either of these two functions.

$f_1(x)$ has an infinite number of maxima and minima which occur when x belongs to a certain set E_1.

$f_2(x)$ is a continuous function which vanishes at the points of E_1 and which, in the interval (α, β) of two consecutive points of E_1 is equal to

$$\frac{\beta - \alpha}{2} f_1 \left[\frac{2}{\beta - \alpha}\left(x - \frac{\alpha + \beta}{2}\right)\right]$$

$f_2(x)$ has the same total variation as that of $f_1(x)$ because, in (α, β), the total variation of $f_2(x)$ is $\frac{\beta-\alpha}{2} V$.

The function $f_1 + \frac{1}{2^2} f_2$ has an infinite number of maxima and minima, in each interval (α, β). Indeed, if $f_1 = a$, it is of unbounded variation in (α, β) and this implies the consequence that it has an infinite number of maxima and minima. It is because, if a function has only a finite number of maxima and minima, it is sufficient to calculate the number ν relative to a division $x_1, x_2, \cdots, x_p$ at which appears the abscissa of all the maxima and minima, for calculating the total variation V, which is therefore finite. If $f_1 = b$, f_1 has a bounded derivative in (α, β), while the derivative of f_2 takes all the positive and negative values, hence again the existence of an infinite number of maxima and minima. Let E_2 be the set of values of x for which $f_1 + \frac{1}{2^2} f_2$ is maximum or minimum.

By operating on $E_1 + E_2$, as well as on E_1, we will form f_3, resulting in $f_1 + \frac{1}{2^2} f_2 + \frac{1}{3^2} f_3$ and E_3.[6]

By continuing so, we define the different terms of the series

$$f(x) = f_1 + \frac{1}{2^2} f_2 + \frac{1}{3^2} f_3 + \cdots$$

which is uniformly convergent, because $|f_i|$ is smaller than 1.

The continuous function $f(x)$ has a maxima and minima in every interval. In an arbitrary interval (l, m), indeed, provided that n is large enough, there are more than two points in E_n. Let us suppose that there are three consecutive points r, s, t in E_n, with f equal to the sum $s_n = f_1 + \frac{1}{2^2} f_2 + \frac{1}{3^2} f_3 + \cdots + \frac{1}{n^2} f_n$ for these three points, f will have a maximum or a minimum, at least, between r and t, depending on whether s corresponds to a maximum or a minimum.

The function f admits all the maxima and minima of s_n, therefore f is of unbounded variation in any interval if $f_1 = a$. On contrary if $f_1 = b$, the total variation of s_n being finite and smaller than $V(1 + \frac{1}{2^2} + \cdots + \frac{1}{n^2})$, f is of bounded variation in any interval (see Sect. 5.1).

Let us focus now on the discontinuous functions of bounded variation.

Here is a property of singular points, which is easy to prove directly, and which follows immediately from the construction of the most general function of bounded variation starting from two increasing functions: *all points of discontinuity of a function of bounded variation are of first kind.*

Let x_0 be a point of discontinuity; the quantity

$$s_g(x_0) = f(x_0) - f(x_0 - 0)$$

is the left jump of the function at x_0;

$$s_d(x_0) = f(x_0 + 0) - f(x_0)$$

is right jump of the function at x_0; finally

$$s(x_0) = f(x_0 + 0) - f(x_0 - 0)$$

is the jump at point x_0, This being said, let us consider the *jump function of* $f(x)$

---

[6] To be absolutely rigorous, it would be necessary to show that the sum of the lengths of the intervals contiguous to $E_1 + E_2$, intervals which play the role of (α, β), is equal to 2, like the sum of the differences $\beta - \alpha$. It is almost obvious and follows, if we want, from the fact that $E_1 + E_2$ is of exterior extent zero.

$$\varphi(x) = \sum_{a \le x_i < x} s_d(x_i) + \sum_{a < x_i \le x} s_g(x_i),$$

where each of the series contains all the x_i which satisfy the inequality placed below the corresponding sign $\sum$. We easily see that these two series are absolutely convergent and that, if we set

$$f(x) = \varphi(x) + \psi(x),$$

$\psi(x)$ is the continuous function of bounded variation; the total variation of f being the sum of those of φ and ψ.

The most general discontinuous function which is of bounded variation is therefore obtained either by taking the difference of two discontinuous increasing functions or by adding to a continuous function of bounded variation a jump function $\varphi(x)$. This second method shows that we could construct functions with bounded variation by choosing at will the countable set of points of discontinuity, and even the jumps to the right and to the left s_d and s_g, provided that the series $\sum s_d(x)$, $\sum s_g(x)$ are absolutely convergent.

For example, the set of points of discontinuity could be the set of rational numbers, the jumps being, when x is written $\frac{a}{b}$ in the irreducible form,

$$s_d = (-1)^a \frac{1}{a^2 b^2}, \quad s_g = (-1)^b \frac{1}{a^3 b^3}.$$

Let us show that, for calculating the total variation of a discontinuous function f, it is sufficient to take the limit of numbers ν provided by a sequence of divisions $D_1, D_2, \cdots$ in intervals of length tending to zero and such that any point of discontinuity of f being a point of division of D_i for a certain value of i. It is sufficient to prove this for the jump functions φ. Now, if ξ is a point of discontinuity of φ and belongs to $D_i, D_{i+1}, \cdots$, there exist in division D_{i+p} an interval $(\xi, \xi + h_p)$ of origin ξ, whose contribution to ν_{i+p} tends to $|s_d(\xi)|$ when p increases indefinitely. Thus, in ν_n, there are terms which tend to the first K terms of $\sum |s_d(x_i)| + \sum |s_g(x_i)|$, for n increasing to infinity. Therefore, the limit of ν_n cannot be smaller than the total variation of φ, but since it cannot be greater either, the ν_n does indeed tend to the total variation of φ.

We can also use the divisions D_i satisfying the indicated conditions but obtained with the help of reducible sets of points and no longer only with the help of a finite number of points. It is only necessary to define with precision what we mean by the number u; this would be, if we denote by $\delta_i = (a_i, b_i)$ the different contiguous intervals to the set E of division points and by $x_1, x_2, \cdots$ the points of this set,

$$u = \sum |f(b_i - 0) - f(a_i + 0)| + \sum |s_g(x_i)| + \sum |s_d(x_i)|.$$

It is clear that the first term tends to the total variation of the continuous part ψ of f and the second term ends up with containing the absolute value of all the jumps of f, thus providing the total variation of φ.

When at any point $f(x)$ is contained between $f(x+0)$ and $f(x-0)$ we can take for u the quantity u'

$$u' = \sum |f(b_i - 0) - f(a_i + 0)| + \sum |f(x_i + 0) - f(x_i - 0)|;$$

but, in the general case, this quantity would give a very small limit. When f is of bounded variation, the series constituting u' is convergent, therefore we could arrange, as per our choice, the terms of

$$\sum [f(b_i - 0) - f(a_i + 0)] + \sum [f(x_i + 0) - f(x_i - 0)].$$

Let us group those provided by δ_i and the x_i interior to an interval δ' contiguous to the derived set E' of E. It is clear that the sum of these terms is $\sum[f(\beta' - 0) - f(\alpha' + 0)]$ if δ' is the interval (α', β'). Therefore, by such groupings, we transform the sum related to E into a similar sum related to E'; then into the sum related to E'', etc. And finally we conclude that, if (a, b) is the considered interval, we have, for f of bounded variation,

$$f(b) - f(a) = \sum [f(b_i - 0) - f(a_i + 0)] + \sum [f(x_i + 0) - f(x_i - 0)].$$

5.2 The Rectifiable Curves

Let C be a curve defined on (a, b)

$$x = x(t), y = y(t), z = z(t).$$

Let us consider a polygon P inscribed in this curve, with its vertices, in the order they appear on P, corresponding to increasing values of t,[7] $a, \alpha_1, \alpha_2, \cdots, \alpha_p, b$. We can consider P as a curve defined on (a, b) with the help of functions $\xi(t), \eta(t), \zeta(t)$ equal to $x = x(t)$, $y = y(t)$, $z = z(t)$ for the values $a, t_1, t_2, \cdots, t_p, b$ of t.

This being said, let two sequences of polygons inscribed in C, P_i and Π_j, be chosen such that the maximum of the differences $t_k - t_{k-1}$ tends to zero as $\frac{1}{i}$ on one hand, and as $\frac{1}{j}$ on the other hand. The length of a polygon is, by definition, the sum of the lengths of these edges; we would compare the length s_i of P_i with that of σ_j of Π_j.

Let us suppose that the two consecutive vertices m_1, m_2 of P_i correspond to $t = \theta_1$ and $t = \theta_2$. The points μ_1, μ_2 of Π_j which correspond to these values of t tend to m_1, m_2, when j increases to infinity. The smallest limit, for j infinite, of the arc length $\mu_1\mu_2$ is therefore at least equal to the length of the edge m_1m_2. But, this being true for each edge, the smallest limit of σ_j is at least equal to s_i. And since we could permute P_i and Π_j, the lengths s_i and σ_j tend to the same limit when i and j increase to infinity, and they are always smaller than their limit.

---

[7] When we speak of a polygon inscribed in a curve, we always assume this last condition satisfied.

When the maximum length of the edge of a polygon inscribed in a curve tends towards zero, the length of this polygon tends towards the upper limit of the lengths of polygons inscribed in the curve. It is this limit that we call the length of the curve.

A curve is said to be *rectifiable* if it is of finite length. The study of the rectifiable curves was initiated by Ludwig Scheeffer,[8] then continued by Jordan[9] to whom we owe the following result:

For a curve to be rectifiable, it is necessary and sufficient, that the functions $x(t)$, $y(t)$, $z(t)$ which define it be of bounded variation.

Indeed, any edge of a polygon inscribed in the curve has a length at least equal to each of the projections $\delta_x, \delta_y, \delta_z$ of this edge onto the axes, and of length at most equal to $\delta_x + \delta_y + \delta_z$. But the sum of the projections δ_x is the variation ν_x of the function $x(t)$ for the values of t corresponding to the vertices.[10] The length of the polygon is, therefore, greater than ν_x; it is, similarly, greater than ν_y or ν_z, but it is less than $\nu_x + \nu_y + \nu_z$; the property is proved.

Moreover the *arc length from t_0 to t $(t > t_0)$ of a rectifiable curve is a continuous non-decreasing function of t*, since the increment of this arc, in an arbitrary interval, is contained between the increments of ν_x and $\nu_x + \nu_y + \nu_z$.

To calculate the length of a curve, we could use the polygons having an infinite vertices corresponding to the values of t forming a reducible set; because the initial reasoning applies to these polygons.

A rectifiable plane curve is quadrable, because if we divide it into n pieces of equal lengths of $\frac{s}{n}$, each of them can be enclosed in a circle of radius $\frac{s}{2n}$, and the sum $\frac{\pi s^2}{4n}$ of areas of these circles tend to zero as $\frac{1}{n}$.

Let us suppose that $x(t)$, $y(t)$, $z(t)$ have the integrable derivatives; then $|x'(t)|$, $|y'(t)|$, $|z'(t)|$ are also integrable, because we can write

$$x'^2 = u, \quad |x'| = +\sqrt{u},$$

and if we square or if we take the arithmetic square root of an integrable function, we still have an integrable function.

If l, m, n, L, M, N are the lower and upper limits of $|x'|$, $|y'|$, $|z'|$ in an interval (t_1, t_2), the sums such as $\sum(t_2 - t_1)(L - l)$, extended to any division of (a, b) into partial intervals, tends to zero when the intervals used tend to zero.

The chord (t_1, t_2) has a length δ which satisfies the inequality

$$a = (t_2 - t_1)\sqrt{l^2 + m^2 + n^2} \leq \delta \leq (t_2 - t_1)\sqrt{L^2 + M^2 + N^2} = A.$$

---

[8] *Allgemeine Untersuchungen über Rectification der Curven (Acta Mathematica, t. V).*

[9] *Cours d'Analyse*, t.1, 2nd edition. Scheeffer and Jordan have also examined the case where $x(t)$, $y(t)$, $z(t)$ are not continuous.

[10] The curve $x = x(t)$, $y = 0$, $z = 0$, which is used in the argument, is called the *projection on ox of the given curve; the projection of xoy* is $x = x(t)$, $y = y(t)$, $z = 0$.

Therefore, an inscribed polygon has a length contained between the corresponding sums $\sum a, \sum A$. If the length of the edges of polygon tends to zero, $\sum a$ and $\sum A$ tend to the same limit; because we have

$$\sum A - \sum a \leq \sum (t_2 - t_1) \cdot (L - l) + \sum (t_2 - t_1) \cdot (M - m) \\ + \sum (t_2 - t_1) \cdot (N - n).$$

The limit of $\sum a$ and $\sum A$ is the length of the curve. But, since the integral $\int_a^b \sqrt{x'^2 + y'^2 + z'^2}\, dt$, which exists according to our hypothesis, is also always contained between $\sum a$ and $\sum A$, we can conclude that, if x', y', z' *exist and are integrable, the arc length over* (a, b) *is*

$$\int_a^b \sqrt{x'^2 + y'^2 + z'^2}\, dt.$$

The previous reasoning also shows that if $f'(x)$ exists without being integrable, and we will see that it is possible, the length of the curve $y = f(x)$ is included between the integrals by lower and upper sums of $\sqrt{1 + f'^2}$.

We will obtain the generalisation of this proposition to the curves $x(t)$, $y(t)$, $z(t)$, as well as a result related to the case where $\sqrt{x'^2 + y'^2 + z'^2}$ is a derivative, using the following considerations.

We suppose that x', y', z' exist; then, from point x_0, y_0, z_0, t_0, whatever it may be, as the origin, we can draw a chord whose length $\sqrt{\delta x_0^2 + \delta y_0^2 + \delta z_0^2}$ differs at most by $\varepsilon \delta t_0$, from the quantity $\delta t_0 \sqrt{x_0'^2 + y_0'^2 + z_0'^2}$; and we could even subject δt_0 to be smaller than a certain predetermined quantity λ.

The curve being defined on (a, b), from point $a = t_1$ as origin, we could draw a chord satisfying the indicated conditions; it corresponds to (t_1, t_2). Form t_2 we could draw a new chord which corresponds to (t_2, t_3) and so on. If, after a finite number of operations, we arrive at b, the construction is thus achieved. Otherwise, t_n has a limit point t_ω starting from which, and taking it as origin, we can draw a chord $(t_\omega, t_{\omega+1})$, then from $t_{\omega+1}$ we draw $(t_{\omega+1}, t_{\omega+2})$ and so on. If we do not attain b, we approach a limit point $t_{2\omega}$, starting from which we operate similarly, as we did starting from t_ω.

We thus have the intervals whose origins t_α have for indices the different finite and transfinite numbers α. It is necessary to show that we arrive at b after having exhausted the sequence of transfinite numbers, that is, with the help of a countably infinite number of intervals $(t_\alpha, t_{\alpha+1})$. This is quite obvious, because there are no more than $\frac{b-a}{\eta}$ intervals of length greater than η, and all the intervals, being greater in length than one of the numbers $1, \frac{1}{2}, \frac{1}{3}, \cdots$, form a finite or countable set.

The set of values $t_1, t_2, \cdots$ is reducible, since it is closed and countable; therefore we can use the chords drawn for evaluating the length of the curve. The sum of the length of these chords differ from the sum

$$I = \sum (t_{\alpha+1} - t_\alpha)\sqrt{x'^2(t_\alpha) + y'^2(t_\alpha) + z'^2(t_\alpha)},$$

by at most

$$\varepsilon \sum (t_{\alpha+1} - t_\alpha) = \varepsilon(b - a)$$

If ε and λ simultaneously tend towards zero, $\varepsilon(b - a)$ tends to zero, the sum of the lengths of chords tend to the length s of the curve, I therefore tends to s. But, according to the form of I, we can write, if $\sqrt{x'^2 + y'^2 + z'^2}$ is bounded,

$$\underline{\int_a^b \sqrt{x'^2 + y'^2 + z'^2}\, dt} \leq s \leq \overline{\int_a^b \sqrt{x'^2 + y'^2 + z'^2}\, dt}.$$

Let us now suppose that $\sqrt{x'^2 + y'^2 + z'^2}$*, bounded or not, be the derivative of a function* $\sigma(t)$. If we have chosen each interval $(t_\alpha, t_{\alpha+1})$ in a manner which satisfies, not only the previously mentioned conditions, but also, what is possible, the inequality

$$\varepsilon \delta t_\alpha \geq |\delta\sigma(t_\alpha) - \delta t_\alpha \sqrt{x'^2(t_\alpha) + y'^2(t_\alpha) + z'^2(t_\alpha)}|$$

I tends to the increment $\sigma(b) - \sigma(a)$ of $\sigma(t)$ in (a, b) when ε and λ simultaneously tend towards zero. We therefore have

$$s = \sigma(b) - \sigma(a).$$

The arc length is the increment of the function σ.

I draw your attention to the construction used in the previous proof.

I assume that a method for constructing one or more intervals starting from an arbitrary point t_0 has been provided. I will say that an interval (a, b) has been covered, starting from a, by a chain of intervals chosen from those defined by the given method, when we have constructed, by this method, an interval (t_1, t_2) with origin $t_1 = a$, then an interval (t_2, t_3) with origin t_2 and so on, and if necessary, an interval $(t_\omega, t_{\omega+1})$ whose origin is the limit of $t_1, t_2, \cdots$, and so on. It has been proved that in this way we necessarily arrive at b after a finite or countably infinite number of operations, so that the constructed chain will indeed cover the entire (a, b).[11]

---

[11] When the given method associates multiple intervals with the same origin t_α, one must choose among these intervals the one that will be designated as $(t_\alpha, t_{\alpha+1})$. This choice can be made arbitrarily if the need to choose arises only a finite number of times. If it arises infinitely many times, to avoid the difficulties that arise from the expression "choose infinitely many times," it is better to eliminate the choice by indicating the rule according to which $(t_\alpha, t_{\alpha+1})$ will be determined among all possible intervals. In the previous proof each interval $(t_\alpha, t_{\alpha+1})$ can be required to be the largest one which satisfies the imposed conditions; moreover, within the set of these intervals, there is indeed one interval larger than all the others.

In the final Note we find a detailed study of use of these chains of intervals.

The Search for Primitive Functions 6

6.1 The Indefinite Integral

Let $f(x)$ be a bounded integrable function defined on (a, b); the function

$$F(x) = \int_a^x f(x)\,dx + K$$

is *the indefinite integral of* $f(x)$.

By applying the mean value theorem, we see that *the indefinite integral of* $f(x)$ *is a continuous function of bounded variation*[1] *and it admits* $f(x)$ *as its derivative at all points where* $f(x)$ *is continuous.*

What happens at the α if $f(x)$ is not continuous there? In this case, the derivative may still be equal to $f(\alpha)$, which is the case for $\alpha = 0$ if $f(x)$ is zero for all x except when x is the inverse of an integer, and equal to 1 when x is the inverse of an integer; there may be a derivative different from $f(\alpha)$, which is the case for $\alpha = 0$ when $f(x)$ is zero everywhere except for $x = 0$; there may not be any derivative, which is the case for $\alpha = 0$ when $f(x) = \cos \mathfrak{L}|x|$ for $x \neq 0$ and $f(0) = 0$.[2]

Thus, integration can lead to functions that do not have a derivative everywhere. This consequence was indicated by Riemann, who drew attention to the indefinite integral of the function

$$f(x) = \sum \frac{(nx)}{n^2}.$$

[1] I leave it to the reader to show that the total variation of $F(x)$ on (a, b) is exactly equal to $|\int_a^b |f(x)|\,dx|$. This proposition had also been proved incidentally (Sect. 5.1) as a particular case of the result related to the length of a curve $x(t)$, $y(t)$, $z(t)$, when $x'(t)$, $y'(t)$, $z'(t)$ are integrable. Indeed, it is enough to consider the curve $x(t) \equiv f(t)$, $y(t) \equiv z(t) \equiv 0$.

[2] Then the indefinite integral is $\frac{x}{2}(\sin \mathfrak{L}|x| + \cos \mathfrak{L}|x|)$.

R. Jain, *Lebesgue's Theory of Integration*,
https://doi.org/10.1007/978-981-96-1169-0_6

This[3] indefinite integral $F(x)$ admits as its derivative $f(x)$ when x is not of the form $\frac{2p+1}{2n}$.

Let us suppose $\alpha = \frac{2p+1}{2n}$ and β increases towards α. We just saw that $f(\beta)$ tends to $f(\alpha) + \frac{\pi^2}{16n^2}$. Therefore, according to the mean value theorem, the same holds true for $\frac{F(\beta)-F(\alpha)}{\beta-\alpha}$.

On the contrary, this ratio would tend to $f(\alpha) - \frac{\pi^2}{16n^2}$ if β decreases towards α; therefore $F(x)$ does not have the derivative for the values of the form $\frac{2p+1}{2n}$.

This is the first known example of a function for which it is not clearly legitimate to say that it generally admits a derivative. Many functions were known, such as Cauchy's function, for example $+\sqrt{x^2}$, which did not have derivatives at certain points. However, these points were exceptional, they never formed an everywhere dense set. In Riemann's example, on the contrary, there are points without a derivative in every interval. The principle of condensation of singularities will give us as many examples as we want, of functions similar to that of Riemann; if the numbers a_p are all rational numbers, $\int \sum \frac{\cos \mathfrak{L}|x-a_p|}{p^2} dx$ is one of these functions.

The integration produces functions which do not always have a derivative. By an entirely different method, Weierstrass constructed a function that is nowhere differentiable[4]. It is obvious that the integration cannot yield such functions: *The points at which an indefinite integral does not have a derivative form a set of measure zero*, since these points belong to the set of points of discontinuity of the integrated function $f(x)$. Thus, the points without derivatives are still, in some sense, exceptional.

When a function $f(x)$ is bounded, but not integrable, we can attach to it *the two indefinite integrals by upper and lower sums*

$$\overline{F(x)} = \overline{\int_a^x f(x)\,dx} + K, \quad \underline{F(x)} = \underline{\int_a^x f(x)\,dx} + K.$$

These two functions are continuous, of bounded variation, and admits f as its derivative at all points where f is continuous.[5]

The concept of an indefinite integral is related to an important generalisation of the definite integral.

If a function $f(x)$ defined on (a, b) is non-integrable over (a, b), but integrable over any interval (α, β) interior to (a, b), we can hope to define an integral of $f(x)$ over (a, b) by assuming, in principle, the continuity of the indefinite integral and by applying the Cauchy's methods.

[3] See Sect. 3.1. The indefinite integral is obtained by integrating term-by-term.

[4] See *Journal de Crelle,* Vol. 79, or JORDAN, *Cours d'Analyse,* 2nd edition, t.1, p. 316.

[5] The property regarding the set of points without a derivative is also true for the integrals by lower and upper sums; later we would see, that it belongs to all functions of bounded variation.

We easily see that the assumed conditions are never realised if $f(x)$ is bounded. However, if $f(x)$ is unbounded, Cauchy's method can lead us to a definite number; it will be so in particular, around a and b, $|f(x)|$ is less than a function of definite infinitesimal order at infinity, less than 1.[6]

Regarding Riemann integration, all reasoning applied about Cauchy integration and the Cauchy–Dirichlet procedures can be repeated; I will not dwell on this point.[7]

6.2 The Derivative Numbers

The integration applies to the functions that are not derivative function. A function zero everywhere, except at $x = 0$, is not a derivative function, since its primitive function, if it exists, would be continuous, constant for positive and negative x, therefore always constant and however its derivative would not be zero for $x = 0$. This shows that the notions of indefinite integral and primitive function are different.

It seems that it has long been admitted that the first of these notions includes the second, and as a result, the integration always allows to resolve the problem of the search for primitive functions. In any case, instead of dealing with this problem, we have study what services could the integration provide in solving the problems, generalisations in the various senses of the problem of primitive functions.

For the study of this problem it will be useful for us to know a few properties of derivative numbers.

Let $f(x)$ be a continuous function,[8] let us consider the ratio

$$r[f(x), x_0, x_0 + h] = \frac{f(x_0 + h) - f(x_0)}{h};$$

and let h tend to 0. If we subject h to take only the negative values, the smallest and the largest of the limits of the ratio are two *derivative numbers to the left* at point x_0. These two numbers, which have been defined and studied by P. Du Bois Reymond and Dini, are still called the *anterior oscillatory extremes*. The smallest limit is the *left-hand inferior derivative number*,

6 In a more general sense, we can apply all standard theorems given related to the existence of an integral when the quantity placed under the integration sign becomes infinite at a point.

7 These questions are related to a generalisation of the integral given by Jordan in Volume II of the second edition of his *Cours d'Analyse*. If generalisations in the text allow us to define the integral of $f(x)$ in any interval contiguous to a closed set E, Jordan calls the *integral of* $f(x)$ the sum of the integrals over the intervals contiguous to E. For the integral of a sum to be the sum of the integrals, it is necessary to add that the outer extent of E must be zero. These questions are connected to the work of Harnack *(Math. Ann., Bd XXI, XXIV)*, Holder *(Math. Ann., Bd XXIV)*, de la Vallée Poussin *(J. de Liouville, serie* 4, *vol. VIII)*, Stolz *(Wiener Berichte, Bd $CVII$)*, Moore *(Trans. Amer. Math. Soc., vol. II)*.

8 We can also consider the case of discontinuous functions, but the definitions of the text would be sufficient for us.

the greatest limit is the *left-hand superior derivative number*. By giving to h the positive values, we define the *two right-hand derivative numbers* or *posterior extreme oscillators*.

These four numbers, which are not necessarily finite, are denoted

$$\lambda_g, \Lambda_g, \lambda_d, \Lambda_d;$$

If we want to emphasise the function f and the value x_0 in question, we write $\lambda_g f(x_0)$, $\Lambda_g f(x_0)$.[9]

The geometric significance of these numbers is simple. Let $y = f(x)$ be a curve, let us consider an arc AB of this curve corresponding to the interval $(x_0, x_0 + h)$; let us assume it to be positive. All the straight lines joining A to any point of AB are all the lines of a certain angle XAY. Let h tend to zero, the angle XAY varies in such a manner that, for the value of h, it contains all the angles corresponding to the values less than h.

This would be enough for us to conclude the existence of the right limits $\xi A, \eta A$ for XA and YA. The angular coefficients of these two right limits are the derivative numbers to right.

We could make figure for the curve $y = x \sin \frac{1}{x}$; for $x = 0$ the two inferior derivative numbers are -1 and the two superior derivative numbers are $+1$. For this curve the angle XAY is fixed. On the contrary, it is true for the function

$$y = x \sin \frac{1}{x} + x^2 \sin \frac{1}{x},$$

which admit the same derivative numbers as the previous one for $x = 0$. The derivative numbers could replace, in some studies, the ordinary derivatives. In the study of the variation of a function for example: if all the four derivative numbers are positive, the function is increasing; if the two posterior derivative numbers are positive and the two anterior numbers negative, the function admits a minimum for $x = x_0$; if the two right-hand derivative numbers are of opposite signs, the function is neither increasing nor decreasing to the right of $x = x_0$, but if one of the two is zero we cannot say anything anymore.

When $\Lambda_d = \lambda_d$, we say that the function admits a *right-hand derivative* equal to Λ_d; if $\Lambda_g = \lambda_g$, the value of Λ_g is the *left-hand derivative*.

If $\Lambda_d = \lambda_d = \Lambda_g = \lambda_g$, the function has a derivative equal to Λ_d. This definition is identical to the classical definition, except in the case where $\Lambda_d = \pm\infty$.[10]

Let us apply this definition to the integral. The mean value theorem gives

$$l \leq r[F(x), \alpha, \beta] \leq L,$$

if F is one of the three indefinite integrals and if l and L are the lower and upper limits of f in (α, β); we can even suppose that α is excluded from the interval (α, β).

If β tends towards α by values less than α, we see that *the left-hand superior derivative number at $x = \alpha$ of one of the indefinite integrals of a bounded function $f(x)$ is at most*

9 We sometimes also use the notation D_-, D^-, D_+, D^+ or d_-, D_-, d_+, D_+.

10 With this definition $\sqrt[3]{x}$ admits a definite derivative, ∞, for $x = 0$.

equal to the limit superior of $f(x)$ to the left of α and the left-hand inferior derivative number of $f(x)$[11] *is at most equal to the limit inferior of $f(x)$, to the left of α. Here, we are talking about limit inferior and limit superior at α, excluding the point α* (see Sect. 3.1).

Let us suppose that $f(\alpha - 0)$ exists, then the two limits of $f(x)$ to the left of α are $f(\alpha - 0)$, therefore; when $f(\alpha - 0)$ *exists, one of any indefinite integrals of the bounded function $f(x)$ admits, for $x = \alpha$, a left-hand derivative equal to $f(\alpha - 0)$.*

We reason in the same way for the derivative numbers and the right-hand derivative.

Riemann's function $\sum \frac{(nx)}{n^2}$, which admits points of discontinuity of first kind only, leads to an indefinite integral which has, at any point, a well-defined right-hand derivative and a left-hand derivative. In short, it is the existence of these right-hand and left-hand derivatives which has been demonstrated in Sect. 6.1.

If $f(\alpha - 0)$ and $f(\alpha + 0)$ exist and are equal, the integral of $f(x)$ admits the common value of $f(\alpha - 0)$ and $f(\alpha + 0)$ as its derivative when $x = \alpha$, regardless of the value of $f(\alpha)$.

There exists a proposition for the derivative numbers similar to the theorem of finite increment[12]:

If L and l are the upper and lower limits of any one of the four derivative numbers of the function $f(x)$ in (a, b), we have

$$l \leq r[f(x), a, b] \leq L.$$

I suppose that l and L are related to Λ_d; the other cases reduce to this one, because we obviously have:

$$\lambda_d[f(x)] = -\Lambda_d[-f(x)],$$
$$\lambda_g[f(x)] = -\Lambda_g[-f(x)],$$
$$\Lambda_g[f(x)] = +\Lambda_d[f(-x)].$$

We, therefore, assume that l and L are limits of Λ_d; to demonstrate the second inequality, it suffices to prove that there exist values of Λ_d at least equal to

---

[11] **Translator's Note:** The original text states 'left-hand inferior derivative number of $f(x)$'. However, given the context and standard mathematical conventions, M. Lebesgue may have intended to refer to the 'left-hand inferior derivative number of one of the indefinite integrals of $f(x)$' ensuring consistency with the earlier reference to the 'left-hand superior derivative number of one of the indefinite integrals of $f(x)$'.

[12] We know that this theorem is stated as follows:

If a function $f(x)$ is continuous in the interval (a, b), and admits a well-determined derivative for each value of x interior to (a, b), there exists a number ξ in this interval such that

$$f(b) - f(a) = f'(\xi)[b - a].$$

This statement does not assume that $f'(x)$ to be bounded or even finite, but if $f'(x)$ is infinite, this must be $+\infty$, and not $\pm\infty$.

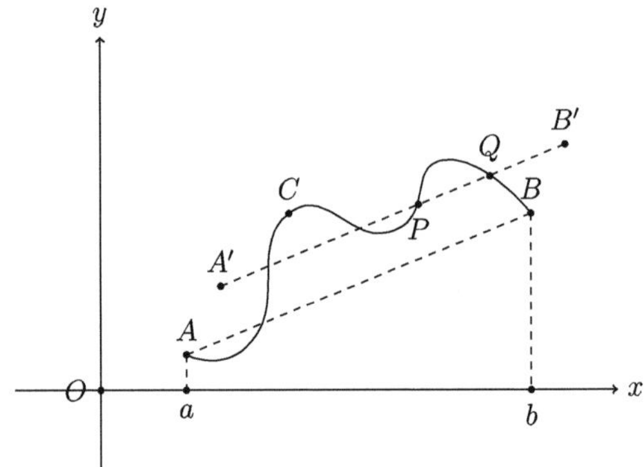

Fig. 6.1 Curve Δ

$$r[f(x), a, b].$$

For this, I adopt the geometric language because it appears to me more expressive; we can translate it easily, if we want, in analytic language. The property is obvious if the curve C which $f(x)$ represents reduces itself to the chord AB joining its extremities (Fig. 6.1).

If this is not the case and if there exist points of the curve C above AB (i.e. on the edge of $y = +\infty$), I displace the line AB parallel to itself at $A'B'$ in a manner that it cuts C.

Above $A'B'$ there are arcs of C, let PQ be one of them. At point P of $A'B'$, Λ_d and λ_d are obviously greater than or at least equal to the angular coefficient of PQ, i.e., $r[f(x), a, b]$ and the property is proved in this case.

Finally, if C does not have a point above AB (Fig. 6.2), I displace AB parallel to itself towards $y = -\infty$, and let $A'B'$ be the last position in which it has common points with C. If P is one of any of these points, at this point Λ_d and λ_d are at least equal to $r[f(x), a, b]$. Hence, the property is proved in all the cases. In both the cases shown in the figure, we have reasoned on an arc P_p, of origin P, and located above parallel to AB drawn through P. Furthermore, for the sake of what follows, we took P to be different from A.

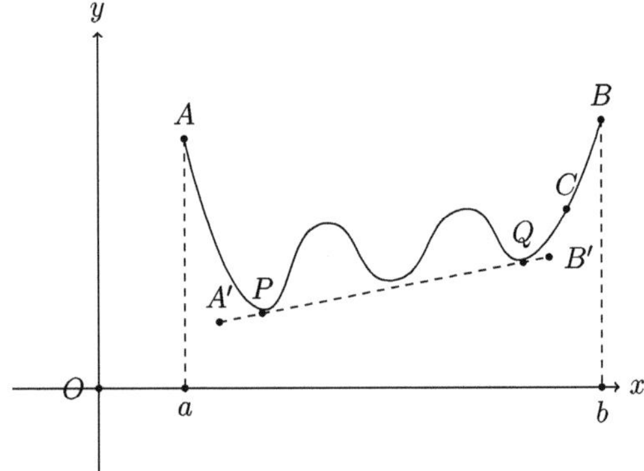

Fig. 6.2 Curve Λ

From the previous theorem, it follows that *the four derivative numbers have the same upper limit and the same lower limit in any interval.*

Indeed, let us compare the upper limits L and L' of Λ_d and λ_g. Since Λ_d has for limit L and Λ_d is the limit of ratio $r[f(x), \alpha, \beta]$, where α and β belong to the considered interval (a, b), we can find α and β in (a, b) such that $r[f(x), \alpha, \beta]$ is greater than $L - \varepsilon$. The maximum of λ_g in (α, β), and therefore in (a, b), is, as a result, at least equal to $L - \varepsilon$. This is enough to prove that L and L' are equal.

The common value of L and L' is at the same time upper limit of the ratio $r[f(x), \alpha, \beta]$.

The property stated for the upper and lower limits in an interval implies the same property for the limit superior and the limit inferior at a point. In particular, if for one of the derivative numbers these two limits are equal, the same holds true for the others. This can be stated as: *If at a point x_0, one of the derivative numbers is continuous, the same holds true for the other three derivative numbers and furthermore, the function admits a derivative for $x = x_0$.*

Here is another obvious consequence: *if the four derivative numbers are bounded, they admit the same upper integral and the same lower integral; if one them is integrable, then so are the others and they have the same integral.*

In the case of derivatives, the Rolls[13] theorem is a particular case of the theorem of finite increment; in the case of derivative numbers the theorem, similar to the Rolle's theorem, can be stated as: *If the continuous function $f(x)$ vanishes at a and b, the limits of the derivative numbers in (a, b) are, either all zero, or all different from zero and of opposite signs.*

This statement is justified by noting that if $f(x)$ is not constant, $r[f(x), \alpha, \beta]$ takes positive and negative values.

We can therefore say: *if the continuous function $f(x)$, not constant in (a, b), vanishes at a and b, there exist points, interior to the interval (a, b), for which the two right-hand or left-hand derivative numbers are positive and non-zero and the other points where they are negative and non-zero.* If in Figs. 6.1 and 6.2, we suppose $bB > aA$, the two derivative numbers at point P, interior to (a, b), are in fact different from zero and positive.

The converse can be stated in the following form: *if we know that the two right-hand or left-hand derivative numbers of $f(x)$ are never both different from zero and are of the same sign, then $f(x)$ is a constant.*[14]

Among the continuous functions it is necessary to note that *the functions of bounded derivative numbers* possess many properties of the differentiable functions. This class of functions include the indefinite integral of bounded functions. The functions of bounded derivative numbers are those for which we always have

$$|r[f(x), \alpha, \beta]| < M,$$

[13] This theorem is stated as follows:

If a continuous function $f(x)$ vanishes at a and b, and admits at points interior to (a, b) a definite derivative, with definite magnitude and sign, finite or infinite, this derivative vanishes in (a, b).

[14] This property corresponds to the following: If derivative of a continuous function is zero at every x in (a, b), the function is constant.

where M is a fixed number. This inequality, known as *condition of Lipschitz,* appears in almost all the arguments on the existence of the solutions of the differential equations. This shows the practical importance of the functions of bounded derivative numbers.

We would revisit in chapter 9 the study of these functions; for now it would suffice to indicate an immediate property:

A function of bounded derivative numbers and less than M in absolute value, is of bounded variation, and its total variation is at most $M\delta$ in an interval of extent δ.

Let now a rectifiable curve be

$$x = x(t), y = y(t), z = z(t),$$

defined on (a, b), and let $s(t)$ be its arc from a to t.

The equation $s(t) = s$ can be solved in t when s is in the interval $[0, s(b)]$; it admits a unique solution, except in the case where $x(t)$, $y(t)$, $z(t)$ would be constant simultaneously in an interval. Except in this case, $t(s)$ is a well-determined increasing function,

$$x = x[t(s)], y = y[t(s)], z = z[t(s)]$$

represents the given curve and these functions of s are the continuous functions of derivative numbers at most equal to 1.

The study of rectifiable curves, and as a result, of the functions of bounded variation, is therefore linked, initially, to the study of functions of bounded derivative numbers. We would have the opportunity of using this remark.

Furthermore, there exist continuous functions of bounded variation and unbounded derivative numbers, the function $x^2 sin\frac{1}{x}$ is such an example.

The functions we have just seen are defined on the whole interval, but it is clear that the concept of derivative and the derivative numbers extend immediately to a function defined only for points of a set, or considered only for points of a set. The function $\chi(x)$ (Sect. 3.1) is discontinuous at every point; however it admits a derivative equal to zero, in the set of rational numbers at rational points x and a derivative equal to zero, in the set of irrational numbers at irrational points x.

6.3 Functions Determined by One of Their Derivative Numbers

Let us return to the search for the function primitives. The problem:

A. *To find a function whose derivative is a given function,*

does not admit, in general, a solution. Therefore, we replace it by two others:

B. *To determine if a given function is a derivative function.*

C. *To find a function when its derivative is known.*

To these problems, correspond the following:

A'. *To find a function whose right-hand superior derivative number (or one of the other derivative numbers) is given.*
B'. *To determine, if a given function is right-hand superior derivative number of an unknown function*
C'. *To find a function when its right-hand superior derivative number is known.*

We would first clarify the indetermination of the solution of problem C' by showing that *a function is determined, up to an additive constant, when we know the finite value of one of its derivative numbers for each value of the variable.*

Let, in fact, two functions $f_1(x)$ and $f_2(x)$ have at each point the same right-hand superior derivative number. We have, by hypothesis,

$$\Lambda_d f_1(x) = \Lambda_d f_2(x)$$

and also

$$\lambda_d[-f_2(x)] = -\Lambda_d f_2(x),$$

as we see by referring to the geometric or analytic definition of the derivative numbers. This definition also provides the inequality

$$\lambda_d[f_1(x) - f_2(x)] \leq \Lambda_d f_1(x) + \lambda_d[-f_2(x)] \leq \Lambda_d[f_1(x) - f_2(x)].$$

in which the middle term is zero.

Therefore, the function $f_1(x) - f_2(x)$ never has its two right-hand derivative numbers, different from zero and of the same sign; it is constant.

Our proposition is proved. The proof does not assume that the function has bounded derivative numbers, but it does assume that the given derivative number is finite; otherwise, the middle term, in the inequality we have used, does not make any sense.

Moreover, the proposition is no longer necessarily true when the known value of the derivative number, or of the derivative, is not finite everywhere. Specifically, we may cite two continuous functions $f_1(x)$ and $f_2(x)$, *having at any point the same derivative, of definite sign and magnitude but not finite everywhere, and their difference is not a constant.*

To this end, let us again consider the set Z of Sect. 3.2 and function $\xi(x)$ of Sect. 5.1. $\xi(x)$ is continuous, increasing; it has a zero derivative in each interval contiguous to Z, it has positive or zero derivative numbers at the points of Z. Therefore, if we find a continuous function $f_1(x)$ having a definite derivative at every point, this derivative being $+\infty$ at the points of Z, the continuous function $f_2(x) = f_1(x) + \xi(x)$ has, at every point, the same derivative as that of $f_1(x)$ without differing from it by a constant.

To construct $f_1(x)$, let us arrange the intervals contiguous to Z in an infinite sequence, with their lengths forming, for example, the following sequence

$$\delta_1 = \frac{1}{3}, \quad \delta_2 = \delta_3 = \frac{1}{3^2}, \quad \delta_4 = \delta_5 = \delta_6 = \delta_7 = \frac{1}{3^3}, \ldots$$

if this order has been adopted, the series $\Sigma(\delta_i)^k$ will be convergent for k smaller than one and sufficiently close to one, in a precise manner, as soon as $\frac{2}{3^k}$ is smaller than one. Let us assume that k is chosen in this way and now let us define $f_1(x)$.

If x is a point of Z, we will take

$$f_1(x) = 2^{1-k} \sum_0^x (\delta_n)^k,$$

the summation being extended to the intervals δ_n included between 0 and x. If (α, β) is an interval contiguous to Z, we would take

$$f_1(x) = f_1(\alpha) + (x - \alpha)^k, \quad \text{for} \quad \alpha \le x \le \frac{\alpha + \beta}{2}$$

and

$$f_1(x) = f_1(\beta) - (\beta - x)^k, \quad \text{for} \quad \frac{\alpha + \beta}{2} \le x \le \beta.$$

It is clear that $f_1(x)$ is continuous at point $\frac{\alpha+\beta}{2}$, because

$$f_1(\beta) - f_1(\alpha) = 2^{1-k}(\beta - \alpha)^k = 2\left(\frac{\beta - \alpha}{2}\right)^k;$$

and it admits a definite derivative.

The function $f_1(x)$ is therefore continuous in (0, 1), monotonically increasing and it admits a definite and finite derivative at all the points not belonging to Z. Now, we will show that at points of Z it has a derivative equal to $+\infty$.

Let x_0 and x_1, $x_0 < x_1$, be two points of Z. We have:

$$\begin{aligned} f_1(x_1) - f_1(x_0) &= 2^{1-k} \sum_{x_0}^{x_1} (\delta_n)^k > 2^{1-k} \left(\sum_{x_0}^{x_1} \delta_n\right)^k > \left(\sum_{x_0}^{x_1} \delta_n\right)^k \\ &= (x_1 - x_0)^k, \end{aligned}$$

because, since k is smaller than one, we have

$$\sum_{x_0}^{x_1} (\delta_n)^k > \left(\sum_{x_0}^{x_1} \delta_n\right)^k.$$

However, in an interval (α, β) contiguous to Z, we obviously have

$$f_1(x) - f_1(\alpha) \ge (x - \alpha)^k;$$

therefore, for any x greater than the value x_0, belonging to Z we have

$$f_1(x) - f_1(x_0) \geq (x - x_0)^k.$$

And as a result $f_1(x)$ has, at point x_0, right-hand derivative, at least equal to that of $(x - x_0)^k$ at this point, therefore equal to $+\infty$.

Similarly, we will prove that $f_1(x)$ has a left-hand derivative equal to $+\infty$ at the points of Z. Therefore $f_1(x)$ does indeed fulfil the imposed conditions; the functions $f_1(x)$ and $f_2(x)$ show us that a function is not determined up to an additive constant by the knowledge of its derivative at every point, when it is not finite everywhere.[15]

The functions encountered in Analysis, and that are not bounded generally take infinite values at some points. We can also say that it is rare to be able to prove that a function is always finite without proving, thereby, that it is bounded. Therefore, the study of the problem C and C' was hardly of interest except in case where the given function, as a derivative or as a derivative number, is bounded if one could not establish cases where the solution functions to the problems C and C' remain determined, up to an additive constant, although the given function may not be finite everywhere. It is this what we are going to do:

If the finite number $\Lambda_d f(x)$ is given for all the values of the variable, except for the points of a set E, the *continuous function* $f(x)$ is determined, up to an additive constant, in every interval which does not contain points of E in its interior; therefore, the same holds for every interval if E is reducible, as we have seen by revisiting the reasoning used in chapter 1 in the context of research of Cauchy and Dirichlet.

We would have a similar result whenever we know a solution set of one of the following problems:

D. *In what set of points is it sufficient to know the finite derivative of a function for which this function is determined up to an additive constant?*

D′. *In what set of points is it sufficient to know the finite value of the superior derivative number to the right of a function for which this function is determined up to an additive constant?*

We have just quoted a family of sets responding to the question: the sets complementary to the reducible sets; meaning that, the ones which we obtain by removing the points of a reducible set from the considered interval. We owe Ludwig Scheeffer a more general solution:

A function is determined, up to an additive constant, when we know for each value of x, except, may be, for those of a countable set D, the finite value of the right-hand superior derivative number of this function.

---

[15] The previous example is due to M. Hans Hahn (*Mon. Math. Phys.*, 1905). It was constructed in response to a question that I had asked in the first edition of this book.

Since then other examples have been given, in particular by M. Denjoy. See also M. Ruziewicz (*Fund. Mat.*, *t.*1).

Let $f_1(x)$ and $f_2(x)$ be the two functions having, in general, the same finite right-hand superior derivative number we will show that we always have

$$f_1(x) - f_2(x) = f_1(a) - f_2(a),$$

and to this end, we show the equality

$$f_1(b) - f_2(b) = f_1(a) - f_2(a) - H, \tag{6.1}$$

where H is non-zero, is impossible. It is enough to consider the case where H is positive, since the other case would reduce to this one by interchanging f_1 and f_2. Similarly, we can assume $b > a$.

Let us consider the function

$$\phi_c = c(x - a) + f_1(x) - f_2(x) - f_1(a) + f_2(a) + \frac{H}{2},$$

in which c is a constant such that

$$0 < c < \frac{H}{2(b-a)}.$$

Then

$$\phi_c(a) = \frac{H}{2} > 0, \quad \phi_c(b) = c(b-a) - \frac{H}{2} < 0;$$

the function ϕ_c being continuous and vanishes between a and b; let x_c be the greatest value included between a and b at which ϕ_c vanishes. We obviously have

$$\Lambda_d \phi_c(x_c) \leq 0.$$

Hence, we can conclude that x_c is a point of D.

In fact, we have shown (Sect. 6.3), that for any point not belonging to D, we have

$$\Lambda_d[f_1(x) - f_2(x)] \geq 0;$$

therefore at these points we have

$$\Lambda_d \phi_c(x) \geq c > 0.$$

To each value of c of the interval $\left[0, \frac{H}{2(b-a)}\right]$ thus, corresponds a point x_c of D. But, if c and c_1 are different, then x_c and x_{c_1} are as well; because the equality

$$\phi_c(x_c) = \phi_{c_1}(x_{c_1}) = 0$$

implies

$$c(x_c - a) = c_1(x_{c_1} - a)$$

and x_c is different from a.

Therefore, for the equality (6.1) to be possible, it is necessary that D has the cardinality of the continuum.[16]

A consequence of this property, pointed out by Ludwig Scheeffer, is that a function is determined when we know its finite derivative, for all the values of irrationals. However, a function is not determined when we know, for each rational value of x, the finite value of its derivative. To prove this, let $x_1, x_2, \ldots$ be positive rationals numbers. Let us construct an interval δ_1 of incommensurable length, having x_1, as mid-point. Let x_{α_2} be the first of the x_i not belonging to δ_1; let us construct an interval δ_2 of incommensurable length, centred at x_{α_2} and not overlapping with δ_1. If $x_{\alpha 3}$ is the first of x_i which belongs neither to δ_1 nor to δ_2, x_{α_3} is the mid-point of an interval of incommensurable length that overlaps neither with δ_1 nor with δ_2, and so on.

The function $f(x)$, equal to the sum of the lengths of the intervals δ and the parts of the interval δ, contained between 0 and x, is a continuous increasing function of x, which admits $+1$ as derivative for all the rational values of x. However, this function is not necessarily of the form x+const., since $f(+\infty) - f(0)$ is the sum of the lengths of δ, a sum that can have any desired positive value.

The continuous function $f(x) - 1$ is not constant and in any interval there exist points where its derivative is zero.

It was in the context of studying functions whose derivatives vanish in every interval, that Ludwig Scheeffer embarked on his research on the determination of a function by its derivative numbers.

As functions whose derivative vanishes in every interval, we can still cite the function $\phi(x)$ (Sect. 3.3) and the function $\xi(x)$ (Sect. 5.1).

The previous proof, inspired from certain methods of the proof of the theorem of finite increment or from the Taylor's formula, is quite artificial, and here is another:

The two functions f_1 and f_2 having the same Λ_d at any point, except, possibly, at the points of D, the function $f(x) = f_1 - f_2$ has, at any point not belonging to D , a Λ_d positive or zero and a λ_d negative or zero. If α is such a point, let us associate it with the largest interval $(\alpha, \alpha + h)$ such that we have

$$f(\alpha + h) - f(\alpha) < \varepsilon h.$$

Let us suppose that the points of D are arranged in a simply infinite sequence, $x_1, x_2, \ldots$. Let us associate x_n with the largest interval (x_n, x'_n) such that we have

$$f(x'_n) - f(x_n) < \frac{\varepsilon}{2^n}.$$

[16] The previous proof is, almost exactly, that of L. Scheeffer. I have also followed his statement, but it is worth noting that the proof only assumes that D does not have the cardinality of the continuum, which does not perhaps mean that D is countable.

Each point of (a, b) is now the origin of an interval δ attached to this point; we could cover (a, b), starting from a, with the help of a chain of intervals δ (Sect. 5.2). Let us use these intervals for calculating $f(b) - f(a)$, we find that this quantity is at most equal to

$$\varepsilon \sum h + \varepsilon \sum \frac{1}{2^n} \leq \varepsilon(b - a - 1);$$

but, ε is arbitrary, therefore $f(b) = f(a)$; and, since this reasoning could be used for any subset of (a, b), the function $f(x)$ is constant.

This method of proof leads to another result. Let us suppose that the exceptional set is, no longer a countable set D, but a set E of measure zero. This means that the points of E could be covered with the help of an infinite countable number of intervals d whose sum of lengths is as small as we want.

To a point α, not belonging to E, let us associate an interval $(\alpha, \alpha + h)$, or δ, as it has been done above. To a point α of E we now associate, as the interval δ, the interval δ_1, whose origin is α and whose extremity is the extremity of the interval d containing α.

We recover (a, b) from a with the help of a chain of intervals δ and δ_1; this chain gives, as upper limit of the increment $f(b) - f(a)$ of $f(x)$ in (a, b), the number $\varepsilon \sum h$ increased by the sum of increments of $f(x)$ in the intervals δ_1. The sum λ of the lengths of the δ_1 is smaller than the sum relative to d, therefore it is as small as we want. In general, this does not allow us to conclude that the corresponding sum of increments of $f(x)$ is as small as we want. However, if f_1 and f_2 have derivative numbers less than M in absolute value, this sum is smaller than $2M\lambda$. Thus:

A function, of bounded derivative numbers, is determined, up to an additive constant, when we know its right-hand superior derivative number, for any value of x, except at those of a set of measure zero.

This statement does not provide us with any information relating to the indetermination of the problem C' when the given derivative numbers are not bounded.

However, let us notice that it is only for the points of E for which we needed to know that the upper limit of the derivative numbers of $f(x)$ is of the form $2M$, without having the necessity to know M any longer; this would lead us to the following statement which contains all the previous ones:

A continuous function is determined, up to an additive constant, when we know that it has a finite right-hand superior derivative number at any point, except, possibly, at points of a countable set D, and when we know this finite derivative number, except at most at the points of a set E of measure zero.

Let $f = f_1 - f_2$ always be the difference of two functions fitting the given data and let us cover (a, b), starting from a, with a chain of intervals. Those intervals which have for origin the points of D, or not belonging either to D or E, are chosen as described and give in $f(b) - f(a)$ a contribution equal to $\varepsilon(b - a + 1)$ at most. Let us partition E into sets E_i, where each set E_i consists of points where right-hand superior derivative number of f satisfies the inequality

$$i - 1 \leq |\Lambda_d f| < i;$$

and encloses E_i, which is of measure zero, in the intervals d_i of total length $\frac{\varepsilon}{2^i i}$. An interval $(x, x+\delta)$ of the chain having for origin a point of E_i would be taken interior to d_i and such that

$$\left| \frac{f(x+\delta) - f(x)}{\delta} \right| < i.$$

Then the intervals of the chain having for origin the points of E_i have a contribution to $f(b) - f(a)$, at most, equal to their length multiplied by i, therefore less than $\frac{\varepsilon}{2^i i} \times i = \frac{\varepsilon}{2^i}$. And, as a result, we find

$$f(b) - f(a) < \varepsilon(b - a + 2).$$

We keep only one restriction: the right-hand superior derivative number is finite.

This restriction is moreover necessary: The function $\xi(x)$ (Sect. 5.1) is not a constant, although it has its derivative zero everywhere, except, possibly, at the points of Z, which is of measure zero.

The previous theorem could be advantageously transformed; for these transformations I would use a useful generalisation of the notion of the lower limit and upper limit which is due to M. Baire.[17]

Let $f(x)$ be a function; the upper limit of $f(x)$, in an interval (a, b), is a number L such that the set $E(f > m)$, formed from the points x of (a, b) such that $f(x)$ be greater than m, exists as soon as m is smaller than L, while it does not contain any point for $m > L$; the lower limit of $f(x)$ in the interval (a, b) can be defined similarly.

Similarly there exists a number L_1 such that the set $E(f > m)$ is countable for $m > L_1$ and is uncountable for $m < L_1$. This number L_1 is called by M.Baire *the upper limit of $f(x)$ in (a, b), when we neglect the countable sets.*

This example would be enough to understand what we mean by the limit superior or limit inferior, in an interval or at a point, of a function when we neglect the countable sets, or the non-dense sets, or the sets of measure zero. If, by neglecting certain sets, we obtain the equal limit inferior and limit superior, we can say that,up to these sets, the function is continuous.

Given these definitions, here are the two propositions which I have in view: *The limit superior and inferior of a derivative number are the same, whether or not we neglect the countable sets.*

The limit superior and inferior of a finite derivative number are the same, whether or not we neglect the countable sets.

I show, for example, the first of these two propositions. If the limits superior L and L_1 of a derivative number $\Lambda\phi(x)$, obtained by taking into account and then by ignoring the countable sets, are equal, and if K is a finite number contained between L and L_1, the derivative number $\Lambda[\phi(x) - Kx]$ is negative, except for the points of a countable set for which it is positive.

---

17 *These Sur les fonctions de variables relles (Annali di Matematica,* 1900*).*

However, it is sufficient to revisit, by modifying it slightly, one of the two reasonings that led us to the theorem of Scheeffer to see that it is impossible.

6.4 Search for Function Whose Derivative Number is Known

We will try to solve the problems B' and C' in the case where the function, given as Λ_d, is bounded.

Let us divide the positive interval (a, b) into partial intervals. In (α, β) the lower and upper limits of $f(x)$ are l and L, therefore, if F is the sought function such that $\Lambda_d F(x) = f(x)$, we have

$$(\beta - \alpha)l \leq F(\beta) - F(\alpha) \leq (\beta - \alpha)L.$$

If we add similar inequalities, relative to the partial intervals, we have, by letting these intervals tend towards zero,

$$\underline{\int_a^b} \Lambda_d f(x)\, dx \leq F(b) - F(a) \leq \overline{\int_a^b} \Lambda_d f(x)\, dx.$$

From this inequality it follows, in particular, that: *if one of the derivative numbers of a function $f(x)$ is integrable, in which case the three others are as well and have the same integral, its indefinite integral is of the form $f(x)$+const.*. More specifically: *when a derivative is integrable, there is an identity between its primitive functions and its indefinite integrals.*

These statements would obviously apply to the cases where the given function becomes infinite in the neighbourhood of the points of reducible sets, on the condition that the generalisation of the integral mentioned in Sect. 6.1 is used.

If we take into account the theorems stated at the end of the previous section, we see that if we know the derivative numbers everywhere, except for, at the values of a countable set,—or if we know it everywhere, except for, at the values of a set of measure zero, and moreover if we also know that it is bounded everywhere, we could still apply the previous theorems on the condition of extending the integral which appears there, to the set in which we know the derivative number.

To this observation is linked another one, which is more important. The case in which we can solve problem C' is the one where the given derivative number is integrable. This derivative number then has points of continuity; at these points, the integral of this derivative number has a derivative equal to the given derivative number, and we know the derivative of the unknown function everywhere, except at the points of a set of measure zero. It would be sufficient to use the known values of the derivative to obtain the function. The solved case of problem C' therefore is actually reduced to the problem C.

The previous reasoning allows us to answer the questions B and B' in an important case: the one where the given function is integrable. To determine, for example, if a given integrable function $f(x)$ is an exact derivative, we will form its indefinite integral $F(x)$, then we would find if we have

$$f(x) = \lim_{h \to 0} \frac{F(x+h) - F(x)}{h}.$$

We, therefore, have a regular method of calculation allowing to determine if f is or is not an exact derivative. It is true that it is necessary to investigate if a certain expression does or does not have the known limit $f(x)$. However, a derivative is, by definition, a limit, it is unlikely that we could replace the indicated method of calculation by another, in which we do not use the limits.

We have found a necessary and sufficient condition for an integrable function to be a derivative. It does not take the usual form of such conditions. Most often, a necessary and sufficient condition for the existence of a fact A is stated as the existence of a property B that always accompanies fact A and is always accompanied by it. However, to avoid a mere tautology, it is necessary that we know a regular method of calculation allowing to determine if we have or not, the property B. It is this method which has been directly provided for the case at hand.

If we want to state the necessary and sufficient condition found in the usual form, we could, as Darboux called it, the mean value in (a, b) of an integrable function $f(x)$ the quantity $\frac{1}{b-a}\int_a^b f(x)\,dx$; then we call the mean value at a point x_0, if it exists, the mean value in $(x_0 - h, x_0 + k)$, when the positive numbers h and k tend to zero; and we have the following statement:

For an integrable function to be a derivative function, it is necessary and sufficient that it has, at any point, a definite mean value and, that it be equal to its mean value everywhere.

6.5 Riemannian Integration Considered as an Inverse Operation of the Differentiation

We have seen that we have generalised the problem of finding primitive functions in various ways. Let us now search if one of these generalisations allows us to consider the integration in the sense of Riemann as the inverse problem of differentiation.

If we recall that an indefinite integral admits as derivative the integrated function at all the points where the integrated function is continuous, we are led to pose, along with M. Volterra, the following problem: *Find a continuous function which admits a given bounded function $f(x)$ as its derivative at all points where $f(x)$ is continuous.*[18]

[18] In reality, M. Volterra is looking for the functions which admit $f(x)$ as its derivative at all points which are neither points of discontinuity of $f(x)$, nor limit points of such discontinuities. Moreover, M. Volterra assumes implicitly that the functions he is looking for have bounded derivative numbers. For these two reasons the results he obtains differ from those in the text. Moreover, any function is obviously a solution of M. Volterra's problem, if the points of discontinuity of $f(x)$ form an everywhere dense set. In contrast, only very specific functions satisfy the statement in the text.

This problem is always possible, because both the lower and upper integral of $f(x)$ can be used to address the question. However, in general, it is indeterminate, meaning not all its solutions can be expressed in a formula of the form $F(x)$+const. When $f(x)$ is not integrable, the problem is always indeterminate. If $f(x)$ is integrable, the problem may be determinate; for example when the set of points of discontinuity is reducible. However, it may also be indeterminate, as is the case when the set of points of discontinuity contains a perfect set E. We have learned (Sect. 2.2) how to construct a continuous function that is not constant everywhere but is constant in every interval contiguous to E. When added to a function that solves the proposed problem, this function provides a new solution to the problem.

Thus, our problem encompasses as a particular case the problem of indefinite Riemannian integration, but it is more general than the previous problem.

Now, let us propose to *find a function of bounded derivative numbers which admits a bounded function $f(x)$ as derivative at all points where $f(x)$ is continuous.*

This new problem is always possible and still admits two integrals of $f(x)$ for solution. However, if $f(x)$ is integrable, it is determined, because the derivative of the sought function of bounded derivative numbers is known everywhere, except at the points a set of measure zero. This problem is, therefore, determined only for integrable functions; and when it is determined, its solution is the indefinite integral of $f(x)$.

We could thus, in a certain sense, consider Riemannian integration as the inverse operation of the differentiation.

Definite Integration Using Primitive Functions 7

7.1 Direct Search for Primitive Functions

We have obtained the theorems that theoretically allow, in the extended cases, to determine if a given function is a derivative function and, if it is so, to find its primitive function. In practice, only one of these theorems is commonly used: every continuous function is a derivative function. As for the actual calculation of the primitive functions, it is never done by means of a definite integral,[1] but with the help of a method known by the name of integration by parts and integration by substitution. These two methods can be applied, whether the functions are continuous or not.

We can also use the following theorem: *A uniformly convergent series of derivative functions represent a derivative function.*

Its primitive function is obtained by taking the sum of the primitive functions of the terms of the given series, the constants being so chosen that the series obtained is convergent for one of the values of the variable.

---

[1] However, it is sometimes possible to practically perform the search for a primitive function with the help of definite integrals. We find an example of such a search in *l'introduction a l'etude des fonctions d'une variable reelle* of J. Tannery, p. 284.

Moreover, the mathematicians, especially those who used the method of indivisibles, have performed a number of quadratures by applying what would later become the definition of the definite integral. After the invention of differential calculus and the introduction of the notions of derivative and primitive function, the quadratures performed earlier formed the first elements of the table of derivatives and the primitive functions. Currently, this table is primarily presented in the form of a table of derivatives; we see that the historical order is exactly the reverse of that adopted in our courses.

R. Jain, *Lebesgue's Theory of Integration*,
https://doi.org/10.1007/978-981-96-1169-0_7

Let

$$f = u_1 + u_2 + \cdots = u_1 + \cdots + u_n + r_n = s_n + r_n,$$
$$F = U_1 + U_2 + \cdots = U_1 + \cdots U_n + R_n = S_n + R_n,$$

be the given series and the series of the primitive functions, which is, by hypothesis, convergent for a certain value x_0.

Let us choose n sufficiently large, for which we have, for any positive p,

$$|s_{n+p}(x) - s_n(x)| < \varepsilon;$$

the theorem of finite increment gives, if (a, b) is the considered interval,

$$\begin{aligned}|S_{n+p}(x) - S_n(x)| &< \varepsilon|x - x_0| + |S_{n+p}(x_0) - S_n(x_0)| \\ &\leq \varepsilon(b - a) + |S_{n+p}(x_0) - S_n(x_0)|.\end{aligned}$$

This inequality shows that the series F is uniformly convergent in (a, b), since it is convergent for x_0.

Let us evaluate the ratio

$$r[F(x), x, x + h] = \frac{F(x + h) - F(x)}{h} = \Lambda F(x),$$
$$\Lambda F = \Lambda S_n + \Lambda R_n = \Lambda S_n + \lim_{p\to\infty} \Lambda(S_{n+p} - S_n).$$

The quantity $\Lambda(S_{n+p} - S_n)$ is less than ε in absolute value, according to the theorem of finite increment. Therefore, if we let h tend to zero, any of the limits of ΛF differs from the limit $s_n(x)$ of ΛS_n by at most ε. Since ε is arbitrary, it is thus shown that $F(x)$ admits $f(x)$ for its derivative.

This theorem also allows to use the principle of condensation of singularities in the construction of the derivative functions.

When a derivative function is given by a series of non-negative derivative functions, we can take the primitive functions term-by-term on the condition of choosing the constants in a manner that the obtained series is convergent.

To demonstrate this, I will retain the earlier notation, and to simplify the language, I assume that the series F is convergent for the origin of the interval (a, b) under consideration, and that $U_1, U_2, \ldots$, vanish for $x = a$. Let $\mathfrak{F}$ be one of the primitive functions of f which vanishes at $x = a$. It needs to be shown that $F = \mathfrak{F}$.

All the U_i are positive, therefore S_n increases with n. However, since f is at least equal to s_n, $\mathfrak{F}(x)$ is at least equal to $S_n(x)$, and $S_n(x)$ converges to a limit $F(x)$, at most equal to $\mathfrak{F}(x)$.

The same reasoning when applied to the positive interval $(x, x + h)$ shows that $\mathfrak{F}(x + h) - \mathfrak{F}(x)$ is at least equal to $F(x + h) - F(x)$, and as a result, $f(x)$, the derivative of $\mathfrak{F}(x)$, is at least equal to $\Lambda_d F(x)$.

On the other hand $F(x+h) - F(x)$ is greater than $S_n(x+h) - S_n(x)$, therefore $\lambda_d F(x)$ is at least equal to the derivative $s_n(x)$ of $S_n(x)$, and, since n is arbitrary, $\lambda_d F(x)$ is at least equal to $f(x)$.

Therefore, $F(x)$ has a right derivative equal to $f(x)$; by a similar reasoning in the negative interval $(x, x-h)$, we see that $F(x)$ also admits $f(x)$ for the left derivative. Hence, the theorem is proved.

We can also say: *if derivative functions f_n increase to a derivative function f, their primitive functions converge to the primitive function of $f(x)$, if the constants are suitably chosen.*

Indeed, we can write

$$f = f_1 + (f_2 - f_1) + (f_3 - f_2) + \cdots,$$

and all the terms, which are derivative functions, are positive, with possible exception of the first term.

The theorem is still true if, instead of considering the functions increasing with the integer n, we consider the derivative functions $f(x, \alpha)$ increasing with the parameter α and converging to a derivative function f when α tends to α_0.

Finally, it must be noted that it is necessary to know that the function f, whether a limit or a sum, is a derivative function, in order to have the right to apply the previous theorem: the function

$$f(x, \alpha) = -e^{-\alpha x^2}$$

increase to the function $f(x)$ zero everywhere except for $x = 0$ where it is equal to -1, when α increases to infinity. However, $f(x, \alpha)$ is a derivative function whereas $f(x)$ is not.

These two properties will allow us to perform the search for the primitive functions in the extended cases.

First of all, when a function is the sum of a uniformly convergent series of derivative functions, it is a derivative function for which we know how to find the primitive functions. This is an important theoretical application.

Let a continuous function $f(x)$ be defined on (a, b). Let us mark the points $a = x_0, x_1, x_2, \ldots, x_n = b$ taken sufficiently close to each other so that, in (x_i, x_{i+1}), the oscillation of f is smaller than ε.

In the curve $y = f(x)$, let us inscribe the polygon line $y = \varphi(x)$ where vertices have the abscissa $x_0, x_1, x_2, \ldots, x_n$, and $f(x)$ and $\varphi(x)$ differ by less than ε. This means that $\varphi(x)$ converges uniformly to $f(x)$, when ε tends towards zero. Therefore, it will be sufficient to show that $\varphi(x)$ is a derivative function, for us to assert the same for $f(x)$. However, $\varphi(x)$ being in (x_i, x_{i+1}), the polynomial of first degree

$$\varphi(x) = f(x_i) + (x - x_i)\frac{f(x_{i+1}) - f(x_i)}{x_{i+1} - x_i},$$

is the derivative of the continuous function which, in (x_i, x_{i+1}), is defined by

$$\Phi(x) = \Phi(x_i) + (x - x_i)f(x_i) + \frac{(x - x_i)^2}{2}\frac{f(x_{i+1}) - f(x_i)}{x_{i+1} - x_i}.$$

It is shown that *any continuous function is a derivative function, and this is done without resorting to integration.*[2]

When we are able to represent a function in the form of a series of derivative functions, all having the same sign, we will have a systematic method of calculation that allows us to determine if f is an exact derivative, since the primitive function of f cannot be other than the sum of the primitive functions of the terms of the given series (compare Sect. 6.4).

Thus, the two theorems on the primitive functions of series allow us, in certain cases, with respect to the determination of the primitive functions, to accomplish what the theorems on the integration enable us to do for integrable functions.

I leave aside similar remarks related to the search for a function whose derivative number is a given function. I will now outline some properties of derivative functions which will sometimes immediately allow us to recognise that a given function is not a derivative function.

7.2 Properties of Derivative Functions

A derivative function cannot pass from one value to another without taking all the intermediate values. Indeed, let us assume that we have $f'(a) = A,\ f'(b) = B$, and let C be a number included between A and B. We can take $h > 0$ sufficiently small for $r[f(x), a, a+h]$ to be included between A and C and $r[f(x), b-h, b]$ to be included between B and C. The function $r[f(x), x, x+h]$ is, h being fixed, a continuous function of x. When x varies from a to $b-h$, it passes from one value included between A and C to a value included between C and B, therefore for some value x_0 of $(a, b-h)$ we have $r[f(x), x_0, x_0+h] = C$. The theorem of finite increment shows that in (x_0, x_0+h) there exists a value c such that $f'(c) = C$.[3]

Therefore, the derivative functions possess one of the properties of the continuous functions. Darboux, in his Memoire *Sur les fonctions discontinues*,[4] emphasised a lot on this property. We have taken, in France, the habit of defining a continuous function as the ones which cannot pass from one value to another without passing through all the intermediate

[2] We might be tempted, to apply the theorem on the uniformly convergent series of by relying on a proposition attributed to Weierstrass: every continuous function can be represented by a uniformly convergent series of polynomials. However, to make this method suitable for the purpose, it is necessary to establish Weierstrass's theorem without using the integration. The proof I gave in *Bulletin des Sciences Mathématiques* de 1898, in a Note *Sur l'approximation des fonctions,* satisfies this condition. In another Note: *Remarques sur la définition de l'Intégrale*, published in the same Bulletin in 1905, I used a different idea than what we have just used here.

[3] This does not assume that $f'(x)$ is finite, but only that $f'(x)$ is always well-defined in magnitude and sign.

[4] *Annales de l'École Normale,* 1875.

values, and this definition was considered as equivalent to that of Cauchy. Darboux, who constructed in his Memoire, a discontinuous derivative function in the Cauchy's sense, was able to show that the two definition of continuity were very different.[5]

It is easy to cite the discontinuous functions which do not pass from one value to another without taking, at least once, each intermediary value. This is the case of the function equal to $\sin\frac{1}{x}$ for $x \neq 0$ and to any value in the interval $(-1, +1)$ for $x = 0$.

It is very curious that a function can possess this property, which has been taken for the definition of continuity, and yet be discontinuous at every point. To construct such a function, I write the number x, taken between 0 and 1, in a system of numeration, the decimal system, for example

$$x = \frac{a_1}{10} + \frac{a_2}{10^2} + \frac{a_3}{10^3} + \cdots$$

Let us consider the sequence of digits of odd rank $a_1, a_3, a_5, \ldots$. If it is not periodic, we will take $\varphi(x) = 0$; if it is periodic, and if the first period starts at a_{2n-1}, then we will take

$$\varphi(x) = \frac{a_{2n}}{10} + \frac{a_{2n+2}}{10^2} + \frac{a_{2n+4}}{10^3} + \frac{a_{2n+6}}{10^4} + \cdots$$

It is obvious that the function $\varphi(x)$ thus defined takes all the values in the interval $(0, 1)$ no matter how small the interval is. Therefore, $\varphi(x)$ is discontinuous at every point. Moreover, $\varphi(x)$ does not take values outside of $(0, 1)$. So it does not pass from one value a to another value b without taking all the values of $(0, 1)$, and, *consequently*, all the values between a and b.

It is necessary to note that, with the definition criticized by Darboux, the sum of two continuous functions is no longer necessarily a continuous function. In fact, if

$$f_1(x) = \sin\frac{1}{x} \quad \text{for} \quad x \neq 0 \quad \text{and} \quad f_1(0) = 1,$$

and if

$$f_2(x) = -\sin\frac{1}{x} \quad \text{for} \quad x \neq 0 \quad \text{and} \quad f_2(0) = 1.$$

the two functions f_1 and f_2 cannot pass from one value to another without taking all the intermediary values and this does not hold for $f_1 + f_2$, since

$$f_1 + f_2(x) = 0 \quad \text{for} \quad x \neq 0 \quad \text{and} \quad f_1(0) + f_2(0) = 2.$$

The sum of two derivative functions being a derivative function, it is necessary, according to the previous remark, to state as a new property that the sum of two derivative functions cannot pass from one value to another without taking all the intermediary values. We can also

[5] Allow me to point out that in 1903 we still taught in high schools of Paris the definition criticized by Darboux in 1875. This is all the more surprising since the property which is stated in Cauchy's definition is the one which appears directly in almost all the proofs, while the property of continuous and derivative functions is hardly used except in the theorem of substitutions and its consequences.

say that the difference of the two derivative functions cannot change sign without vanishing, which, if we think about the geometric representation, we can state thus: *Two derivative functions cannot cross each other without meeting.*

Here is an example of the application of this property. Let $\psi(x)$ be a function equal to the function $\varphi(x)$ (Sect. 7.2) when $\varphi(x)$ is not equal to x, and equal to 0 when $\varphi(x) = x$. $\psi(x)$, as $\varphi(x)$, cannot pass from one value to another without passing through all the intermediary values, therefore the first theorem does not permit to ascertain that $\psi(x)$ is not a derivative function; but, since $\psi(x)$ crosses the continuous function x in any interval and meets, however, only for $x = 0$, the second property shows that $\psi(x)$ is not a derivative.

Before investigating if the function $\varphi(x)$ is a derivative, I will show, how an important special case of Scheeffer's theorem is deduced immediately from the theorem of Darboux.

Let us assume that the derivative of a function $f(x)$ is always well defined in magnitude and sign (we do not suppose it to be finite). Then, if it is not always equal to a given number A, the set of values of x for which $f(x)$ is different from A has the cardinality of the continuum. In fact, either $f'(x)$ is constant and the property is proved, or $f'(x)$ takes two values B and C, and in that case, it also takes all the values included between B and C, which are all different from A, except may be one. The set of these values of $f'(x)$, different from A has the cardinality of the continuum, and the same holds for the set of corresponding values of x.

With this in mind, if $f(x)$ always has a derivative, and if this derivative is zero, except may be for a countable set of values of x, we can assert that it is always zero. This is the theorem of Scheeffer in a particular case.

Let us return to the function $\varphi(x)$. Is it a derivative? The previous two theorems do not seem to easily provide an answer to this question. One method is to apply a previously proved theorem; a bounded derivative function has the same maximum whether or not we neglect the sets of measure zero.[6] It is not difficult to show that $\varphi(x)$ is non-zero only for set of values of x of measure zero (see Sect. 8.2), $\varphi(x)$ is, therefore, not a derivative function.

This result may be obtained in a totally different manner. *A derivative cannot be discontinuous at every point, and $\varphi(x)$ is discontinuous at every point.*

This property of derivative functions follows from a theorem due to M.R. Baire. $f'(x)$ is the limit, for $h = 0$, of the function $r[f(x), x, x+h]$, continuous at x when h is constant; therefore, it is a *function of first class*, that is, a limit function of continuous functions. But M.Baire showed that if we consider a function of class one on any perfect set, there exists points where it is continuous on this perfect set; this is expressed by saying that it is *point wise discontinuous on any perfect set.*[7]

[6] I recall that this theorem was obtained without using integration.

[7] It would be more appropriate to say that a function of class *one* is either continuous or at the most point-wise discontinuous on each perfect set and it is indeed point-wise discontinuous on certain perfect sets.

We will show this proposition and its converse in Chap. 10.

7.3 The Integral Deduced from the Primitive Functions

In a lot of cases, we can, without resorting to integration, determine whether a given function is a derivative, and we could also hope to find the primitive function of a given derivative without integration. Previously, we resolved these questions by using the definite integral; one might wonder if, conversely, we could define define the integral with the help of primitive functions. This is the method of Duhamel and Serret.[8] For these authors *a function* $f(x)$ *has an integral in* (a, b) *when it admits a primitive function* $\mathfrak{F}(x)$ *in* (a, b). *This integral* I_a^b *is, by definition,*

$$I_a^b f(x)\,dx = \mathfrak{F}(b) - \mathfrak{F}(a).$$

This definition is not equivalent to Riemann's definition. On one hand, we know that there exist functions that are integrable in Riemann's sense, which are not the derivative functions; on the other hand, as we will see, there exist derivative functions which are not integrable in Riemann's sense.

The first example of such a function is due to M. Volterra (*Giornale de Battaglini,* 1881); here is how we obtain it:

Let E be a non-dense perfect set which is not an integrable group (Sect. 4.1). Let (a, b) be an interval contiguous to E, let us consider the function

$$\varphi(x, a) = (x - a)^2 \sin \frac{1}{x - a};$$

its derivative vanishes an infinite number of times between a and b, let $a + c$ be the greatest value of x not greater than $\frac{a+b}{2}$ at which φ' vanishes. Given this, we define a function $F(x)$ by the following conditions: it is zero at the points of E; in any interval (a, b) contiguous to E, it is equal to $\varphi(x, a)$ from a to $a + c$; from $a + c$ to $b - c$, the function F is constant and equal to $\varphi(a + c, a)$; from $b - c$ to b, F is equal to $-\varphi(x, b)$.

This function $F(x)$ is obviously continuous. It has a derivative; this is obvious for the points which do not belong to E; let x_0 be a point of E, the ratio $r[F(x), x_0, x_0 + h]$ is zero if $x_0 + h$ is point of E. If $x_0 + h$ is not a point of E, it belongs to an interval contiguous to E, let α be one of the extremities of this interval which is in $(x_0, x_0 + h)$; we obviously have

$$|r[F(x), x_0, x_0 + h]| = \left| \frac{F(x_0 + h)}{h} \right| \leq \frac{(x_0 + h - \alpha)^2}{|h|} \leq |h|,$$

therefore $F(x)$ has a zero derivative at all the points of E.

[8] In reality, Duhamel and Serret hardly considered anything other than continuous functions. For these functions, according to what has been mentioned earlier, their definition is equivalent to that of Cauchy; it is only a matter of differences in exposition at that point.

The derivative F' of F is bounded, because the derivative of $x^2 \sin \frac{1}{x}$, which is zero for $x = 0$, and which, for x different from zero, is equal to

$$2x \sin \frac{1}{x} - \cos \frac{1}{x},$$

is bounded. However, this derivative F' is not integrable, in Riemann's sense, because at all points of E, the maximum of F' is $+1$ and its minimum is -1, as is the case at point $x = a$ for the function $\varphi'(x, a)$; but E is not an integrable group, by hypothesis.

By a suitable application of the principle of condensation of singularities, we obtain a derivative function which is not integrable over any interval, however small it may be.[9]

Duhamel's definition applies, therefore, to the bounded functions to which Riemann's definition does not apply. Furthermore, Duhamel's definition applies to the unbounded functions because there exist unbounded derivatives, but still always finite derivatives of $x^2 \sin \frac{1}{x^2}$, for example.

To the definition of Duhamel and Serret, we can apply the generalisation used by Cauchy and Dirichlet. I will not dwell on this generalisation, nor, for the time being, on the following one, which contain as a particular case the definition of Riemann and Duhamel for the bounded functions: *A bounded function $f(x)$ is said to be summable, if there exists a function of bounded derivative number $F(x)$ such that $F(x)$ admits $f(x)$ as its derivative except for a set of values of x of measure zero. The integral in (a, b) is then, by definition, $F(b) - F(a)$.*[10]

Let us adopt the definition of Duhamel and Serret without generalisation. The integral of Duhamel (Integral D) possesses some of the properties of Riemann's integral.

We have

$$I_a^b + I_b^c + I_c^a = 0.$$

The sum of two D integrable functions is a D integrable function and has an integral that is sum of the integrals; but the product of two D integrable functions is not necessarily D integrable.[11]

A uniformly convergent series of D integrable functions is a D integrable function and the integration can be performed term-by-term; this is the proposition of Sect. 7.1. From the proposition in Sect. 7.1 we deduce that if the D integrable functions, $f_n(x)$, increase to a D integrable function $f(x)$, the integral of f_n increases to that of f, for a positive interval of integration.

[9] M. Kopke has constructed differentiable functions of bounded derivative which vanish in every interval. These derivatives are obviously not integrable.

[10] Compare with Sect. 6.5, where, as soon as f is given, we know at what points we may not necessarily have $F'(x) = f(x)$; here on the contrary, we do not know.

Different functions $F(x)$ corresponding to the same function $f(x)$ differ only by an additive constant.

We will encounter these bounded summable functions in the following chapters.

[11] For example the product $x\left(x^2 \sin \frac{1}{x}\right)$ is not D integrable.

The similar proposition for Riemann's integral is true. We would base its proof on the one provided in Sect. 7.1.

Let us keep the notation of Sect. 7.1. $f, u_1, u_2, \ldots,$ are now the positive integrable functions. $\mathfrak{F}, U_1, U_2, \ldots,$ are their indefinite integrals which vanish at the origin a of the considered interval.

We obviously have $f \geq s_n$, hence $\mathfrak{F} \geq S_n$, and since the S_n are increasing the series of U is convergent. The increment of $\mathfrak{F}$, in an arbitrary positive interval, is at least equal to that of S_n, and therefore, to that of F; F is a bounded derivative number. To show that $\mathfrak{F} = F$, it is sufficient to show that these two functions have the same derivative everywhere, except for a set of values of x of measure zero.

At any point where $f, u_1, u_2, \ldots$ are all continuous, $\mathfrak{F}, U_1, U_2, \ldots$ have the derivatives, and the reasoning of Sect. 7.1 shows that at these points F has same derivative as that of $\mathfrak{F}$. But the points where f is not continuous form a set of measure zero $E(f)$, the points of discontinuity of u_i form a set of measure zero $E(u_i)$; the union of all these sets give a set of measure zero E. And we have $F' = \mathfrak{F}'$, except, possibly, at the points of E.

Hence, we deduce the theorem:

When the integrable functions f_n increase to an integrable function f, the integral of f_n converges to that of f.[12]

Now, we should ask ourselves what services the integral in the sense of Duhamel and Serret could provide.

These integrals cannot provide any service in the search for primitive functions, since they presuppose this search has been performed. However, the integrals in the sense of Riemann are used primarily to calculate the limit of sums.

The reasoning of Sect. 6.3 shows that a D integral is a limit of sum. Therefore, we can hope to use these integrals for the calculation of the limit of sums. We have seen (Sect. 5.2) that this is indeed possible, as it has been shown that the length of a curve is the D integral of $\sqrt{x'^2 + y'^2 + z'^2}$, whenever this integral exists.[13]

---

[12] We can note that this property remains true when dealing with so-called summable functions, which are integrable according to the generalisation mentioned in Sect. 7.3.

[13] I can only mention another application of the D integrals: when a bounded derivative function admits a trigonometric expansion, the coefficients of its expansion are given by well-known formulae of Euler and Fourier, with the integrals appearing in these formulae being D integrable.

Indeed, there exist bounded derivative functions which are not Riemann integrable, but still admit trigonometric expansion. For the proof of these properties, we can refer to a Memoire *Sur les series trigonometriques* that I published in the *Annales de l'École Normale*(November 1903).

The new studies on the integral are, however necessary, because we have still not solved the problem of finding primitive functions; moreover, for calculating the length of a curves with tangents, both integrations are insufficient.[14]

Further, I would like to add that if the two integrations we have studied generally seem sufficient, this is solely because, almost always, we deliberately restrict ourselves to the consideration of continuous functions, and often even to the consideration of analytic functions.

---

[14] It is easy to see that $\sqrt{1+\left[\left(x^2 \sin \frac{1}{x}\right)'\right]^2}$ is not an exact derivative.

Starting from there, we show without any difficulty that the quantity $\sqrt{1+F'(x)^2}$, where F is non-integrable derivative function of M. Volterra, neither integrable in the sense of Riemann nor in the sense of Duhamel.

Therefore, the curve $y = F(x)$ cannot be rectified, by either of the two methods used.

For the application mentioned in the previous Note, both integrations are also insufficient. This becomes obvious when considering the sum of non-integrable derivative that can be represented trigonometrically, and a non-derivative function that can be represented trigonometrically.

8 The Definite Integral of Summable Functions

8.1 The Problem of Integration

The classical applications of integration of continuous functions, as well as the previous applications of integration in the sense of Riemann or in the sense of Duhamel and Serret, are sufficient to highlight the role of certain simple properties resulting from all the definitions of the integral we have already studied. They also serve to convince us that these properties must necessarily belong to the integral, if we want some analogy to be there between this integral and the integral of the continuous functions.

This is why *we propose to associate with any bounded function*[1] $f(x)$*, defined on a positive, negative, or null, finite interval* (a, b)*, a finite number,* $\int_a^b f(x)\,dx$ *which we call the integral of* $f(x)$ *in* (a, b) *and which satisfies the following conditions:*

1. *For any* a, b, h *we have*

$$\int_a^b f(x)\,dx = \int_{a+h}^{b+h} f(x-h)\,dx;$$

2. *For any* a, b, c *we have*

$$\int_a^b f(x)\,dx + \int_b^c f(x)\,dx + \int_c^a f(x)\,dx = 0;$$

3. $$\int_a^b [f(x) + \varphi(x)]\,dx = \int_a^b f(x)\,dx + \int_a^b \varphi(x)\,dx;$$

4. *If we have* $f \geq 0$ *and* $b > a$*, we also have*

[1] The word *bounded* is necessary if we want the integral to be always *finite*.

R. Jain, *Lebesgue's Theory of Integration*,
https://doi.org/10.1007/978-981-96-1169-0_8

$$\int_a^b f(x)\,dx \geq 0;$$

5. *We have*

$$\int_0^1 1 \times dx = 1;$$

6. *If, when the index n increases, $f_n(x)$ increases to $f(x)$, the integral of $f_n(x)$ tends towards that of $f(x)$.*

The significance, necessity and consequences of the first five conditions of this *problem of integration* are quite obvious; we will not dwell on them.

The condition 6 has a special role. It neither has the same nature of simplicity as the first five, nor the same trait of necessity.[2] Moreover, while it is easy to construct the numbers satisfying any four of first five conditions, without satisfying all five, showing that all the five conditions are independent, we do not know if the six conditions of the problem of integration are independent or not.[3]

By stating the six conditions of the problem of integration, we define the integral. This definition belongs to the class of what can be called *descriptive* definitions; in these definitions, we state the characteristic properties of the 'being' that we want to define. In the *constructive* definition, we state the operations that are necessary to obtain the 'being' that we want to define. Constructive definitions are more often used in Analysis; however, we sometimes use descriptive definitions[4] as well. The definition of integral, according to Riemann, is constructive, while the definition of the primitive functions is descriptive.

[2] It appears so insignificant that it is generally unknown, even for the case where f and f_n are integrable in the sense of Riemann or even continuous. However, it could be the case that some of its consequences have, on the contrary, a very high degree of necessity. To prepare for the introduction of this condition 6, I worked in Sects. 7.1 and 7.3 with integration and with search for the primitive function of the limit of an increasing sequence of the functions.

[3] The answer to this question does not matter for the application, but it is of interest from the viewpoint of principles. If it was shown that this sixth condition is independent of the other five, it would be necessary to try to replace it with a simpler sixth condition and especially to explore whether, among the system of numbers which satisfy only the first five conditions, there are any as useful as the one being studied.

Just before the second edition, M. St. Banach studied the question posed in the first edition, and concluded that the six conditions of the problem of integration are independent (*Fund. Math., t. IV*).

[4] The use of these descriptive definitions is indispensable for the initial stages of a science when we want to construct that science in a purely logical and abstract manner. *See* the thesis of M. J. Drach *(Annales de l'École Normale,* 1898*)* and the Memoire of M. Hilbert *Sur les fondements de la Géométrie (Annales de l'École Normale,* 1900*)*. Such a definition is then called axiomatic, since it enumerates the necessary axioms. Thus, it is self-sufficient and forms a complete whole.

In contrast, the descriptive definitions introduced in the course of development of a theory, such as the definition of the integral, do not claim to enumerate all the axioms on which it relies. They do not form a complete whole and cannot be isolated from the exposition of the rest of the theory.

When we state a constructive definition, it is necessary to show that the operations mentioned in this definition are possible. Similarly, a descriptive definition is also subject to certain conditions: it is necessary that the stated conditions be compatible.[5] The general method used so far to show that the conditions are compatible is the following: we choose in a class of 'beings' previously defined, 'beings' which possess all the stated properties. This class of 'beings' is typically the class of integers[6]; and it is assumed that the descriptive definition of these integers does not contain any contradictions.

It is also necessary to study the nature of the indetermination of the 'beings' that we want to define. Let us assume, for example, that we have shown the impossibility of the existence of the two different classes of 'beings' satisfying the specified conditions, and additionally, we have shown the compatibility of these conditions by choosing a class of 'beings' that satisfy them. This class of 'beings' will be the only one defined, so that the constructive definition used to make the choice is exactly equivalent to the given descriptive definition.

We would find a constructive definition equivalent to the descriptive definition of the integral.[7]

First, we will easily show, relying on the conditions 3 and 4, that we have condition S

$$\int_a^b kf(x)\,dx = k\int_a^b f(x)\,dx \tag{S}$$

where k is a constant. This being said, let $f(x)$ be any function. We denote by $E[\alpha < f(x) < \beta]$ the set of values of x for which we have $\alpha < f(x) < \beta$, and by $E[f(x) = \alpha]$ the set of values of x for which $f(x) = \alpha$; and we will use similar other notations.

Let (l, L) be a positive interval containing in its interior the interval of the variation of $f(x)$[8]; let us partition this interval into partial intervals with the help of the numbers

---

[5] Meaning thereby, none of their consequences are of the form: A is not A. It is also necessary, as I have already said, to investigate if the conditions are independent.

[6] *See* the Memoires already cited by M. Hilbert. This is because we can show the compatibility of the stated conditions in the descriptive definitions of the early stages of the Geometry with the help of the system of integers that it is legitimate to say that the Geometry can be entirely constructed starting from idea of number. From the viewpoint of the arithmatisation of the science, the main interest of the precise definition of the integral, as posed by Cauchy, is that it reduces the various notions of magnitude involved in geometry (area, volume, length of the curves, etc.) to that of the length of a segment, i.e., the difference of the two numbers. This definition of Cauchy completes the work of Descartes who, by the use of coordinates, reduced all the geometries to that of a straight line.

[7] By placing ourselves in the same point of view, we can say that the work presented in this book has the primary goal of seeking a constructive definition, equivalent to the descriptive definition of primitive functions.

[8] In other words, the lower and upper limits of $f(x)$ are included between l and L but not equal to $\{l, L\}$; with the notations of the text, there would be no need to change anything if l and L were equal to the limits of $f(x)$; however, at other times, we would need to take some precautions for the extreme values of the indices which appear in the summation. Here, we could extend the first summation only from 0 to $n - 1$ and the second from 1 to n.

$$l_0 = l < l_1 < l_2 < \cdots < l_n = L,$$

let us suppose that $l_{i+1} - l_i$ is never greater than ε.

Let us denote by ψ_i $(i = 0, 1, 2, \dots, n)$ the function equal to 1 when x belongs to $E[f(x) = l_i]$, or to $E[l_i < f(x) < l_{i+1}]$, and zero for the other points; let us denote by Ψ_i $(i = 0, 1, 2, \dots, n)$ the function equal to 1 when x belongs to $E[l_{i-1} < f(x) < l_i]$ or to $E[f(x) = l_i]$ and zero for the other points. We obviously have

$$\varphi(x) = \sum_{i=0}^{i=n} l_i \psi_i(x) \le f(x) \le \sum_{i=0}^{i=n} l_i \Psi_i(x) = \Phi(x).$$

When we know how to integrate the functions ψ which take only the values 0 and 1, we will deduce, thanks to the conditions 3 and S, the integrals of $\varphi(x)$ and $\Phi(x)$, which include the integral of $f(x)$ (conditions 3 and 4).[9]

Moreover, $\varphi(x)$ and $\Phi(x)$ differ from $f(x)$ by ε at most, and therefore converge uniformly to $f(x)$ when ε tends towards zero; it is easy to conclude that their integrals converge to that of $f(x)$.

Indeed, if the lower and upper limits of $g(x)$ are m and M, according to 3 and 4, $\int_a^b g(x)\,dx$ is included between

$$\int_a^b m\,dx = m \int_a^b dx \quad \text{and} \quad \int_a^b M\,dx = M \int_a^b dx;$$

now, by making

$$g(x) = f(x) - \varphi(x),$$

we have

$$\varepsilon \ge M \ge m \ge 0,$$

therefore, the integral of $g(x)$ is smaller in modulus than $\varepsilon \int_a^b dx$, a quantity which tends towards zero with ε.

To be able to calculate the integral of any function, it is sufficient to know how to calculate the integral of functions ψ which take only the values 0 *and* 1.

It should be noted that we have incidentally shown the possibility of term-by-term integration of uniformly convergent series, if the problem of integration is possible.

The quantity, $\int_a^b dx$ which appears in the previous proof, is easily calculated; by using 1, 2 and 5, we see that it is equal to $b - a$.

If the function is included between l and L, its integral in (a, b) is included between $l(b - a)$ and $L(b - a)$; this is the mean value theorem.

---

[9] Here, we suppose, for some time, the problem of integration is possible.

If we apply this theorem after having decomposed (a, b) into partial intervals, we find that $\int_a^b f(x)\,dx$ is included between the sums which are used to define the integrals by lower and upper sums; *the integral is therefore included between the integrals by lower and upper sums*. In particular, if the problem of integration is possible for integrable functions in the sense of Riemann, then for these functions it does not admit, a solution other than the Riemann's integral.

8.2 The Measure of the Sets

Let us now concentrate on functions ψ which take only the values 0 and 1. Such a function is entirely defined by the set $E[\psi(x) = 1]$ of values where it is different from 0; the integral of such a function, in a positive interval, is a positive number or zero which we can consider as attached to the subset of the set $E[\psi(x) = 1]$ included in the interval of integration. If we translate in the geometric language, the conditions of the problem of integration of functions ψ, we have a new problem, *the problem of the measure of sets*.

To state it, I recall that the two sets of points on a line are considered equal if, by the displacement of one of the two, we can make them coincide. Moreover, a set E is considered the sum of sets e if every point of E belongs to at least one of the sets e.[10] Here is the question to be resolved:

We propose to associate with each bounded set E, formed from points of ox, a positive number or zero, m(E), that we call the measure of E, and which satisfies the following conditions:

1′ *Two equal sets have the same measure;*
2′ *The measure of the sum of a finite or a countably infinite number of sets, which are pairwise disjoint, is the sum of their measures;*
3′ *The measure of the set of all points of* $(0, 1)$ *is* 1.

The condition 3′ replaces the condition 5; the condition 2′ results from the application of conditions 3 and 6 to the series

$$\psi = \psi_1 + \psi_2 + \cdots,$$

in which all the terms and the sum are the functions ψ; as for condition 1′ it is equivalent to condition 1. However, an explanation is necessary. There are two types of equal sets: those which can be made to coincide by a translation along the ox and those which can be made to coincide by a rotation of π around a point of ox. Condition 1′ applies only to the former. I did not put this restriction in the statement because, in the subsequent reasoning, we can

[10] With our definition, the sets e could, therefore, have common points.

restrict ourselves to using only translations as displacements, and yet we will still always obtain equal measures[11] for two equal sets in either case.

A simple consequence of conditions $1', 2', 3'$ is that any positive interval (a, b) has a measure equal to its length $b - a$, whether or not the extremities make a part of the interval.[12]

If we refer to Chap. 3, we immediately see that, if the problem of the measure is possible, we have

$$e_i(E) \leq m(E) \leq e_e(E);$$

for the J measurable sets, the problem of the measure is possible in more than one way and measure corresponds to extent in the sense of Jordan.

Now, let E be an arbitrary set. We can enclose its points in a finite or countably infinite number of non-intersecting intervals. The measure of the set of points of these intervals is, according to $2'$, the sum of the lengths of these intervals. This sum is an upper limit of the measure of E. The set of these sums has a lower limit denoted by $m_e(E)$, the *exterior measure* of E, and we obviously have

$$m(E) \leq m_e(E) \leq e_e(E).$$

Let $C_{AB}(E)$ be the complement of E with respect to AB, which means, the set of points not making a part of E but making a part of a segment AB of ox containing E. We must have

$$m(E) + m[C_{AB}(E)] = m(AB),$$

therefore

$$m(E) = m(AB) - m[C_{AB}(E)] \geq m(AB) - m_e[C_{AB}(E)];$$

the lower limit thus found for $m(E)$, a limit which is necessarily positive or zero, is called the *interior measure* of E, denoted as $m_i(E)$; it is obviously greater or at least equal to the interior extent of E.

For comparing the two numbers m_e, m_i, we use a theorem due to M. Borel:

---

[11] All the conditions required for the integration problem concerning the functions ψ have been stated. However, we are afraid that this is not sufficient for integral of arbitrary functions—one defined as soon as the integral of functions ψ are—will also satisfy these conditions. In the sequel, it is shown that these fears are not justified.

We can now show this without using the values of the integral of functions ψ, and we could also show that, if we remove the words *or a countably infinite* from $2'$, we obtain a new problem of measure that corresponds exactly to the problem of integration posed with conditions 1, 2, 3, 4, 5 without condition 6.

[12] This has already been expressed by the equality $\int_a^b dx = b - a$.

If we have a family of intervals Δ *such that every point of an interval* (a, b)*, including a and b, be interior*[13] *to at least one of the* Δ*, then there exists a family consisting of a finite number of the intervals of* Δ *which has the same property [every point of* (a, b) *is interior to one of them].*

Let (α, β) be one of the intervals of Δ containing a, the property to be shown is obvious for the interval (a, x), if x is included between α and β. I mean that this interval can be covered with the help of a finite number of intervals of Δ, which I express by saying that the point x is reached. It needs to be shown that b is reached. If x is reached, all the points of (a, x) are reached; if x is not reached, then none of the points of (x, b) are reached. Therefore, if b is not reached, there must be a first 'unreached' point, or a last reached point. Let x_0 be this point. It is interior to an interval of Δ, (α_1, β_1). Let x_1 be a point of (α_1, x_0), x_2 be a point of (x_0, β_1); x_1 is reached by hypothesis, and the finite number of intervals of Δ which are used to reach it, plus the interval (α_1, β_1), allows us to cover $x_2 > x_0$. Therefore, x_0 is neither the last reached point, nor the first 'unreached' point; hence, b is reached.[14]

From the theorem of M. Borel it follows that *if we cover the whole interval* (a, b) *with the help of a countably infinite number of intervals of* Δ*, the sum of the lengths of these intervals is at least equal to the length of the interval* (a, b).[15] Indeed, we can also cover (a, b) with the help of a finite number of intervals of Δ and the theorem, being obviously true when we consider only these intervals in a finite number, is all the more true, when we consider all the intervals of Δ.

Let us now go back to the set E and its complement $C_{AB}(E)$. Let us enclose the first in a countably infinite number of intervals α, and the second in intervals β, we have

---

13 'Interior' is taken in a narrow sense which excludes the extremities.

14 M. Borel has given, in his thesis and in his *Leçons sur la théorie des fonctions,* two proofs of this theorem. These proofs essentially assume that the set of intervals Δ is countable, which is sufficient in some application. However, there is some interest in demonstrating the theorem as stated in text. For example, for the applications I made in my thesis using the theorem of M. Borel, it was necessary to show for a set of intervals Δ having the cardinality of the continuum.

We have deduced from theorem, as stated in the text, a pretty beautiful proof of the uniformity of the continuity.

Let $f(x)$ be a continuous function at all points of (a, b), including a and b; by definition, each point of (a, b) is interior to an interval of Δ in which the oscillation of $f(x)$ is less than ε. With the help of a finite number of them, we can cover (a, b); let l be the length of the smallest interval used; in any interval of length l the oscillation of f is at most 2ε, since such an interval intersects at most two intervals of Δ; hence, the continuity is uniform.

This application of the completed theorem makes it quite clear, it seems to me, that we can use it anywhere in the theory of the functions. Since the time when the first edition of this book appeared the theorem of M. Borel has been studied a lot; it has given rise to the attributions of priority which did not seem to me justified in any way, as I had the opportunity to express in an analysis of the book of M. and Ms. Young, cited in Sect. 4.1 (*Bull. des Sc. math.,* 1907). For the relationship between the proofs in the text and the first demonstration of M. Borel, see the final Note.

15 If, as I assumed in the proof, we admit that every point of (a, b) is interior to one of the Δ, we can replace *at least equal* by *greater*.

$$m(\alpha) + m(\beta) \geq m(AB),$$

since AB is covered by the intervals α and β. Hence, we deduce

$$m_e(E) + m_e[C_{AB}(E)] \geq m(AB),$$
$$m_e(E) \geq m(AB) - m_e[C_{AB}(E)],$$
$$m_e(E) \geq m_i(E).$$

The interior measure is never greater than the exterior measure.

The sets whose two measures, exterior and interior, are equal are called *measurable* and their measure is the common value of m_e and m_i.[16] It remains to find out if this measure satisfies the conditions $1'$, $2'$, $3'$. This is obvious for $1'$ and $3'$. It remains to study condition $2'$.[17] During the course of our verification we will use this obvious fact that the subset (E, I) of a set E, which is contained in an interval I, is certainly measurable whenever E is measurable.

Let $E_1, E_2, \ldots$ be the measurable sets, in a finite or countable number, which are pairwise disjoint, and let E be their sum set.

From the definition of the measurable sets, it follows that we can enclose E_i in a countably infinite number of intervals α_1 and $C_{AB}(E_i)$ in the intervals β_i in a manner that the measure

---

[16] It is only for these sets that we will study the problem of measure. I do not know if we can define, or even if there exist sets other than measurable sets. If they exist, what is stated in the text is not sufficient to affirm either that the problem of measure is possible or it is impossible for these sets. On the subject of possibility and determination of the problem of the measure for all sets, see the work of M. Banach, cited in Sect. 8.1. As for the question of the existence of the non-measurable sets, there has not been much progress since the first edition of this book. However, the existence of such sets is certain for those who accept a certain mode of reasoning based on what we call the *axiom of Zermelo*. By this reasoning, we indeed arrive, at the conclusion: *there exist* non-measurable sets; but this affirmation must not be considered contradicted if we came to show that no man will ever be able to *name* a non-measurable set!

On these questions we could consult the second edition of *Leçons sur la théorie des fonctions* of M. Emile Borel.

[17] The geometric definition of the measure not only allows us to compare two equal sets, but also two similar sets. The ratio of the measure of two similar sets with a ratio k is $|k|$, a condition that could have been imposed a priori; it corresponds to condition S_1 for the the problem of integration

$$\int_a^b f(x)\,dx = k \int_{\frac{a}{k}}^{\frac{b}{k}} f(kx)\,dx \qquad (S_1)$$

The conditions S (Sect. 8.1) and S_1 constitute what we can call *the condition of similitude*, they reveal how should an integral transform under a transformation

$$x_1 = kx, \quad f_1(x) = kf(x).$$

Maybe we could replace condition 6 by some other conditions of this nature.

of the common subsets of α_i and β_i is equal to ε_i. The positive numbers, ε_i, are chosen in such a way that the series $\sum \varepsilon_i$ is convergent and has the sum ε.

Let α'_2, β'_2 be the subsets of α_2, β_2 which are contained in the intervals β_1; let α'_3, β'_3 be the subsets of α_3, β_3 which are contained in β'_2 and so on. E_i is enclosed in α'_i. Therefore, E is enclosed in $\alpha_1 + \alpha'_2 + \cdots$, and hence, its exterior measure is at most equal to the sum $m(\alpha_1) + m(\alpha'_2) + m(\alpha'_3) + \cdots = s$; let us evaluate this sum. We obviously have

$$m(\alpha_1) + m(\beta_1) = m(AB) + \varepsilon_1,$$
$$m(\alpha'_2) + m(\beta'_2) \leq m(\beta_1) + \varepsilon_2,$$
$$m(\alpha'_i) + m(\beta'_i) \leq m(\beta'_{i-1}) + \varepsilon_i \quad (i > 2),$$

hence, by addition,

$$m(\alpha_1) + m(\alpha'_2) + m(\alpha'_3) + \cdots + m(\alpha'_i) \leq m(AB) + \varepsilon_1 + \varepsilon_2 + \cdots + \varepsilon_i,$$

and this is sufficient for showing that the series s is convergent; moreover we have

$$m(E_i) \leq m(\alpha'_i) \leq m(\alpha_i) \leq m(E_i) + \varepsilon_i,$$

therefore, s is included between $\sum m(E_i)$ and $\sum m(E_i) + \varepsilon$, This gives

$$m_e(E) \leq \sum m(E_i).$$

The complement of $E, C_{AB}(E)$, can be enclosed in β'_i. Now, β'_i has, in common with $\alpha_1 + \alpha'_2 + \alpha'_3 + \cdots$, the intervals $\alpha'_{i+1} + \alpha'_{i+2} + \cdots$, along with a subset of the intervals common to α_1, β_1, a subset of those common to $\alpha_2, \beta_2, \ldots$ and a subset of those common to $\alpha_i, \beta_i, \beta'_i$. Therefore, its measure is at most equal to

$$[m(AB) - s] + \varepsilon_1 + \varepsilon_2 + \cdots + \varepsilon_i + m(\alpha'_{i+1}) + m(\alpha'_{i+2}) + \cdots,$$

and, as a result,

$$m_e[C_{AB}(E)] \leq m(AB) - \sum m(E_i),$$

that is,

$$m(AB) - m_e[C_{AB}(E)] \geq \sum m(E_i),$$

or

$$m_i(E) \geq \sum m(E_i).$$

the lower and upper limits found, respectively, for $m_e(E)$ and $m_i(E)$ show that these two quantities are equal. Therefore, the set E is measurable and of measure $\sum m(E_i)$, and hence the condition $2'$ is indeed verified.

The set of the measurable sets contains the set of J measurable sets, but it is much wider, as we will see. Indeed, we can perform the following two operations on the measurable sets without leaving the set of measurable sets:

I. Make the sum of countably infinite number of sets;
II. Take the subset common to all the sets of a family containing a finite or countably infinite number of sets.

To show this, let us first note that the second operation does not differ essentially from the first one, because if E is a subset common to $E_1, E_2, \ldots, C(E)$ is the sum of $C(E_1), C(E_2), \ldots$. Therefore, it is enough to concentrate on the first operation; Let

$$E = E_1 + E_2 + E_3 + \cdots$$

If E_i' is the set of points of E_i not making a part of $E_1 + E_2 + \cdots + E_{i-1}$, we have

$$E = E_1 + E_2' + E_3' + \cdots,$$

the terms of the sum being pairwise disjoint. Now, it is easy to see that E_2' is measurable. Indeed, let us enclose E_1 in the intervals α_1, $C(E_1)$ in the intervals β_1, E_2 in α_2, $C(E_2)$ in β_2 and let ε_1 and ε_2 be the lengths of the subsets common to α_1 and β_1 on one hand and, to α_2 and β_2 on the other. If α_2' and β_2' are the subsets of α_2 and β_2 common to β_1, E_2' can be enclosed in α_2' and $C(E_2')$ in $\alpha_1 + \beta_2'$ and the subsets common to these two systems of intervals have a measure at most equal to $\varepsilon_1 + \varepsilon_2$, therefore E_2' is measurable. Hence, it follows that

$$E_1 + E_2 = E_1 + E_2'$$

is measurable, and therefore the set E_3', which is a subset of E_3 and does not belong to the measurable set $E_1 + E_2$, is also measurable and so on. All the E_i' are measurable, and consequently, E[18] is measurable. An interval being a measurable set, by applying the operations I and II a finite or a countably infinite number of times starting from intervals, we obtain measurable sets; these are the ones which M. Borel called *measurable sets;* let us call them *B measurable sets.* These are the most important of the measurable sets. For an arbitrary set, while can only assert the existence of the two numbers m_e, m_i, without being able to specify the sequence of operations needed to calculate them, it is easy to obtain the measure of a B measurable set by following step-by-step construction of this set. We will make use of the property $2'$ whenever we use operation I; when we use the operation II, we will employ a theorem with an immediate proof:

---

[18] If E_1 contains E_2, we can speak of their difference $E_1 - E_2$. This difference is measurable if both E_1 and E_2 are measurable because it is the subset common to E_1 and $C(E_2)$.

The measure of the subset common to the sets $E_1, E_2, \ldots$ is the limit of $m(E_i)$ if each set E_i contains all those with a higher index.[19]

The closed sets are B measurable, since they are complements of the sets formed of points interior to a finite or a countably infinite number of intervals. Let E be such a set, the measure of its complement is obviously the interior extent of this complement, therefore the measure of a close set is its exterior extent. Hence, it follows that the property which we have used: *a closed set of measure zero is an integrable group* (Sect. 3.2).

As an application of these theoretical considerations, let us calculate the measure of a set E of points of $(0, 1)$ such that the sequence of their odd decimal digits is periodic (Sect. 7.2).

Let

$$x = \frac{a_1}{10} + \frac{a_2}{10^2} + \frac{a_3}{10^3} + \cdots$$

such a number, written as

$$x = y + \frac{a_2}{10^2} + \frac{a_4}{10^4} + \frac{a_6}{10^6} + \cdots = y + z;$$

where y is rational. The set of numbers y is countable. To each rational number there corresponds a set of numbers x that have same measure as the set of the numbers z whose digits of odd rank are all zero. To show that E is measurable and of measure zero, it is, therefore, sufficient to show that the set of numbers z possesses this property.

Now, this set is obtained by removing from the interval $(0, 1)$ the interval $\left(\frac{1}{10}, 1\right)$. Then from $\left(0, \frac{1}{10}\right)$ we remove the intervals $\left(\frac{p}{10^2} + \frac{1}{10^3}, \frac{p+1}{10^2}\right)$, where p is an integer smaller than 10. After that, from each remaining interval $\left(\frac{p}{10^2}, \frac{p}{10^2} + \frac{1}{10^3}\right)$, we remove the intervals $\left(\frac{p}{10^2} + \frac{q}{10^4} + \frac{1}{10^5}, \frac{p}{10^2} + \frac{q+1}{10^4}\right)$, and so on. In each operation, we remove $\frac{9}{10}$ of the remaining intervals. The set of z is therefore B measurable and of measure zero.

---

[19] M. Borel has indicated (note 1, p. 48 of *Leçons sur la théorie des fonctions*) the principles which guided us in the theory of the measure. When we seek to construct the sets on which the application of these principles permits to attach a measure, we are immediately led to the class of the B measurable sets. This class has, hitherto, sufficed in practice. The primary advantage of reasoning about measurable sets and not just on B measurable sets, is not that we consider a larger class of sets, but that we start from the fundamental property of the sets to which we can attach a measure, rather than from a construction process that is perpetually evolving. As a result, as we have seen above, it was possible to show quite simply the compatibility of the conditions of problem of the measure of all measurable sets, whereas this proof has not been achieved when considering only the B measurable sets.

Among the B measurable sets, it seems that M. Borel initially considered only those obtained by performing only a finite number of operations I and II; following the links I made between the B measurable sets and the functions considered by M. Baire, the use of the definition in the text became more widespread. On these questions and on the existence of the non B measurable sets, see a Memoire that I published in the *Journal de Mathematique*, in 1905, and the book of M. de la Vallée Poussin, cited in Sect. 7.1. For more recent research, see the collection of *Fundamenta* and a Memoire of M. M. Lusin and Sierpinski (*Journ. de Math.*, 1923.)

8.3 The Measurable Functions

For the previous considerations to allow us to attach an integral to a function $f(x)$, it is necessary that, however small ε may be, we can find the numbers l_i (Sect. 8.1) such that the corresponding functions ψ_i, as well as the functions Ψ_i are associated with the measurable sets. Let us suppose that the sets corresponding to ψ_i are measurable, and let α and β be any two numbers. For a given value of ε there corresponds a certain system of numbers l_i; let l_p be the smallest of those which are included between α and β, and l_{p+q} be the largest. The set

$$E(\psi_p = 1) + E(\psi_{p+1} = 1) + \cdots + E(\psi_{p+q} = 1) = E(\varepsilon)$$

is measurable; but when we give to ε a sequence of decreasing values tending towards zero $\varepsilon_1, \varepsilon_2, \ldots,$ we have

$$E[\alpha < f(x) < \beta] = E(\varepsilon_1) + E(\varepsilon_2) + \cdots,$$

therefore $E[\alpha < f(x) < \beta]$ is measurable.

We will say that a bounded or unbounded function is measurable if, for any α and β, the set $E[\alpha < f(x) < \beta]$ is measurable. When this is the case, the set $E[f(x) = \alpha]$ is also measurable, because it is the subset common to the sets $E[\alpha - h < f(x) < \alpha + h]$ when h tends towards zero. Similarly, we would see that, for a function to be measurable, it is necessary and sufficient that the set $E[\alpha < f(x)]$ is measurable for any α; or equivalently, that the sets $E[\alpha \leq f(x)]$ are measurable. We would also see that $f(x)$ is measurable if, and only if, for each value of α, there exists a set, which we will denote by $E[\alpha \leq f(x), f(x) < \alpha]$, contained in $E[\alpha \leq f(x)]$ and containing $E[\alpha < f(x)]$ which is measurable.

The sum of two measurable functions is a measurable function. Let f_1 and f_2 be two measurable functions. For any number ε, there corresponds a division of their interval of variation, whether it is finite or infinite, with the help of the numbers l_i such that $l_{i+1} - l_i$ is at most equal to ε. Now, let us consider the sets E_{ij} of values of x, such that we have both

$$l_i < f_1(x), \quad l_j < f_2(x) \quad (l_i + l_j > \alpha);$$

E_{ij} is measurable as a subset common to $E[l_i < f_1(x)]$ and to $E[l_j < f_2(x)]$.

The sum $E(\varepsilon)$ of sets E_{ij} is measurable, since each of them is measurable; and if we give to ε the values ε_i tending towards zero, we have

$$E[\alpha < f_1 + f_2] = E(\varepsilon_1) + E(\varepsilon_2) + \cdots,$$

therefore $f_1 + f_2$ is a measurable function.

Similarly, we will prove that we can perform all the operations discussed regarding integrable functions (Sect. 3.3) on measurable functions without ceasing to obtain measurable functions. Moreover, *the limit of a convergent sequence of measurable functions is a measurable function;* if f_n tends towards f, we can construct a set $E[\alpha \leq f(x), \alpha < f(x)]$, by taking the sum of sets E_n; where E_n is the subset common to the sets $E[f_n(x) < \alpha]$, $E[f_{n+1}(x) < \alpha], \ldots$ and all these sets are measurable, if the functions f_n are measurable.

Let us apply these results; the two functions $f = \text{const.}$, and $f = x$ are obviously measurable, therefore any polynomial is measurable; Any function which is the limit of polynomials is also measurable. Therefore, according to a theorem of Weierstrass, any continuous function is measurable. The discontinuous functions which are limits of continuous functions, that M. Baire called *functions of first class*, are measurable. The functions which are not of first class and which are limits of functions of first class (M. Baire called them *functions of second class*) are measurable functions.

Let us again note that the functions thus formed step by step are B measurable, meaning that the corresponding sets are B measurable; these are the functions we will only encounter.

We can often show that a function is measurable by using the following property: if, when disregarding a set of values of x of measure zero, the function $f(x)$ is continuous, it is measurable. This is because the limit points of a set $E[\alpha \leq f(x)]$ which do not make a subset of this set necessarily belong to the neglected set of measure zero, therefore they form a set of measure zero. The set $E[\alpha \leq f(x)]$, being closed except for a set of measure zero, is measurable. In particular, this shows that any Riemann integrable function is measurable; we also see that the function $\chi(x)$ of Dirichlet, which is not integrable, is measurable.

8.4 Constructive Definition of the Integral

Now, let us define the integral of a bounded measurable function by assuming the interval of integration (a, b) is positive. We know that, if it is a function ψ, this integral is

$$m[E(\psi = 1)],$$

and, if we are dealing with an arbitrary function $f(x)$, the integral must be the common limit of the integrals of φ and Φ (Sect. 8.1) when the maximum of $l_{i+1} - l_i$ tends towards zero. According to the conditions of the integration problem, these integrals are

$$\sigma = \sum_{i=0}^{i=n-1} l_i \langle m\{E[f(x) = l_i]\} + m\{E[l_i < f(x) < l_{i+1}]\}\rangle,$$

$$\Sigma = \sum_{i=1}^{i=n} l_i \langle m\{E[l_{i-1} < f(x) < l_i]\} + m\{E[f(x) = l_i]\}\rangle.$$

We already know that these two numbers differ by less than $\varepsilon(b-a)$ because $\Phi - \varphi$ is smaller than ε. If we let ε tend towards zero, by interspersing the new numbers between the l_i, then σ increases, Σ decreases, $\Sigma - \sigma$ tends towards zero; therefore σ and Σ have the same limit.

Let $\sigma_1, \Sigma_1; \sigma_2, \Sigma_2; \ldots$ be the sum obtained by this method. Let $\sigma_1', \Sigma_1'; \sigma_2', \Sigma_2'; \ldots$ be the sums obtained by letting ε tend towards zero in another manner[20]; let σ_1'', Σ_1'' be the sums obtained by combining the numbers l_i yielding σ_1, Σ_1 and σ_1', Σ_1'; let σ_2'', Σ_2'' be the ones obtained by combining the l_i yielding $\sigma_2, \Sigma_2, \sigma_1', \Sigma_1'; \sigma_2', \Sigma_2'$; and so on. We obviously have

$$\sigma_i \leq \sigma_i'' \leq \Sigma_i'' \leq \Sigma_i,$$
$$\sigma_i' \leq \sigma_i'' \leq \Sigma_i'' \leq \Sigma_i';$$

the second of these inequalities shows that σ_i' and Σ_i' have the same limit as σ_i'' and Σ_i'' because, we know that σ_i'' and Σ_i'' have a limit and $\Sigma_i' - \sigma_i'$ tends towards zero. The first one shows that this limit is also that of σ_i and Σ_i.

The value of the limit is, that is, the integral, is therefore independent of how the maximum of $l_{i+1} - l_i$ tends towards zero.

Let us complete this definition by setting

$$\int_a^b f(x)\,dx = -\int_b^a f(x)\,dx.$$

It remains to see if the integral satisfies the conditions of the integration[21] problem.

First, let us get rid of the conditions 1, 2, 4, 5. There is nothing to say about the latter. The condition 4 follows from the fact that σ, for example, is positive when f is positive or zero.

The condition 1 follows, for the case of a positive interval, from the fact that the sums σ formed for the two integrals $\int_a^b f(x)\,dx$, $\int_{a+h}^{b+h} f(x-h)\,dx$ using the numbers l_i, are identical. And hence we pass on to the case of negative intervals.

To verify the condition 2, it is obviously sufficient to examine the case $a < c < b$; then if we use the same l_i for calculating the approximate values $\sigma_a^c, \sigma_c^b, \sigma_a^b$ of the integrals

$$\int_a^c f(x)\,dx, \int_c^b f(x)\,dx, \int_a^b f(x)\,dx,$$

obviously, we have

$$\sigma_a^b = \sigma_a^c + \sigma_c^b;$$

because we have the similar equalities between the corresponding measures of the sets which appear in the three sums σ.

Therefore, there remain the conditions 3 and 6 that need to be verified for a positive interval (a, b). To do this, let us first prove that, in such an interval, we have

---

[20] The l_i that yield σ_p' and Σ_p' do not necessarily contain those that yielded σ_{p-1}' and Σ_{p-1}', while the l_i that yield σ_p and Σ_p include the l_i related to σ_{p-1} and Σ_{p-1}.

[21] In the case where non-measurable functions exist, it is necessary to add that we confine ourselves to the consideration of only measurable functions.

$$\int_a^b f\,dx \leq \int_a^b g\,dx \leq \int_a^b f\,dx + \eta(b-a),$$

when we have

$$f \leq g \leq f + \eta.$$

Let us use the same numbers chosen appropriately, l, L, l_i to calculate the approximate values $\sigma(f)$ and $\sigma(g)$ of two integrals. Let us denote by F_i and G_i the two sets

$$E[l_i \leq f(x) < l_{i+1}], \quad E[l_i \leq g(u) < l_{i+1}],$$

we have

$$\sigma(f) = \sum_{i=0}^{i=n-1} l_i m(F_i) = l_{n-1} m(F_0 + F_1 + \cdots + F_{n-1}) - \sum_{i=0}^{i=n-2} (l_{i+1} - l_i) m(F_0 + F_1 + \cdots + F_i),$$

$$\sigma(g) = \sum_{i=0}^{i=n-1} l_i m(G_i) = l_{n-1} m(G_0 + G_1 + \cdots + G_{n-1}) - \sum_{i=0}^{i=n-2} (l_{i+1} - l_i) m(G_0 + G_1 \cdots + G_i).$$

Now, because $f \leq g$, $F_0 + F_1 + \cdots + F_i$ contains $G_0 + G_1 + \cdots + G_i$ and, for $i = n-1$, these two sets are identical. Therefore, the first terms of $\sigma(f)$ and $\sigma(g)$ are equal and the other terms are less in $\sigma(g)$ than in $\sigma(f)$, in absolute value, which proves the first inequality

$$\int_a^b f\,dx \leq \int_a^b g\,dx.$$

Since we have

$$g \leq f + \eta,$$

therefore, we have

$$\int_a^b g\,dx \leq \int_a^b (f + \eta)\,dx;$$

to calculate approximately the last integral, let us use the numbers $l + \eta$, $L + \eta$, $l_i + \eta$; it is clear that we find

$$\begin{aligned}\sigma(f+\eta) &= \sum_{i=0}^{i=n-1} (l_i+\eta)m(F_i)\\ &= \sum_{i=0}^{i=n-1} l_i m(F_i) + \eta \sum_{i=0}^{i=n-1} m(F_i) = \sigma(f) + (b-a)\eta.\end{aligned}$$

Hence

$$\int_a^b g\,dx \le \int_a^b (f+\eta)\,dx = \int_a^b f\,dx + \eta(b-a).$$

We can again say that, if two functions differ by less than ζ the integrals differ by less than $\zeta(b-a)$. Because, saying that f and g differ by less than ζ, means that g is included between $f-\zeta$ and $f+\zeta$; therefore $\int_a^b g\,dx$ is included between

$$\int_a^b f\,dx - \zeta(b-a) \quad \text{and} \quad \int_a^b f\,dx + \zeta(b-a).$$

This being said, let f and φ be two measurable and bounded functions in the positive interval (a, b); we have learned (Sect. 8.1) to associate with them the functions f_1 and φ_1 taking only the finite number of values and differing, respectively, from f and φ by less than ε.

Then $f_1 + \varphi_1$ differs from $f + \varphi$ by less than 2ε. Therefore, we have

$$\begin{aligned}\left|\int_a^b f\,dx - \int_a^b f_1\,dx\right| &< \varepsilon(b-a),\\ \left|\int_a^b \varphi\,dx - \int_a^b \varphi_1\,dx\right| &< \varepsilon(b-a),\\ \left|\int_a^b (f+\varphi)\,dx - \int_a^b (f_1+\varphi_1)\,dx\right| &< 2\varepsilon(b-a).\end{aligned}$$

The equality to be shown,

$$\int_a^b (f+\varphi)\,dx = \int_a^b f\,dx + \int_a^b \varphi\,dx,$$

will, therefore, follow from this:

$$\int_a^b (f_1+\varphi_1)\,dx = \int_a^b f_1\,dx + \int_a^b \varphi_1\,dx.$$

Now, let us suppose that f_1 takes only the values $a_1, a_2, a_3, \ldots$ at the respective points of the sets $E_1, E_2, \ldots$; and that φ_1 takes only the values $b_1, b_2, \ldots$ at the points of $e_1, e_2, \ldots$. And let E_{ij} be the set of points common to E_i and e_j, we have

$$\begin{aligned}\int_a^b (f_1+\varphi_1)\,dx &= \sum_{i,j}(a_i+b_j)m(E_{ij})\\ &= \sum_i a_i\left[\sum_j m(E_{ij})\right]+\sum_j b_j\left[\sum_i m(E_{ij})\right]\\ &= \sum_i a_i m(E_i)+\sum_j b_j m(e_j)\\ &= \int_a^b f_1\,dx+\int_a^b \varphi_1\,dx.\end{aligned}$$

The condition 3 is, therefore, fulfilled indeed. The condition 6 is also satisfied, as shown by the following property:

If the measurable $f_n(x)$, bounded for all n and x, have a limit $f(x)$, then the integral of $f_n(x)$ tends towards integral of $f(x)$.

Indeed, we know that $f(x)$ is integrable; let us evaluate

$$\int_a^b [f(x)-f_n(x)]\,dx.$$

If we always have $|f_n(x)| < M$ and if $f - f_n$ is less than ε in E_n, where $f - f_n$ is less than the function equal to ε in E_n and M in $C(E_n)$, then the integral is at most equal in modulus to

$$\varepsilon m(E_n)+Mm[C(E_n)].$$

But ε is arbitrary, and $m[C(E_n)]$ tends towards zero with $\frac{1}{n}$, since there is no point common to all the E_n, therefore

$$\int_a^b (f-f_n)\,dx$$

tends towards zero. The property is proved.[22]

Another form of this theorem is the following:

If all the remainder of the series of measurable and bounded functions are, in absolute value, less than a fixed number M, the series is term-by-term integrable.

The previous definitions and the results can be extended to some unbounded functions. Let $f(x)$ be an unbounded measurable function. Let us choose infinitely many numbers $\ldots, l_{-2}, l_{-1}, l_0, l_1, l_2, \ldots$, spread out from $-\infty$ to $+\infty$ such that $l_{i+1}-l_i$ is always less than ε. We can form two series, infinite in both directions,

[22] M. Osgood, in a Memoire of *American Journal* 1897 : *On the non uniform convergence,* has shown a particular case of this theorem in which f and f_n are continuous. The method of M. Osgood is totally different from the one presented in the text.

$$\sigma = \sum_{-\infty}^{+\infty} l_i m\{E[l_i \leq f(x) < l_{i+1}]\},$$

$$\Sigma = \sum_{-\infty}^{+\infty} l_i m\{E[l_{i-1} < f(x) \leq l_i]\}.$$

By revisiting the previous arguments, we immediately see that if one of them is convergent, and consequently absolutely convergent since all the terms with sufficiently small indices are negative or zero and all those with sufficiently large indices are positive or zero, then the other is also convergent. Under these conditions, both σ and Σ converge to a well determined limit, when the maximum of $l_{i+1} - l_i$ tends towards zero in any manner. This limit is, by definition, the integral of $f(x)$ in the positive interval of integration. Hence, we can extend it to a negative interval as done previously.

We call *summable functions* those functions to which the completed constructive definition of the integral applies.[23] Any bounded measurable function is summable.[24] If f is a summable function, $|f|$ is also a summable function and, if the interval of integration (a, b) is positive, we have

$$\left|\int_a^b f\,dx\right| \leq \int_a^b |f|\,dx,$$

because the approximate values of the two members are $|\sigma|$ and the sum of the absolute values of the terms of σ. If f is summable and if $f_{M,N}$ denotes the function equal to f when f is included between $-M$ and $+N$ and zero otherwise, then as M and N tend towards $+\infty$, we have,

$$\int_a^b f\,dx = \lim \int_a^b f_{M,N}\,dx,$$

because the approximate values of the two members are σ and the limit of the contribution to σ of terms with indices ranging between $-m$ and n, when m and n increase indefinitely.[25] We would have the same equality if we has subjected $f_{M,N}$ to be respectively equal to $-M$ and $+N$, instead of being zero, when f is respectively less than $-M$ or greater than $+N$.

[23] I depart here from the language used in my thesis, where I called functions that I now refer to as *measurable* functions the *summable functions*. With the conventions of the text, which are now adopted by all, the word *summable* plays in the theory of integration, the same role as the word *integrable* in Riemannian integration.

When a function is measurable without being summable, it does not have integral. However, when it concerns a function that is always positive or bounded from below (or always negative function or bounded from above) we may still say that it has an infinite integral, for the reasons that become clear immediately.

[24] The bounded summable functions considered here, are the same which have been discussed at the end of the previous chapter, this will become apparent in Chap. 9.

[25] The converse is exact, that is to say that if $\int_a^b f_{M,N}\,dx$ tends towards a definite limit *when M and N tend towards $+\infty$ independently and in an arbitrary manner*, $f(x)$ is summable and its integral is the considered limit.

We do not know any bounded non-summable function, on the contrary, it is easy to cite unbounded non-summable functions. The function, zero for $x = 0$ and equal to

$$\left(x^2 \sin\frac{1}{x^2}\right)' = 2x \sin\frac{1}{x^2} - \frac{2}{x}\cos\frac{1}{x^2}$$

is an example; however, this function can be integrated by the methods of Cauchy and Dirichlet developed in Chap. 1. We could, in some cases, apply these methods to non-summable functions to define their integrals. We will return to this generalisation, but for the time being let us limit ourselves to the summable functions.

But we are going to give this concept a new extension: let us suppose that a function $f(x)$ is given, or is considered only at the points of a set E; we can still form sets $E[l_i \leq f(x) < l_{i+1}]$ for any l_i and l_{i+1}, but now these sets will all be contained in E. If these sets are measurable, regardless of the choice of l_i and l_{i+1}, we will say that f is measurable in E. Let us note that this implies that E itself is measurable because E is the sum of different sets $E[l_i \leq f(x) < l_{i+1}]$ associated with a choice of numbers l_i.

If f is measurable in E, we can therefore form σ and Σ. If one of these sums is convergent, in which case both must be, f is said to be summable over E, *and the integral of f extended to E, $\int_E f\,dx$,* is the common limit to which σ and Σ converge, when we vary the choices of l_i such that the maximum of $l_{i+1} - l_i$ tends towards zero.

In short, nothing changes in the definition and we could have simply stated that

$$\int_E f\,dx = \int_a^b (f)\,dx;$$

where, (a, b) is a positive interval containing E and (f) is a function, equal to f at points of E and zero at the other points. This second definition will allow us to conclude very easily from the properties of integrals in the intervals, the properties of integrals in the sets.

We have deviated from the problem of integration as we initially posed it in the beginning of the chapter; how should we modify it to provide a descriptive definition of the integrals of summable functions in the measurable sets? To answer this question, let us first note that we still have a proposition analogous to the one in problem number 6 of the integration problem. To be more precise, *if, as the index n increases, the functions $f_n(x)$, which are summable over a measurable set E, monotonically increase to the function $f(x)$, which is also summable over E, then the integral of $f_n(x)$ over E monotonically increases to the integral of $f(x)$.*

Indeed, we can assume that the f_n are non-negative otherwise we would reason on the functions $f_n - f_1$. Since the f_n are positive or zero, the same holds for f. Let us assume

$$f = g + h, \quad f_n = g_n + h_n,$$

g being equal to f when f is less than an arbitrarily chosen positive number N, and equal to N otherwise; g_n being equal to f_n when g is equal to f, and equal to N otherwise. g is the limit of g_n, h_n is not greater than h.

We have

$$\int f\,dx = \int g\,dx + \int h\,dx, \quad \int f_n\,dx = \int g_n\,dx + \int h_n\,dx;$$

to be truthful, this is only entirely clear when it concerns integrals extended to an interval, but by converting f to (f), we can always assume that it is so.

We had said, a moment ago, that if we let N increase indefinitely, then $\int g\,dx$ converges to $\int f\,dx$, therefore $\int h\,dx$ converges to zero. Let us take N sufficiently large, so that $\int h\,dx$ is less than ε; the same will apply to $\int h_n\,dx$, a fortiori. However, according to the condition 6, for the bounded measurable functions g and g_n, $\int g_n\,dx$ converges to $\int g\,dx$. Therefore, we have

$$\begin{aligned}\lim \int f_n\,dx &= \lim \int g_n\,dx + \int h_n\,dx \\ &\leq \lim \int g_n\,dx + \varepsilon = \int g\,dx + \varepsilon \leq \int f\,dx + \varepsilon,\end{aligned}$$

and the proposition follows immediately.

This case of term-by-term integration of the sequence can, as previously mentioned, be transformed into the case of term-by-term integration of series: *a convergent series of non-negative functions all of whose terms and the sum are summable over a set E, is term-by-term integrable over E.*[26]

Let us apply this to the following particular case: Let f be a summable function in a measurable set E; let us partition E in a finite or a countably infinite number of measurable sets E_i which are pairwise disjoint, and let f_i be the function equal to f in E_i and zero outside of E_i. The previous proposition on the integration of series can be applied to

$$f = f_1 + f_2 + \cdots,$$

by assuming f to be always positive or zero, now

$$\int_E f_i\,dx = \int_{E_i} f\,dx,$$

therefore

$$\int_E f\,dx = \int_{E_1} f\,dx + \int_{E_2} f\,dx + \cdots$$

---

[26] This is the case of integration known as *monotone sequences or series*. A series is called monotone if the sequence of sums s_n of this series is monotone, that is, it is either non-decreasing or non-increasing.

If f is sometimes negative, we will arrive at the same result by applying the above formula to $|f| + f$, then to $|f| - f$ and by adding.

By combining this result with the previous one, we can say: *the integral* $\int_E f(x)\,dx$, *extended to a measurable set* E, *of a summable function* $f(x)$ *in* E, *is a number satisfying the following properties:*

1. *If* E_h *is the set of the values of* x *such that* $x + h$ *belongs to* E, *we have*

$$\int_E f(x)\,dx = \int_{E_h} f(x-h)\,dx;$$

2. *If* E *is the sum of finite or countably infinite, pairwise disjoint, measurable sets* $E_1, E_2, \ldots$, *we have*

$$\int_E f(x)\,dx = \int_{E_1} f(x)\,dx + \int_{E_2} f(x)\,dx + \cdots;$$

3.

$$\int_E [f(x) + \varphi(x)]\,dx = \int_E f(x)\,dx + \int_E \varphi(x)\,dx,$$

$\varphi(x)$ *is assumed to be, like* $f(x)$, *summable over* E;

4. *If we have* $f \geq 0$, *then we have*

$$\int_E f(x)\,dx \geq 0;$$

5.

$$\int_0^1 1 \times dx = 1$$

The reader can easily verify that these properties are characteristic of the integral; therefore, we have here a descriptive definition of the integral of a summable function in a measurable set. The new statement of the problem of integration now contains only five conditions, but that does not reveal any significant difference between the new problem and the older one.

In reality, we could have combined the old conditions 3 and 6 into a single statement related to a case of integration of sums or series. Furthermore, let us note that we have already performed (Sect. 8.2), in the context of measure of sets, the transformation of the condition 6, which pertained to a series of functions, into our new condition 2, which pertains to a series of sets.

It is especially important for the effective calculation of the integrals of functions given by a series expansion to know case of term-by-term integration. M. Vitali has written on this subject a very important Memoire, that I can only point out here[27]; I would confine myself to providing a little more extensive case of integration than the previous ones.

A convergent series of summable functions f_n is term-by-term integrable, when all its remainders $r_n(x)$ are, in absolute value, less than a definite summable function $\varphi(x)$

$$|r_n(x)| \leq \varphi(x).$$

Let $f(x)$ be the sum of the series. The obvious inequality

$$|f(x)| = |f_1(x) + r_n(x)| \leq |f_1(x)| + \varphi(x)$$

shows that $f(x)$ is summable. This being true, we partition the interval or the set E in which we integrate, into three measurable, pairwise disjoint sets: the first one, say A, of these sets is formed from points in which $\varphi(x)$ exceeds a number N, the second one, say B_p, is formed of points not belonging to A and in which the remainder r_p is less than ε in modulus and the remaining points form the third set, say C_p. We have

$$\int_E f(x)\,dx = \int_E f_p(x)\,dx + \int_A r_p\,dx + \int_{B_p} r_p\,dx + \int_{C_p} r_p\,dx;$$

$$\left|\int_A r_p\,dx\right| \leq \int_A |r_p|\,dx \leq \int_A \varphi\,dx,$$

Let us assume N to be chosen in a way that $\int_A \varphi\,dx$ is less than ε:

$$\left|\int_{B_p} r_p\,dx\right| \leq \int_{B_p} |r_p|\,dx \leq \int_{B_p} \varepsilon\,dx = \varepsilon m(B_p) \leq \varepsilon m(E);$$

$$\left|\int_{C_p} r_p\,dx\right| \leq \int_{C_p} |r_p|\,dx \leq \int_{C_p} N\,dx = Nm(C_p).$$

Now, when p increases indefinitely, $m(C_p)$ tends to zero, because the set C_p is contained in all those of lower indices, and there are no common points in all the C_p sets, since the series would diverge at such a point. Therefore, for sufficiently large p, we have

$$\left|\int_E f(x)\,dx - \int_E f_p(x)\,dx\right| < \varepsilon + \varepsilon m(E) + \varepsilon_1.$$

The theorem is therefore proved.

Let us note that, in the statement of this theorem, we did not have to assume that the limit $f(x)$ was summable, whereas we had to make this hypothesis in the statement of Sect. 8.4. We can, with M. B. Levi,[28] transform this latter statement so that we no longer

[27] *Rendiconti del Circolo Matematico di Palermo, t.* 23, 1907.

[28] *Reale Ist Lombardo; Rendiconti, t.* 39, 1906.

need to assume anything about the limit $f(x)$. To give this proposition its full scope, let us define what is meant by a measurable function which is not always finite. It is a function which, at every point in the interval or set E under consideration, takes a definite value in magnitude and sign, but not always a finite one. For such a function f, there is generally a set $E(f = +\infty)$ and a set $E(f = -\infty)$. By stating that f is measurable, we express that these two sets are measurable and f is measurable at the set of points where it is finite. We can also state that the set $E[\alpha \leq f(x) < \beta]$, or the set $E[\alpha < f(x)]$, or the set $E[f(x) \leq \beta]$, is measurable regardless of whether α and β are finite or infinite.

The statement made is related to the increasing sequence of measurable functions, such a sequence necessarily has a measurable limit, but it is not necessarily finite everywhere.

Let $f(x)$ be the limit of a sequence of increasing, finite and summable functions $f_n(x)$.

If the sequence of integrals of the functions $f_n(x)$ converges, then f is infinite only at points of a set of measure zero, f is summable over the set of points where it is finite, and the integral of f is the limit of the integrals of f_n.

If the sequence of integrals of f_n tends to infinity, then f is infinite at points of set of non-zero measure, or is not summable over the set of points where it is finite.

f never takes the value $-\infty$ anywhere. Furthermore, we can reason solely on the set of points where f_1, and therefore f are positive. If $E(f = +\infty)$ is of non-zero measure λ, then f_p will be greater than some number N at the points of a set of measure $\frac{\lambda}{2}$; at least for a p large enough, the integral of f_p will be greater than $\frac{\lambda}{2}N$. Since N can be an arbitrary value, the sequence of $\int f_p \, dx$ cannot converge.

Therefore, let us assume that $E(f = +\infty)$ is of measure zero, and let us remove this set from the set of integration, which does not modify the integral of f_n. Now we are dealing with increasing sequences of functions which are always finite and have a limit that is also finite. Therefore, if f is summable, the series of integrals of f_n converges to the integral of f (Sect. 8.4). If f is not summable, it means that the integral of function g, which is equal to f when f is less than N and zero elsewhere, increases indefinitely as N increases indefinitely. And since the limit of $\int f_n \, dx$ exceeds $\int g \, dx$, the sequence of integrals $\int f_n \, dx$ is divergent. This justifies the statement.

8.5 Other Forms of the Definition of Integral

We have just defined the integral, both through its properties and through a construction, and we have obtained methods for calculating the integral of functions given by series expansions. At this point, it would be beneficial to take a step back and summarise what we have achieved.

For the construction of the integral of a bounded measurable function defined on an interval, we have deduced the condition of our original problem of integration:

a. The value of the integral for functions φ taking only two values, zero and a constant;
b. The case of term-by-term integration of monotone series or the series which becomes monotone when we remove the first terms;
c. We have proved that any measurable function is the sum of a term-by-term integrable series of functions φ according to b.

It is clear that this construction of integral could be modified in many ways; it would suffice to: a. Begin with knowledge of the integral of a sufficiently broad class of particular functions; b. and to use a term-by-term integration characteristic of series, that we laid down a priori, and sufficiently general to assert that: c. any measurable function is the sum of a series of the considered nature of functions belonging to the envisaged class. I will[29] examine very rapidly a definition of M. W.-H. Young.

a. He starts from the integral of continuous functions.
b. He admits that any monotone sequence is term-by-term integrable.
   He then obtains the integral of limit functions f_l and f_u, respectively, of increasing and decreasing sequence of continuous functions.
   Then the integral of limit functions f_{lu} and f_{ul}, respectively, of sequence of increasing functions f_u and sequence of decreasing functions f_l;
   then the integral of functions f_{lul}, f_{ulu}, etc.
   Naturally, it must be shown that this method of proceeding does not lead to contradictions. Therefore, M. Young proves that the integration of the monotone sequences, he considers, does indeed yield a convergent sequence of integrals. Furthermore, he shows that for monotone sequences converging to same limit, there correspond sequences of integrals with same limit.
c. He must also define the family of functions thus integrated.

Sticking to the previous definitions,[30] these functions would not encompass all measurable functions, but they would include all B measurable functions, that is, all the measurable functions practically useful to consider. Furthermore, M. Young's method provide, in the

---

[29] We can take this opportunity to mention remarks or the work of M. M. Fubini, F. Riesz, Weyl, Egoroff, Lusin, Borel. See, for example, the work I published in *Annales sc de l'École Normale in* 1918.

[30] M. Young, by a new extension, also covers all measurable functions *See*, for example, *Proc. of the Lond. Math. Soc,* 1910.

The letters l and u which appear in the notation of M. Young, are the initials of the *lower* and *upper;* the functions f_l and f_u are the lower and upper semi continuous functions of M. Baire (Sect. 3.1).

We can also use this method of definition for the unbounded functions and for the functions defined on the sets.

Finally, one can also, either with this definition or with the others, deal with the case where the interval or domain of definition of the function is not entirely at a finite distance but extends indefinitely.

process, a classification of these B measurable functions that is related to M. Baire's classification (Sect. 8.3) and which is very interesting.

If we had started: a. from the integral of continuous functions and b. the case of integration of uniformly bounded sequences (Sect. 8.4), it is the classification of M. Baire that would have presented itself to us.

M. W.-H. Young has also introduced[31] the following property which could be taken for definition of the integral:

Given a bounded measurable function $f(x)$ defined on a finite and positive interval (a, b), let us divide (a, b) in a finite or countably infinite number of measurable and pair wise disjoint sets $E_1, E_2, \ldots$. Let δ_i be the measure of E_i, let l_i and L_i be the lower and upper limits of $f(x)$ in E_i, let us form the sums or series

$$\underline{S} = \sum l_i \delta_i, \quad \overline{S} = \sum L_i \delta_i;$$

and, vary the choice of E_i, let us determine the upper bound m of $\underline{S}$ and the lower bound M of $\overline{S}$. These two bounds are equal to each other and to $\int_a^b f(x)\,dx$.

Indeed, let us calculate the contribution of the points of

$$e_n = E[n\varepsilon \leq f < (n+1)\varepsilon]$$

in $\underline{S}$ and $\overline{S}$. The points of e_n are distributed in some of the E_i; they form the set $e_n^{i_1}$ contained in E_{i_1}, the set $e_n^{i_2}$ contained in E_{i_2}, etc. For all the values of $i_1, i_2, \ldots$ of i the number l_i is at most equal to $(n+1)\varepsilon$, therefore the contribution of $e_n^{i_k}$ is at most $(n+1)\varepsilon \times \text{mes}(e_n^{i_k})$ and that of e_n is at most $(n+1)\varepsilon \times \text{mes}(e_n)$. Therefore, we have

$$\underline{S} \leq \sum (n+1)\varepsilon \times \text{mes}(e_n),$$

and by letting ε tend to zero,

$$\underline{S} \leq \int_a^b f(x)\,dx.$$

But similarly, we will find

$$\overline{S} \geq \int_a^b f(x)\,dx.$$

Hence it follows

$$M \geq \int_a^b f(x)\,dx \geq m.$$

But it is sufficient to take the E_n identical to e_n for the difference $\overline{S} - \underline{S}$ to be at most equal to $\varepsilon(b-a)$. Therefore, $M = m$.

The similarity between the definitions of M. W.-H. Young and that of Riemann (as explained on Sect. 3.2) is obvious. Note that our constructive definition of the integral is also

---

[31] *Proc. Lond. Math. Soc.*, 1905, and *Ph. Trans. London*, 1905.

very similar to that of Riemann; the difference is that, while Riemann divided the intervals of variation of x in small partial intervals, it is the interval of variation of $f(x)$ that we have subdivided.

This approach is necessary and its advantages are obvious. When we form the sum $S = \sum f(\xi_i)(x_{i+1} - x_i)$ for a continuous function $f(x)$, we group the values of x that yield values of $f(x)$ that are very close to each other. It is because these values are close that we can replace them in S by any one of the $f(\xi_i)$. However, if $f(x)$ is discontinuous, there is no longer any reason that the choices of intervals (x_{i+1}, x_i) that become smaller and smaller would lead to the grouping of values of $f(x)$ that are less and less different. And this is why the method of Riemann succeeds only rarely and somehow by chance. Since we want to group the values of $f(x)$ that are close, it is quite clear that we should, as we have done in this chapter, subdivide the interval of variation of $f(x)$ and not the interval of variation of x.

We can still say, by adopting the language of the XVII century-the language of indivisibles-that we are summing up the various indivisibles associated with the given function $f(x)$, i.e., the positive or negative ordinates of the points $[x, y = f(x)]$. To do this, as we have done in algebra when we perform the reduction of similar terms, as in arithmetic when adding the numbers by summing the digits in the units place then the digits at tens place, etc., we have grouped the indivisibles of the same magnitude or nearly the same magnitude.

Now, we shall perform the summation of these indivisibles by grouping all those which are positive and all those which are negative and thus, we will have a definition similar to that in Chap. 3. For this, we will assume that the problem of the measure of sets formed of the points in a plane has been solved. This problem is posed similarly to the case of the straight line, with condition 3′ becoming: *the measure of the set of points whose coordinates satisfy the inequalities*

$$0 \leq x \leq 1, \quad 0 \leq y \leq 1,$$

is 1.

We will easily prove that the measure of a square is its area in the elementary sense of the word. Hence, we will deduce that the measure of an arbitrary set lies between its exterior measure and interior measure, which are defined as in the case of the straight line, with squares replacing the intervals.

To prove that the interior measure never exceeds the exterior measure, it is necessary to show that a square C can be covered with the help of a finite number of squares c_i only if the sum of areas of c_i is at least equal to the area of C, which can be done in an elementary way.[32] Then we must prove the theorem of M. Borel by replacing the word *interval* in its statement with the word *square* or the word *domain.*

---

[32] For this question and everything related to the measure of polygons, we can consult the Note D of the *Géométrie élémentaire* of M. Hadamard with interest.

The proof can be given similarly to the case of the line. However, I would like to mention here how we can use M. Peano's curves and other similar curves (Sect. 4.1).

Let the domain D be such that any point interior to D or on the boundary of D is interior to one of the domains Δ. We can define, with the help of a parameter t varying from 0 to 1, a curve C which fills the domain D and which does not pass through any exterior points.[33] Each domain Δ cuts arc on C corresponding to certain intervals of variation for t, let δ be these intervals. A domain Δ may also have points on its boundary common with C, and these points do not form intervals; we neglect these points and deal only with the intervals. $(0, 1)$ is obviously covered with the δ, and therefore, with a finite number of them, according to the theorem of M. Borel for the case of line. Consequently, D is covered with the Δ in a finite number which correspond to these δ.

Having established this property, the sequence of arguments and definitions continues as in the case of line, with intervals always being replaced with squares. Just as in the case of the line, we define the measurable sets, the B measurable sets, and we show the same properties regarding them.

It is necessary not to confuse the measure of the sets of points in the plane with that of the sets of points on a line. We will distinguish between them when there is doubt by referring to them as *the surface measure* m_s *and linear measure* m_l.[34]

Let us arrive at the definition of the integral.

To any function $f(x)$ let us associate two surface sets

$$E[f(x) > 0,\ f(x) \geq y \geq 0] = E_1[f]$$
$$E[f(x) < 0,\ f(x) \leq y \leq 0] = E_2[f(x)];$$

by analogy with what has been done earlier (Chap. 3, Sect. 4.2), it is natural to call *the integral of the function* f, *the quantity*

$$I = m_s[E_1(f)] - m_s[E_2(f)].$$

Let us study under what conditions this definition is applicable; we will show that it is when the function f is measurable, and only in this case. For this purpose, it will obviously be

[33] To do this, we could establish a bijective and continuous correspondence between the points of a square and those of domain D, then choose the curve C that corresponds to the Peano's curve filling the square. The existence of this correspondence is clear when the curve bounding the domain D is simple, like a polygon, for example. However, the general case requires delicate reasoning. We can refer to, for example, the Thesis of M. Antoine (*Journal de Math.*, 1921).

If we were to envisage the domains other than those bounded by a Jordan's curve, the correspondence may not exist anymore. Furthermore, for such domains, there may not always exist a curve, similar to that of M. Peano, which exactly fills them. A slight modification of reasoning of the text would be necessary in this case; but it is unnecessary to consider it here.

[34] These definitions allow to define the measurable functions of two variables and the double integrals related to these functions. I will not address these questions or others that are related, such as integration by parts and integration under summation sign.

sufficient to show it for the function $\varphi(x)$ equal to $f(x)$ when $f(x)$ is not negative, and zero when $f(x)$ is negative; it is this function $\varphi(x)$ we are going to deal with.

When α decreases, the linear set $E(\varphi \geq \alpha)$ does not loose any point. Hence, we deduce that the lower and upper linear measures $m_{l,i}[E(\varphi \geq \alpha)]$ and $m_{l,e}[E(\varphi \geq \alpha)]$ are the non increasing functions. Moreover, $E(\varphi \geq \alpha)$ is the set of points which belong to all the $E(\varphi \geq \alpha - h)$; hence we deduce that $m_{l,i}[E(\varphi \geq \alpha)]$ and $m_{l,e}[E(\varphi \geq \alpha)]$ are the continuous functions of α at left. This being said, let us suppose that we have

$$m_{l,e}[E(\varphi \geq \alpha)] > m_{l,i}[E(\varphi \geq \alpha)] + \varepsilon,$$

then it will still be the same in a certain interval $(\alpha - h, \alpha)$. Let us consider the subset E of $E_1(\varphi)$ included between $y = \alpha - h$ and $y = \alpha$. Let us enclose the points of E in the squares A, and the points of $C(E)$ in the squares B. We can suppose that A and B have their edges parallel to ox and oy. They have in common the rectangles C, whose sum of areas is at least $m_{s,e}(E) - m_{s,i}(E)$ and differs as little as we want. The intersection of squares A with the straight line $y = K$ is composed of intervals a which enclose $E[\varphi(x) \geq K]$, the intersection of squares B is composed of intervals b which encloses $C\{E[\varphi(x) \geq K]\}$, the intersection of rectangles C is formed by the subsets c common to both a and b; therefore, we have

$$m_l(c) \geq m_{l,e}\{E[\varphi(x) \geq K]\} - m_{l,i}\{E[\varphi(x) \geq K]\};$$

$m_l(c)$ is therefore greater than ε when K varies from $\alpha - h$ to α, and $m_{s,e}(E) - m_{s,i}(E)$ is at least equal to $\varepsilon h.E$ and consequently $E_1(\varphi)$ is measurable only if φ is measurable.

Let us suppose that φ is bounded and measurable and let us partition the interval of variation of φ, using the numbers l_i. Let E be the subset of $E_1(\varphi)$ included between $y = l_{i-1}$ and $y = l_i$, we will evaluate its measure. Let us enclose the points of $E(\varphi \geq l_i)$ in the intervals a and the points of $C[E(\varphi \geq l_i)]$ in the intervals b. Let c be the intervals that are subset of both a and b. Let us consider the set $\mathfrak{A}$ of points whose abscissa are the points of a and whose ordinates are included between l_{i-1} and l_i; Let $\mathfrak{C}$ be the set similarly related to c. The set $\mathfrak{A} - \mathfrak{C}$ being contained in E, we have

$$m_{s,i}(E) \geq m_s(\mathfrak{A}) - m_s(\mathfrak{C}) = (l_i - l_{i-1})[m_l(a) - m_l(c)],$$

Hence, we deduce

$$m_{s,i}(E) \geq (l_i - l_{i-1}) m_l[E(\varphi \geq l_i)].$$

By taking sum of all the similar inequalities, we have

$$m_{s,i}[E_1(\varphi)] \geq \sum l_i m_l[E(l_i \leq \varphi < l_{i+1})] = \sigma.$$

By a similar reasoning, we see that

$$m_{s,e}[E_1(\varphi)] \leq \sum l_i m_l[E(l_{i-1} < \varphi \leq l_i)] = \Sigma.$$

We have shown that both the quantities σ and Σ converge to the same limit when the maximum of $l_{i+1} - l_i$ tends towards zero; therefore, $E_1(\varphi)$ is measurable. The approximate values of σ and Σ found for the measure of $E_1(\varphi)$ leads us to the definition of the integral already given. Therefore, there is an identity between the current geometric definition and the constructive definition studied[35] earlier.

[35] We can say that the reasoning of the text provides an expression of the surface measure based on linear measure. Generalising suitably, these considerations yield the formula which allows for replacing the calculation of a multiple integral by the calculations of successive simple integrals.

The Indefinite Integral of Summable Functions 9

9.1 The Three Indefinite Integrals. The Additive Functions of a Set

We will call an *indefinite integral* of $f(x)$ any one of the functions

$$F(x) = \int_{\alpha}^{x} f(x)\,dx + C$$

C being a constant; the definite integral of $f(x)$ in (a, b) is the increment $F(b) - F(a)$ of the indefinite integral in the interval (a, b).

The indefinite integrals are the continuous functions. If $f(x)$ is a bounded function, this is obvious by virtue of theorem of finite increment. Let us assume, then, $f(x)$ is summable but unbounded; then, we can find N sufficiently large such that the integrals of $f(x)$ over both sets $E(f > N)$ and $E(f < -N)$ are both less than ε in modulus. Let us set $f = f_1 + f_2$, where f_1 is zero on both sets $E(f > N)$ and $E(f < -N)$ and f_2 is zero on the set $E(-N \leq f \leq N)$. Then the indefinite integral of f_1 is a continuous function; and the integral of f_2 in any interval being 2ε at most, around any point x_0, therefore, we can find an interval in which the increment of $f(x)$ is 3ε at most. This proves that $f(x)$ is continuous.

If $f(x)$ is summable then $|f(x)|$ is also summable and, in any interval, the indefinite integral of $f(x)$ undergoes an increment at most equal to that of the indefinite integral of $|f(x)|$ in modulus. Since this latter integral exists and is increasing, therefore it is of bounded variation, *the indefinite integral of $f(x)$ is of bounded variation and its total variation in (a, b) is at most* $\left|\int_a^b |f(x)|\,dx\right|$.

The propositions found in Chap. 5 (Sect. 6.2) related to the limitations of the derivative numbers of $F(x)$ using the maxima and minima of $f(x)$ remain valid; they are similarly

R. Jain, *Lebesgue's Theory of Integration*,
https://doi.org/10.1007/978-981-96-1169-0_9

proved.[1] This leads, very naturally, to the study of the derivation of the indefinite integrals and the search for the primitive functions. But first, we will give a new meaning to the words "indefinite integral".

Where does the name *indefinite integral* come from? It is clear that in the expression "definite integral" and "indefinite integral", *indefinite* does not have the meaning of infinite but rather *non definite,* while *definite* has the meaning *determined.* These two expressions should, therefore, be applicable to the same quantity $\int_\alpha^\beta f(x)\,dx$; this integral will be called *definite* when the interval of integration (α, β) is itself *definite*, that is, *determined* and given and it would be called *indefinite* when (α, β) will be *indefinite*, meaning, non definite, non determined, unknown, or variable. Going back to this primitive sense of denominations, we will, therefore, say that the indefinite integral of $f(x)$ is the function $\Phi(\alpha, \beta)$

$$\Phi(\alpha, \beta) = \int_\alpha^\beta f(x)\,dx = F(\beta) - F(\alpha);$$

this will be a function of two variables or, even better, a function of the interval of integration (α, β). The word "function" here implies, a "correspondence", as in Sect. 3.1. But the task is to associate a number to every interval (α, β). The interval (α, β) serves as the variable or argument our function and the number is the value of the function. The indefinite integral is a function of interval. Any value taken by this function is a definite integral.

The relation that links $\Phi(\alpha, \beta)$ with $F(x)$ allows one to translate any property of $F(x)$ into a property of $\Phi(\alpha, \beta)$ and conversely. And this is why we ordinarily confine ourselves to the consideration of $F(x)$, that we will call, when a doubt may arise, the indefinite integral function of a single variable. In short, these are the properties of $\Phi(\alpha, \beta)$ which we typically study through the intermediary of $F(x)$. Our new language will be more directly suitable to our purpose. However, as we have also considered the integral of $f(x)$ extended to a measurable set E, we should consider the indefinite integral of $f(x)$ as the set function

$$\Psi(E) = \int_E f(x)\,dx;$$

it is implied that the argument E of this function must be measurable and constructed using points from an interval or a set, for points of which the values of $f(x)$ are known. To fix ideas, we will always assume unless otherwise stated, that $f(x)$ is defined on an interval which we will call (a, b); which in reality is not a restriction (Sect. 8.4). Therefore, we will finally consider: a. the indefinite integral function of set, $\Psi(E)$; b. the indefinite integral function of interval, $\Phi(\alpha, \beta) = \Phi(\delta)$, where δ denotes the interval (α, β); c. the indefinite integral function of one variable, $F(x)$. In this chapter we will examine, whether the knowledge of one of these functions implies the knowledge of other two and how the properties of these functions correspond among themselves.

[1] Only now we can make use of the maxima and minima obtained by neglecting sets of measure zero, because if we modify the value of a function at points of such a set, we do not modify the integral of this function.

We will focus on two properties of $\Psi(E)$: the complete additivity and the absolute continuity. We will see later that these properties characterise the functions of a set that are indefinite integrals and consequently summarise and imply all other properties.

A function of a measurable set $\Psi(E)$ *is called additive if* $E_1, E_2, \ldots$ *are pairwise disjoint sets, we have*

$$\Psi(E_1 + E_2 + \cdots) = \Psi(E_1) + \Psi(E_2) + \cdots$$

We can distinguish, with M.de la Vallée Poussin,[2] the case of *restrictive* additivity, in which the previous equality is assured only if the E_i are finite in number, and the case of *complete* additivity, where the equality holds even if there are countably infinite E_i. Only the complete additivity will be important for us.[3] *An indefinite integral is completely additive;* this is the property 2 of Sect. 8.4.

On its own, complete additivity implies many properties. First, let us note that *an additive set function which is unbounded, has some points at which it is not bounded.* By this, it should be understood that there exists a point x such that, however small may be an interval I containing x in its interior, and however large may be a number N, it is possible to find a set E formed from points of I, for which the function $\Psi(E)$, considered on this set, exceeds N in absolute value.

Indeed, if, to every point of (a, b) we could attach an interval I and a number N for which this is impossible, we could cover (a, b) using a finite number, p, of these intervals I and the corresponding numbers N would have an upper bound $\mathfrak{N}$. where E is a set of points of (a, b) that could be considered the sum of sets $E_1, E_2, \ldots, E_p$ located in the considered p intervals, and we will have

$$\begin{aligned} |\Psi(E)| &= |\Psi(E_1 + E_2 + \cdots E_p)| \\ &= |\Psi(E_1) + \Psi(E_2) + \cdots + \Psi(E_p)| \\ &\leq |\Psi(E_1)| + |\Psi(E_2)| + \cdots + |\Psi(E_p)| \\ &\leq p\mathfrak{N}; \end{aligned}$$

therefore, $\Psi(E)$ will be finite.

---

[2] On the functions of set, see the book of M. DE LA VALLÉE POUSSIN, *Integrales de Lebesgue, fonctions d'ensemble, classes de Baire.*

[3] The reader will easily verify that a function $\Psi(E)$, equal to the sum of the lengths of the intervals in which the complement of E is everywhere non-dense, possesses restrictive additivity but not complete additivity.

Hitherto, the restrictive additivity has not been introduced in Analysis. However, there is a great need to distinguish it from complete additivity. Let me explain it with an example. The condition 2 of problem of the measure imposes complete additivity on function $m(E)$. Let us suppose, on the contrary, that we have modified this condition 2 in a way so as to require only restrictive additivity, then the function would no longer be defined, except for the J measurable sets, and this function would have been the extent $e(E)$. Let us note that the function thus obtained, indeed possesses the complete additivity in the domain of the J measurable sets, since it does not differ from $m(E)$, but the function $e(E)$ is not always defined for the sum $E_1 + E_2 + \cdots$, when it is defined for $E_1, E_2, \ldots$

Let x_0 be a point at which $\Psi(E)$ is unbounded and let us consider two intervals $(x_0 - h, x_0)$, $(x_0, x_0 + h)$. It is clear that $\Psi(E)$ is unbounded in one of these two. This being said, to prove that *any completely additive finite function is bounded,* it would be sufficient, therefore, to consider the case where the extremity b of the considered interval (a, b)[4] is a point where a function $\Psi(E)$ is unbounded, and to show that $\Psi(E)$ cannot be both finite and completely additive at the same time. Let $a, a_1, a_2, \ldots$ be a sequence of increasing values tending towards b. Let us denote by $\delta_1, \delta_2, \ldots$ the sets of points defined respectively by

$$a \leq x < a_1, \quad a_1 \leq x < a_2, \quad a_2 \leq x < a_3, \ldots$$

Let $N_1, N_2, \ldots$ be the upper bounds of $|\Psi(E)|$, respectively, for the sets formed of the points of $\delta_1, \delta_2, \ldots$.

If E is a set formed from points of (a_k, b) and if E_i is the subset common to E and δ_i we have, when Ψ is assumed to be completely additive,

$$|\Psi(E)| + |\Psi[\sum(E_i)]| \leq \sum |\Psi(E_i)| \leq \sum_k^\infty N_i,$$

since $\Psi(E)$ is not bounded at point b, the series of the last term is divergent for any k. Moreover, it may contain infinite terms.

Finally, let e_i be a set formed by points of δ_i and for which $|\Psi(e_i)|$ exceeds the smallest, M_i, of the two numbers $N_i - \frac{1}{i}$ and i; the series $\sum M_i$ is divergent. If we partition e_i into those which yield for $\Psi(e_i)$ the positive or zero values, and into those which yield for $\Psi(e_i)$ the negative values, if i' and i'' denote, respectively, the indices of the first and of second, at least one of the series $\sum M_{i'}$ and $\sum M_{i''}$ is divergent. Let us assume that the first of them is divergent. Then we have

$$\Psi(\sum e_{i'}) = \sum \Psi(e_{i'}) > \sum M_{i'} = +\infty,$$

the function Ψ therefore cannot be finite and completely additive. As we are discussing only finite functions of sets, unless expressly stated otherwise, we can simply say: *every completely additive function is bounded.*

A completely additive function[5] *is of bounded variation.* By this we mean that, it is the difference of the two completely additive functions that take only positive or zero values. To show this a few definitions are necessary.

---

[4] If it were a function $\Psi(E)$ defined only for measurable sets formed of points from a given measurable set $\mathcal{E}$, we could extend the definition of $\Psi(E)$ to all measurable sets formed by points of an interval (a, b) containing $\mathcal{E}$, by agreeing that, by definition, for such a set E whose subset common with $\mathcal{E}$ is e, we have $\Psi(E) = \Psi(e)$; and if e does not exist, $\Psi(E) = 0$. Therefore, we can always suppose that it is a function defined for all measurable sets of an interval (a, b).

[5] I will not explain it any longer, that we are dealing with a function taking definite and finite value for each measurable set formed with the points of interval (a, b) or a measurable set $\mathcal{E}$, a case that reduces to the previous one.

Let us consider the values taken by a completely additive function Ψ for the sets e formed with the points of a set E. These values are included between M and $-M$, where M is the upper bound of $|\Psi|$ in entire (a, b). Let $-N(E) \leq 0 \leq +P(E)$ be the smallest interval containing these values of Ψ and zero. These two functions, $N(E)$ and $P(E)$, are completely additive; let us verify this for $P(E)$. Let $E_1, E_2, \ldots$ be pairwise disjoint sets, let $e_1, e_2, \ldots$ be formed from points of $E_1, E_2, \ldots$, we have

$$\Psi(e_1 + e_2 + \cdots) = \Psi(e_1) + \Psi(e_2) + \cdots \leq P(E_1) + P(E_2) + \cdots,$$

hence

$$P(E_1 + E_2 + \cdots) \leq P(E_1) + P(E_2) + \cdots;$$

but we can choose e_i in such a way that $\Psi(E_i)$ exceeds $P(E_i) - \frac{\varepsilon}{2^i}$ if $P(E_i)$ is greater than 0 [if $P(E_i) = 0$, we leave aside this value of i], and then we have

$$\Psi(e_1 + e_2 + \cdots) = \Psi(e_1) + \Psi(e_2) + \cdots > P(E_1) + P(E_2) + \cdots - \varepsilon,$$

or

$$P(E_1 + E_2 + \cdots) > P(E_1) + P(E_2) + \cdots - \varepsilon.$$

The complete additivity of P follows from comparison of these two inequalities.

The function $P(E)$ is called the *total positive variation of* Ψ *in* E, $N(E)$ is its *total negative variation,*

$$V(E) = P(E) + N(E)$$

is its *total variation.* This total variation is also a completely additive function, because it is clear that the sum or the difference of two completely additive functions is a completely additive function. $V(E)$ is the upper bound of $\sum |\Psi(E_i)|$ for the division of E into partial sets E_i.

Let us prove that $\Psi(E)$ *is of bounded variation by showing that we have*

$$\Psi(E) = P(E) - N(E).$$

With the points of E, we can form a set e_1 for which $\Psi(e_1)$ exceeds $\frac{1}{2}P(E)$. With the points of e_1, we can form a set e_2 for which $\Psi(e_2)$ is at most equal to $-\frac{1}{2}N(e_1)$. We have

$$\Psi(e_1 - e_2) \geq \Psi(e_1), \quad N(e_1 - e_2) < \frac{1}{2}N(e_1).$$

With the points of $e_1 - e_2$, we form a set e_3, for which $\Psi(e_3)$ is at most equal to $-\frac{1}{2}N(e_1 - e_2)$, we have

$$\Psi(e_1 - e_2 - e_3) \geq \Psi(e_1 - e_2), \quad N(e_1 - e_2 - e_3) < \frac{1}{2^2}N(e_1).$$

By continuing in this way, we arrive at a set $e_1 - e_2 - e_3 - \cdots$, or E_1. For this set, we have

$$N(E_1) = 0, \quad P(E_1) = \Psi(E_1) \geq \frac{1}{2}P(E), \quad P(E - E_1) \leq \frac{1}{2}P(E).$$

With the points of $E - E_1$, we can similarly form a set E_2 such that

$$\begin{aligned} N(E_2) &= 0, \\ P(E_2) = \Psi(E_2) &\geq \frac{1}{2}P(E - E_1), \\ P(E - E_1 - E_2) &\leq \frac{1}{2^2}P(E). \end{aligned}$$

Then, with the points of $E - E_1 - E_2$, we will form E_3 such that

$$\begin{aligned} N(E_3) &= 0, \\ P(E_3) = \Psi(E_3) &\geq \frac{1}{2}P(E - E_1 - E_2), \\ P(E - E_1 - E_2 - E_3) &\leq \frac{1}{2^3}P(E), \end{aligned}$$

and so on.

It is clear that, for $E_1 + E_2 + \cdots = E^p$, we have

$$N(E^p) = N(E_1) + N(E_2) + \cdots = 0,$$

$$P(E - E^p) \leq P(E - E_1 - E_2 - \cdots - E_k) \leq \frac{1}{2^k}P(E).$$

Therefore

$$P(E - E^p) = 0$$

and, since

$$P(E) = P(E - E^p) + P(E^p),$$

we have

$$P(E^p) = P(E);$$

moreover

$$\Psi(E^p) = P(E^p).$$

In a similar way, we will form E^n such that we have

$$P(E^n) = 0, \quad N(E^n) = N(E); \quad \Psi(E^n) = -N(E^n).$$

Let e be the set of points common to E^p and E^n; then, from the two relations,

$$P(e) \leq P(E^n), \quad N(e) \leq N(E^p).$$

it follows that both non-negative numbers $P(e)$ and $N(e)$ are zero; *consequently* $\Psi(e)$ is zero. Therefore, if we remove e from E^n we do not alter either $P(E^n)$ or $N(E^n)$ or $\Psi(E^n)$.

We can therefore suppose that the two sets E^p and E^n are disjoint. Similarly, we will see that P, N and Ψ are zero for the set of points of E not belonging, either to E^p, or to E^n, so that this set can be added to E^p or to E^n, without changing our relationships. Finally, we see that we can assume E to be divided in two disjoint sets E^p and E^n and such that we have

$$\Psi(E^p) = P(E^p), \quad N(E^p) = 0.$$
$$\Psi(E^n) = -N(E^n), \quad P(E^n) = 0.$$

But

$$E = E^p + E^n,$$

therefore, we have

$$\Psi(E) = \Psi(E^p) + \Psi(E^n) = P(E) - N(E).$$

It is proved that $\Psi(E)$ is of bounded variation.

If, to $P(E$ and $N(E)$ we add a same completely additive and non-negative function $\lambda(E)$, we obtain

$$P_1(E) = \lambda(E) + P(E);$$
$$N_1(E) = \lambda(E) + N(E);$$
$$\Psi(E) = P_1(E) - N_1(E).$$

Therefore, we can express a function of bounded variation in the form of a difference of two non-negative function in an infinite number of ways. Furthermore, the method we have just mentioned is the most general one, that is, if we have

$$\Psi(E) = P_1(E) - N_1(E),$$

where $P_1(E)$ and $N_1(E)$ are two completely additive and non-negative functions, we have

$$P_1(E) - P(E) = N_1(E) - N(E) = \lambda(E),$$

where $\lambda(E)$ is a completely additive and non-negative function. Indeed, let $\lambda(E)$ be the function defined by this double equality. It is absolutely continuous. Let us show that it is non-negative. Let E^p be the set that we have attached to E. We have, since E^p is contained in E,

$$P_1(E) \geq P_1(E^p) = P(E^p) + \lambda(E^p), \quad N_1(E^p) = N(E^p) + \lambda(E^p).$$

But, since $N(E^p)$ is zero and $N_1(E^p)$ should not be negative, $\lambda(E^p) \geq 0$; hence

$$P_1(E) \geq P(E^p) = P(E).$$

therefore

$$\lambda(E) = P_1(E) - P(E)$$

is not negative.

In other words, the *variations* $P(E)$ *and* $N(E)$ *of* $\Psi(E)$ *are, among all the completely additive functions,* $P_1(E)$ *and* $N_1(E)$*, which are non-negative and satisfy the identity*

$$\Psi(E) = P_1(E) - N_1(E),$$

the ones which are the least. This property corresponds exactly to the one in Sect. 5.1 for functions of bounded variation of one variable. However, additionally, we have incidentally seen that $P(E)$ and $-N(E)$ *are two limits, upper and lower of* $\Psi(e)$*, as* e *varies in* E*, and these limits are actually attained*, respectively, for $e = E^p$ and for $e = E^n$. These are the properties that do not have counterparts in functions of one variable.[6] For the previous properties to be entirely proved, it is nevertheless necessary to show that $\Psi(E)$ is necessarily zero in certain sets, because if, for example, $\Psi(E)$ was constantly positive, $-N(E)$ would be constantly zero and would not be the lower limit of $\Psi(E)$, and E^n would not exist. But this is impossible,[7] because *the sets* E *reduced to a point, which give to* $\Psi(E)$ *a non-zero value, form at most a countable infinity.*

In fact, there cannot be infinitely many points, each forming a set E where $\Psi(E)$ exceeds the positive number K. For the sets formed by a countable infinite number of these points, $\Psi(E)$ would be infinite. Then by allowing K to run through a sequence of positive numbers tending towards zero we see that the points for which $\Psi(E)$ has a positive value form a countable set. The same conclusion is applicable to the points for which $\Psi(E)$ is negative.

Each point, constituting on its own a set in which Ψ is not zero is called a *point of discontinuity* of E. Let us form the function $\psi(E)$ equal to the sum of the values taken by Ψ at those points of discontinuity which belong to E. It is clear that $\psi(E)$ is completely additive, and it has for positive and negative variations, the functions $\pi(E)$ and $\nu(E)$, formed in similar way with $P(E)$ and $N(E)$, respectively. Its total variation is the sum $\pi(E) + \nu(E)$, which is the function which can be deduced in the same way from $V(E) = P(E) + N(E)$. $\psi(E)$ is called *the function of jump of* $\Psi(E)$.

The function $\Psi(E) - \psi(E)$ no longer has the points of discontinuity, nor do the functions $P(E) - \pi(E)$, $N(E) - \nu(E)$, $V(E) - \pi(E) - \nu(E)$. Let us show that, for such

[6] However, these properties correspond to the properties of certain functions of one variable since, in a moment, we will define $\Psi(E)$ using a function $F(x)$ of one variable. Nonetheless, these are the properties that we would have hardly thought of, had we not discussed the functions of set. It is because of such properties, there is an interest in considering functions of set.

[7] An exception is made for the case where the set of definition $\mathcal{E}$ would be composed of a finite or countably infinite number of points.

functions, every point x is a *point of continuity,* meaning it can be enclosed in an interval I such that the functions are less than ε in absolute value, for any set formed from points of I.[8] It is sufficient to reason about the largest, in absolute value, of the considered functions, which is

$$V_0(E) = V(E) - \pi(E) - \nu(E).$$

Let us choose in the interval $\delta = (a, b)$, the values a_i increasing towards x_0 and the values b_i decreasing towards x_0; let α_i and β_i be the sets defined, respectively, by

$$a_{i-1} \leq x < a_i, \quad b_i < x \leq b_{i-1}.$$

We have

$$\delta = x_0 + \sum \alpha_i + \sum \beta_i,$$

since the sets of the right-hand side are pairwise disjoint, and $V_0(x_0)$ is zero, therefore, we have

$$V_0(\delta) = \sum V_0(\alpha_i) + \sum V_0(\beta_i).$$

The sums of the right-hand side being convergent, if we take k sufficiently large we have, for $\delta_k = (a_k, b_k)$,

$$V_0(\delta_k) = \sum_{k+1}^{\infty} V_0(\alpha_i) + \sum_{k+1}^{\infty} V_0(\beta_i) < \varepsilon;$$

δ_k would, then, be the interval I that we are looking for.

Thus *a completely additive function, defined on an interval* (a, b), *is continuous at any point if, and only if, it takes a value zero for any set formed of single point.*[9] An indefinite integral is, therefore, continuous.

Now, let us examine what are the properties of $\Phi(\delta)$ and $F(x)$, that correspond to those of $\Psi(E)$, that we have just considered.

When a function $\Psi(E)$ is given, a function of interval is thereby given $\Phi(\delta) = \Psi(E)$. This function is defined for any positive or null interval, meaning, an interval which reduces to a point. Generally, it does not have the same value for an open interval.

$$\alpha < x < \beta,$$

or, for the corresponding closed interval

$$\alpha \leq x \leq \beta;$$

nor for the semi closed intervals

[8] There is a need to show that since we set two definition: the one for the points of continuity, and other for the points of discontinuity.

[9] If such a function takes the values A and B, it takes all the values included between A and B.

$$\alpha < x \leq \beta, \quad \alpha \leq x < \beta.$$

However, there is no need to make this distinction if $\Psi(E)$ is continuous at every point, in which case $\Phi(\delta)$ is zero for any null interval.

To any additive property of Ψ, there corresponds an additive property of Φ, which is stated as: *if an interval* δ *is the sum of disjoint intervals* $\delta_1, \delta_2, \ldots$, *we have*

$$\Phi(\delta) = \Phi(\delta_1) + \Phi(\delta_2) + \cdots$$

If Ψ has restrictive additivity, Φ also has restrictive additivity, that is, the δ_i must be be finite in number. The δ_i could be countably infinite if Ψ has complete additivity, Φ is then called completely additive.

As for the phrase pairwise disjoint, it must be taken in the strict sense if Ψ has points of discontinuity; the constituent intervals can only be without common interior points if Ψ is continuous; for example, in the case of an indefinite integral.

Ψ being assumed completely additive is of bounded variation; Φ *is then of bounded variation, that is, for the pairwise disjoint intervals* $\delta_1, \delta_2, \ldots$, (same remark as above), *the sum* $\sum |\Phi(\delta_i)|$ *remains bounded;* its upper bound, when δ_i are taken from δ, is, indeed, at most the value $V(\delta)$ that the function $V(E)$ takes when $E = \delta$. $\Phi(\delta)$ presents itself as the difference of the two non-negative functions of interval $P(\delta)$ and $N(\delta)$, deduced from $P(\delta)$ and $N(\delta)$.

The points of discontinuity and continuity are defined as above; in short, the previous definitions are applicable. However, we now only consider sets reduced to a closed or open interval, positive or null. Consequently, a property that involves sets which cannot be reduced to intervals does not have an equivalent; For example, the property: any function $\Psi(E)$ of bounded variation attains its upper bound.

Now, let us pass on from a completely additive function of intervals to a function of x. We will denote by $\Phi[a \leq x \leq b]$ and similar notations the values taken by the function Φ for the intervals defined by the inequalities written in the brackets. We can go from $\Phi(\delta)$ to $F(x)$ using either of the following formulae

$$F(X) = \Phi[a \leq x \leq X] + C = C + \Phi[a \leq x \leq b] - \Phi[X < x \leq b],$$
$$F(X) = \Phi[a \leq x < X] + C = C + \Phi[a \leq x \leq b] - \Phi[X \leq x \leq b],$$

in which C denotes a constant. These two formulae are equivalent for all the points X which are points of continuity for Φ; for points of discontinuity, they yield different values for $F(x)$. *The definition of* $F(X)$, *therefore, contains a certain arbitrariness.* We will adopt the first formula; the second choice would yield the results that can be subsequently deduced immediately from the ones we will obtain.

We have

$$F(\beta) - F(\alpha) = \Phi[a \le x \le \beta] - \Phi[a \le x \le \alpha] = \Phi[\alpha < x \le \beta]$$

α is assumed to be less than β.

Therefore if we take arbitrarily $x_1, x_2, \ldots$, such that

$$a < x_1 < x_2 < \cdots < x_p < b,$$

we would have

$$\begin{aligned} &|F(x_1) - F(a)| + |F(x_2) - F(x_1)| + \cdots + |F(b) - F(x_p)| \\ =&|\Phi[a < x \le x_1]| + |\Phi[x_1 < x \le x_2]| + \cdots + |\Phi[x_p < x \le b]|. \end{aligned}$$

And, as the right-hand side is at most equal to $V[a < x \le b]$, *the function $F(x)$ is of bounded variation.*

In the formula of definition of F, let X decrease towards X_0, that is, let us give a sequence of values $X', X'', \ldots$ to X, decreasing towards X_0; we have

$$\begin{aligned} F(X') &= F(X_0) + [F(X') - F(X'')] + [F(X'') - F(X''')] + \cdots \\ &= F(X_0) + \Phi[X'' < x \le X'] + \Phi[X''' < X \le X''] + \cdots \\ F(X'') &= F(X_0) + 0 + [F(X'') - F(X''')] + \cdots \\ &= F(X_0) + 0 + \Phi[X''' < x \le X''] + \cdots \end{aligned}$$

and so on, hence

$$F(X_0 + 0) = F(X_0);$$

the function $F(x)$ is, therefore, continuous from the right.

Similarly, let X increase towards X_0, we have

$$\begin{aligned} F(X_0 - 0) &= \lim_{\substack{X < X_0 \\ X \to X_0}} \{C + \Phi(a \le x \le X)\} = \Phi[a \le x < X_0] + C \\ &= F(X_0) - \Phi(X_0), \end{aligned}$$

the function $F(x)$ *is left discontinuous at points where Φ is discontinuous, and only at these points.*

We thus recover, in particular, this result: *an indefinite integral $F(x)$ is a continuous function of bounded variation.*

From the equalities

$$\begin{aligned} \Phi(X_0) &= F(X_0) - F(X_0 - 0), \\ \Phi[X_0 \le x \le Y_0] &= F(Y_0) - F(X_0 - 0), \end{aligned}$$

which follow from the preceding, it follows that the function Φ is defined by the function F only for intervals, null or non null, that do not have their origin at a when F is known only in (a, b). For F to define Φ in entire $a \leq x \leq b$, let us agree that the formula

$$F(X) = \Phi[a \leq x \leq X] + C$$

will be used only for $a < X \leq b$, and we will set $F(a) = C$.

Then $F(X)$ may be right discontinuous at a and we would have the different formulae for connecting Φ to F depending on whether or not it concerns intervals with origin a.

In this way, we manage to associate with the function of interval Φ a well-determined function of points $F(X)$, whose knowledge would imply that of Φ.

But it is quite obvious that another choice among the possible conventions would have led us to a function $F(X)$ continuous on the left, except, possibly, at b, and with which we would have had

$$\Phi(X_0) = F(X_0 + 0) - F(X_0),$$
$$\Phi[X_0 \leq x \leq Y_0] = F(Y_0 + 0) - F(X_0),$$

except for $Y_0 = b$, in which case the formulae would be different.

Therefore, it is only in a very artificial manner that we have attached to Φ a *definite* function F; if we note that with the two previous conventions we have

$$\Phi(X_0) = F(X_0 + 0) - F(X_0 - 0),$$
$$\Phi[X_0 \leq x \leq Y_0] = F(Y_0 + 0) - F(X_0 - 0).$$

by setting $F(a - 0) = F(a)$, $F(b + 0) = F(b)$, we would be led to consider that any of the functions $F(X)$ of bounded variation satisfying the previous relations are attached to Φ. Two functions $F(X)$ satisfying these conditions will differ, up to an additive constant, only at some of their points of discontinuity. Conversely, if $F(X)$ answers the question, any function of bounded variation, equal to $F(X)$ at all points where they are both continuous, also answers it. *We will often encounter this indetermination of $F(X)$, which one should immediately consider when arriving at conclusions that seem contradictory.*

Let us examine the converse transition from a function of bounded variation $F(X)$ to a function of intervals defined by the formulae

$$\Phi[(X_0 \leq x \leq Y_0] = F(Y_0 + 0) - F(X_0 - 0),$$
$$\Phi[X_0] = F(X_0 + 0) - F(X_0 - 0)$$

and in those which follow from it, for open or half open sets when we want Φ to be additive. We want to prove that the function Φ, thus obtained is completely additive.

Let us consider an interval $\Delta = (l \leq x \leq m)$ and let us divide it by a reducible set of points that belong to E, in the family of the open intervals

$$\delta_i = (l_i < x < m_i)$$

contiguous to E, including points l and m, $x_1, x_2, \ldots$. We would thus have the most general division of an interval into disjoint subsets, except that we could combine δ_i at one or two of its extremities for constituting a semi closed or closed interval. The formula to be shown is, therefore,

$$\Phi(\Delta) = \sum \Phi(\delta_i) + \sum \Phi(x_i),$$

that is

$$F(m+0) - F(l-0) = \sum [F(m_i - 0) - F(l_i + 0)] + \sum [F(x_i + 0) - F(x_i - 0)].$$

Now, this formula follows (Sect. 5.1) from the fact that F is of bounded variation.

If we take the set E suitably, the sum

$$\sum |\Phi(\delta_i)| + \sum |\Phi(x_i)|$$

would yield a value, as close as we want, to the total variation of Φ in Δ. But this sum is written

$$\sum |F(m_i + 0) - F(l_i - 0)| + \sum |F(x_i + 0) - F(x_i - 0)|,$$

quantity that comes as close as we want, to the total variation of the function $F_1(x)$, where $F_1(x)$ is the function deduced from F by modifying the later at its points of discontinuity, except a and b, if those are the points of discontinuity, in a way to obtain a function continuous from the right, except, possibly, at a.

Therefore, we have a relation between the total variation $V(\delta)$ of $\Phi(\delta)$ and the total variation $\nu(X)$ of $F_1(X)$ in $a \leq x \leq X$,

$$V[a \leq x \leq X] = \nu(X).$$

Between the total positive and negative variations of $\Phi(\delta)$, say $P(\delta)$ and $N(\delta)$, and the total positive and negative variations of $F_1(x)$, say $p(x)$ and $n(x)$, we have the relations

$$\Phi(\delta) = P(\delta) - N(\delta), \quad V(\delta) = P(\delta) + N(\delta),$$
$$F_1(X) - F(a) = p(X) - n(X), \quad \nu(X) = p(X) + n(X).$$

Hence

$$P[a \leq x \leq X] = p(X), \quad N[a \leq x \leq X] = n(X),$$

equations which further solidify the relationship between $F(X)$ and $\Phi(\delta)$.

We could easily see that the jump functions of $\Phi(\delta)$, $P(\delta)$, $N(\delta)$, $V(\delta)$, which are defined in the same way as those of $\Psi(E)$, $P(E)$ $N(E)$ $V(E)$, are the interval functions, which can be deduced from the jump functions of $F(X)$ or $F_1(X)$, as well as those of $p(x)$, $n(x)$, and $\nu(x)$.

9.2 The Absolutely Continuous Functions

Having thus studied the transition from $F(x)$ to an interval function $\Phi(\delta)$, let us ask whether we can deduce from completely additive $\Phi(\delta)$, a completely additive set function $\Psi(E)$.

It is clear that $\Psi(E)$ is defined for all closed or open intervals; from absolute additivity, we can deduce the value of $\Psi(E)$ for all B measurable sets, since these sets could be obtained through the summation or differences from the intervals.[10] However, for attaining all B measurable sets, it is necessary for us to bank on the second property of the indefinite integral that we have named the absolute continuity.

A function $\Psi(E)$ is called absolutely continuous if, for any ε positive, we can find a number η such that the following condition is satisfied

$$m(E) \leq \eta \quad \text{implies} \quad |\Psi(E)| \leq \varepsilon.$$

In reality, this property has only been used for additive functions, and it is only for such functions that it deserves to be considered as defining a mode of continuity. Indeed, let Ψ be an additive function and E_1 and E_2 be two sets.

Let

$$E_1 = E + e_1, \quad E_2 = E + e_2,$$

where E is the subset common to E_1 and E_2. The sets E and e_1 on one hand, E and e_2 on the other, are disjoint.

Let us agree to say that two sets E_1 and E_2 are at a distance η[11] if we have

$$m(e_1) \leq \eta, \quad m(e_2) \leq \eta.$$

We have

$$\begin{aligned}|\Psi(E_1) - \Psi(E_2)| &= |[\Psi(E) + \Psi(e_1)] - [\Psi(E) + \Psi(e_2)]| \\ &= |\Psi(e_1) - \Psi(e_2)| \leq |\Psi(e_1)| + |\Psi(e_2)| \leq 2\varepsilon;\end{aligned}$$

thus, to two sets E_1 and E_2, not very distant, correspond values of functions not very different. This is indeed a kind of continuity, and even uniform continuity.[12] This mode of continuity is first noted for indefinite integrals; we have, in fact, the following proposition:

---

[10] However, any B measurable set can be obtained in several ways through additions and subtractions. Therefore, the definition of $\Psi(E)$ would only be complete for B measurable sets if we could prove that it is free from contradictions—and this must be done without using the condition of absolute continuity which is about to be introduced.

[11] M. Borel stated that E_1 and E_2 differ by two η; a better expression in some respects.

[12] A completely additive function of a measurable set, which is continuous in the manner described in the text, that is, with respect to our notion of distance between two sets, *for all measurable sets*, is necessarily uniformly continuous. This mode of uniform continuity is what we call the absolute continuity.

The integral of a summable function $f(x)$, extended to a variable set E, tends towards zero with the measure of E. Indeed, we know that we can choose N in such a way that the integral $\int_a^b \varphi\, dx$, $\int_a^b \varphi_{N,N}\, dx$ differ by less than $\frac{\varepsilon}{2}$, where φ is absolute value of f and $\varphi_{N,N}$ is the function that is deduced from φ, as it has been indicated in (Sect. 8.4). Let, then, E be a set of measure at most equal to $\frac{\varepsilon}{2N}$. Let us divide E into the set e_1 of those points where $\varphi_{N,N}$ is zero and the set e_2 of those points where $\varphi_{N,N}$ is positive.

We have

$$\left|\int_E f(x)\, dx\right| = \left|\int_{e_1} f(x)\, dx + \int_{e_2} f(x)\, dx\right|$$
$$\leq \int_{e_1} \varphi(x)\, dx + \left|\int_{e_2} f(x)\, dx\right| \leq \frac{\varepsilon}{2} + Nm(e_2) \leq \frac{\varepsilon}{2} + N\frac{\varepsilon}{2N} = \varepsilon.$$

An indefinite integral $\Psi(E)$ is, therefore, an absolutely continuous function.

From a completely additive and absolutely continuous function $\Psi(E)$, we will deduce a completely additive and absolutely continuous function of interval $\Psi(\delta)$. This last notion expresses that, *for any pair wise disjoint intervals $\delta_1, \delta_2, \ldots$, the sum $\sum \Phi(\delta_i)$ tends towards zero with $\sum m(\delta_i)$*. We can also state that the sum $\sum |\Phi(\delta_i)|$ tends towards zero because in $\sum \Phi(\delta_i)$ we can conserve only positive terms, providing $\sum'$, or negative terms, providing $-\sum''$, and as $\sum'$ and $\sum''$ must tend towards zero, $\sum |\Phi(\delta_i)| = \sum' + \sum''$ should also tend towards zero. By intervals, without common points, we mean intervals without common interior points, as the absolute continuity of $\Psi(E)$ or $\Phi(\delta)$ obviously implies that these functions will be zero for any interval that reduces to a single point, that is, they are continuous at any point.

If we pass in sequel, from a function $\Phi(\delta)$ having the two considered properties, to a function $F(X)$, where $F(X)$ is of bounded variation and absolutely continuous, that is, *for any system of intervals (α_i, β_i), any pair of which does not have common interior, the sum $\sum[F(\beta_i) - F(\alpha_i)]$ tends towards zero with the sum of the measures of (α_i, β_i)*. Here again, we can, at will, put or not put the sign $|\ \ |$ inside the sign $\sum$. Let us note also that the absolute continuity of $F(X)$ implies, for $F(X)$, the continuity in the ordinary sense and $F(x)$ is of bounded variation.

If, conversely, we start from a function $F(X)$ of bounded variation and absolute continuity, we would deduce a function $\Phi(\delta)$, having the two mentioned properties. Let us now look for a set function $\Psi(E)$ that also has these two properties and reduces to $\Phi(\delta)$ on the intervals. It is clear that if it is a question of calculating $\Psi(E)$ for a set E we could proceed thus: we determine a set A_i of intervals at a distance less than a positive number η_i from E; which is easy, for example, by enclosing E in the intervals. For A_i, we know $\Psi(A_i)$ as equal to the sum of the values of Φ for the various intervals constituting A_i. Then we let η_i tend towards zero and $\Psi(A_i)$ tends towards the sought value $\Psi(E)$.

Therefore, this value $\Psi(E)$ will exist if, by replacing A_i with another set of intervals, B_i, also at a distance less than η_i, from E, $|\Psi(A_i) - \Psi(B_i)|$ tends towards zero with η_i.

However, it is indeed so, since A_i and B_i are obviously at a distance of $2\eta_i$ at most.[13] Finally $\Psi(E)$ itself is calculated using the sums of increments of $F(\beta) - F(\alpha)$, for this reason we will also say that $\Psi(E)$ is *the increment of* $F(X)$ *on the set* E.

Thus, the three families of functions, absolutely continuous and completely additive set functions, interval functions having the same two properties, and functions of a single variable that are absolutely continuous and of bounded variation, are entirely equivalent.

In particular, we see that *the indefinite integral, a function of one variable, of a function* $f(x)$ *completely determines the indefinite integral, a set function, of* $f(x)$. And we have learned to calculate $\int_E f(x)\,dx$, from integral of $f(x)$ in various intervals, by a method similar to the one that allows us to calculate $m(E)$ from the measure of the intervals.

The previous property implies this very important consequence: *two functions* $f_1(x)$ *and* $f_2(x)$*, which have the same integral in every interval, are equal, except at most at points of set of measure zero.* Indeed, by hypothesis, two such functions have the same indefinite integral $F(X)$, and therefore, same indefinite integral $\Psi(E)$. Now, since $E[(f_1 - f_2) \neq 0]$ is the limit, for $\varepsilon > 0$ and tending towards zero, of $E[(f_1 - f_2) > \varepsilon] + E[(f_2 - f_1) > \varepsilon]$, for ε sufficiently small, therefore, one of the two sets just been named, will be of non-zero measure if f_1 and f_2 differ at a set of points of positive measure.[14] And it is clear that, in this set e of non-zero measure, the integral of the function $f_1 - f_2$, either constantly greater than ε, or constantly less than ε,[15] would not be zero. In other words, $\int_e f_1\,dx$ and $\int_e f_2\,dx$ would be different, which is contrary to the hypothesis.

Thus *a function* $f(x)$ *is determined, except at the points of set of measure zero, by knowing one of its indefinite integrals.*

The indetermination that we encounter in this statement is indeed real; because, if we arbitrarily modify $f(x)$ at points of an arbitrarily chosen set of measure zero, we do not alter its indefinite integral. We will later *explore how can we calculate* $f(x)$ *when we know one of its indefinite integral.* However, to give the results obtained the widest possible significance, it would be useful to pursue the study of the set functions for a while.

---

[13] This method of extending a function defined only on the family of interval sets to all measurable sets is related to the proposition that M.Baire called the principle of extension (*see* Baire: *Leçons sur les théories générales de l'Analyse*), whose most well-known application is the definition of the exponential function. It can be stated as follows: *If a function* $f(x)$ *is defined for all rational values of* x *and if it is uniformly continuous over the set of these values, we can extend it, and in a unique way, to any value of* x *such that it remains continuous.*

We could combine this property with the one that serves our purpose into a single statement; we would imitate for that the considerations developed by M. M. Fréchet in his Thesis (*Rendiconti del Circolo Matematico di Palermo,* 1906).

[14] Since f_1 and f_2 are summable, they are measurable, and the set of points where f_1 and f_2 are different is indeed measurable.

[15] **Translator's note:** It should have been $-\varepsilon$ in the original text.

9.3 The Singularities of Non-Absolutely Continuous Functions

Of the two properties mentioned for the indefinite integral function of one variable: being of bounded variation and being absolutely continuous, the first is contained in the second. Indeed, for a function $F(X)$ of unbounded variation, it is possible to choose (Sect. 5.1) a countable system of non-intersecting intervals in such a way that the corresponding series $[F(\beta_i) - F(\alpha_i)]$ diverges. However, such a series, according to the very definition of absolute continuity (Sect. 5.1), is always convergent for an absolutely continuous function.

On the contrary, there exist continuous functions of bounded variation, which are not absolutely continuous; the function $\xi(x)$ of Sect. 5.1 is an example. Indeed, $\xi(x)$ has a total variation equal to 1 in any system of intervals enclosing Z, even though Z is of measure zero.

When considering a function $F(X)$ of bounded variation and assessing to what extent it deviates from absolute continuity, it suffices to take sets of intervals of measure at most equal to η and to form for them the sums $+\sum'$ and $-\sum''$ of positive and negative differences $F(\beta_i) - F(\alpha_i)$. By choosing (α_i, β_i) in any manner, $\sum'$ and $\sum''$ will have two upper limits $M'(\eta)$, $M''(\eta)$ which, as we let η decreases towards zero, decreases towards two limits p_0 and n_0. It is clear that we can assert that $F(X)$ deviates from the absolute continuity: in terms of positive variation by p_0, and in terms of negative variation by n_0, from the point of view of total variation by $p_0 + n_0$.

These numbers n_0 and p_0 could have been defined by applying the previous procedure no longer to $F(X)$, but, respectively, to its positive variation $P(X)$ and its negative variation $N(X)$. In other words we would have replaced $+\sum'$, for example, by the sum of positive variation of $F(X)$ in all the intervals (α_i, β_i). Indeed, by operating in this way, we have $[P(\beta_i) - P(\alpha_i)]$; but, in (α_i, β_i), we can find the non-intersecting intervals $(\alpha_{i,j}, \beta_{i,j})$ such that all the differences $F(\beta_{i,j}) - F(\alpha_{i,j})$ are positive, and their sum, for single variable j, differs as little as we want from $P(\beta_i) - P(\alpha_i)$. We therefore have, by suitably choosing the $(\alpha_{i,j}, \beta_{i,j})$

$$\sum[P(\beta_i) - P(\alpha_i)] - \varepsilon < \sum[F(\beta_{i,j}) - F(\alpha_{i,j})] \leq \sum[P(\beta_i) - P(\alpha_i)]$$

and, as the measure of the set of $(\alpha_{i,j}, \beta_{i,j})$ is at most that of (α_i, β_i), the two methods of defining p_0 are quite equivalent. Let us add that we can obviously require the system of intervals used to contain only a finite number. Obviously, to each method of defining p_0, n_0, and therefore $\nu_0 = p_0 + n_0$, corresponds a different formulation of the condition of absolute continuity.

Let us denote by $P_s(X)$, $N_s(X)$, $V_s(X)$ the numbers p_0, n_0, ν_0 relative to the interval (a, X), it is obvious that in the positive interval (α, X) the numbers p_0, n_0, ν_0 are $P_s(X) - P_s(\alpha)$, $N_s(X) - N_s(\alpha)$, $V_s(X) - V_s(\alpha)$. It is also clear that these three numbers are positive or zero, and are at most equal, respectively, to $P(X) - P(\alpha)$, $N(X) - N(\alpha)$, $V(X) - V(\alpha)$. In other words, the six functions $P_s(X)$, $N_s(X)$, $V_s(X)$; $P(X) - P_s(X)$, $N(X) - N_s(X)$,

$V(X) - V_s(X)$ are non-negative and non-decreasing. These functions are actually defined only in $a < X \leq b$. At point a, we will take them equal to zero and set

$$F_s(X) = P_s(X) - N_s(X).$$

The functions F_s, P_s, N_s, V_s are said to be the *functions of singularities of* F, P, N, V. Here are their characteristic properties: $F_s(X)$ *is the function of singularities of a function of bounded variation* $F(X)$, *the difference* $F(X) - F_s(X)$ *is absolutely continuous and* $F_s(X)$ *is, of all the functions* $G_s(X)$ *such that the difference* $F(X) - G_s(X)$ *is absolutely continuous, the one which has the least total variation, and which vanishes at* $x = a$.

It is obvious, according to the very definition of P_s and N_s, there cannot be a corrective function $G_s(X)$ whose variation is less than $P_s(X)$, $N_s(X)$, $V_s(X)$ in (a, X), and the only function for which these values will attain minima is

$$F_s(X) + \text{const.}$$

But it remains to prove that $F_s(X)$ is a corrective function.

Now we have

$$F(X) - F_s(X) = [P(X) - P_s(X)] - [N(X) - N_s(X)];$$

as the two brackets of the right-hand side are positive or zero, it is, therefore, necessary to prove that the number p_0 relative to the first bracket and the number n_0 relative to the second are zero.

If the number p_0 relative to $P(X) - P_s(X)$ was equal to $\lambda > 0$, that is, we could find finitely many points in (a, b),

$$a = x_0 < x_1 < x_2 < \cdots < x_k = b,$$

such that measure of the set of intervals of even rank (x_{2i-1}, x_{2i}) is less than η and however, the sum

$$\sum\{[P(x_{2i}) - P_s(x_{2i})] - [P(x_{2i-1}) - P_s(x_{2i-1})]\}$$

exceeds λ. This can be further written as

$$\sum[P(x_{2i}) - P(x_{2i-1})] \geq \sum[P_s(x_{2i}) - P_s(x_{2i-1})] + \lambda.$$

On the other hand, we can find in each interval of odd rank (x_{2i}, x_{2i+1}), the intervals (α, β) whose total measure is as small as we want, and which yields a sum

$$\sum[P(\beta) - P(\alpha)]$$

at least equal to $P_s(x_{2i+1}) - P_s(x_{2i})$, according to the very definition of P_s. So that we can assume that the set of (α, β) with respect to all the values of i has a measure less than η and, however, we have

$$\sum[P(\beta) - P(\alpha)] \geq \sum[P_s(x_{2i+1}) - P_s(x_{2i})].$$

Hence, by addition,

$$\sum[P(x_{2i}) - P(x_{2i-1})] + \sum[P(\beta) - P(\alpha)] \geq \lambda + \sum[P_s(x_j) - P_s(x_{j-1})]$$
$$= \lambda + P_s(b);$$

and this is impossible, according to the definition of P_s, since the set of (α, β) and of (x_{2i-1}, x_{2i}) is of measure 2η as small as we want.

The proposition is therefore proved.

Let us set $F = F_s + AC$, where AC represents an absolutely continuous function: *the kernel* of F. The functions F and F_s have the same jumps at every point; therefore, they have the same *function of jumps* S (function φ of Sect. 5.1). If we set

$$F = S + C = F_s + AC = S + C_s + AC,$$

C is the continuous part of F (function ψ of Sect. 5.1 61) and C_s is both the continuous part of F_s and the function of the singularities of the continuous part C of F.[16]

Let us consider a sequence $I_1, I_2, \ldots$ of sets of intervals whose measures tend towards zero and which yield the sums $\sum_1 \Delta V, \sum_2 \Delta V, \ldots$ tending towards the largest possible limit $\nu_0 = V_s(b)$. The similar sums relative to AC, tend towards zero, because of absolute continuity of AC. Therefore, the sums $\sum_1 \Delta V, \sum_2 \Delta V, \ldots$ also tend towards ν_0.

By removing, if necessary, some of the first I, we could assume that the series of measures of I_s is convergent. So, if we denote by I^p the set of intervals $\sum_p^\infty I_k$, the I^p form a sequence possessing all the properties mentioned for the sequence of I^p, and additionally, I^p contains I^{p+1}. Let E_s be the set of points common to all the I^p; it is of measure zero and *for any system of open intervals*[17] *enclosing* E_s, *the sum* $\sum \Delta V_s = \nu_0$.

Indeed, let J be such a system of intervals. For a sufficiently large value of p, I^p is contained in J. Otherwise, as the set K^p, obtained by removing the subsets contained in J from I^p, contains K^{p+1}, there would exist points common to all the K^p, hence to all I^p, and not belonging to J, which is impossible. Therefore, it yields a sum $\sum \Delta V_s$, at least equal to the one yielded by I^p for p very large, and therefore at least equal to ν_0, and exactly equal to ν_0 since no sum $\sum \Delta V_s$ could exceed $\nu_0 = V_s(b)$.

---

[16] The reader can show that the function of jumps is, among all the corrective functions F_c such that $F - F_c$ is continuous, the one which has the least total variation. S and F_s are, therefore, susceptible to similar definitions.

I also leave aside a lot of propositions which may be easily proved, such as this one: the functions of singularities and of jumps of a sum are the sums of the functions of singularities and of jumps of the added functions.

[17] That is, in the strict sense, the points of E_s are interior to the considered intervals. For the construction of E_s, on the contrary, the intervals that formed the I^p were taken closed; that is, the extremities of constituent intervals are considered to make a part of I^p.

When a set is of measure zero and any system of open intervals enclosing it, yields a sum $\sum \Delta V_s$ equal to ν_0, that is, yields a sum $\sum \Delta V$ at least equal to ν_0, this set is called *the set of singularities of* F because, in some sense, all the variation of F_s is concentrated at the points of this set.

Therefore, set E_s that we have just constructed is the set of singularities of F, or, if we want, a set of singularities, because it is clear that the set of singularities is very indeterminate. For example, by adding an arbitrary set of measure zero to E_s, we again have a set of singularities. *Any set of singularities necessarily contains points of discontinuities of* F. *However, these are the only points that it necessarily contains.* Indeed, let x_0 be a point of continuity of F and let us assume that it belongs to E_s. Consider a set L of open intervals enclosing $E_s - x_0$. This set L encloses the subsets E_s^1 and E_s^2 of E_s located in $(a, x_0 - \varepsilon)$ and $(x_0 + \varepsilon, b)$, which are obviously the set of singularities of F in these intervals. Therefore L yields a sum $\sum \Delta V_s$ at least equal to $V_s(x_0 - \varepsilon) + [V_s(b) - V_s(x_0 + \varepsilon)]$. But, by hypothesis, $V_s(x_0 + \varepsilon) - V_s(x_0 - \varepsilon)$ tends towards zero with ε, and therefore, L yields a sum $\sum \Delta V_s$ equal to $V_s(b)$.

Similarly, from the set of singularities, we can remove an arbitrary countably infinite number of points of continuity of F, without it ceasing to be a set of singularities.

Let us consider the function

$$F(x) = \xi(x) + \frac{1}{2^2}\xi(2x) + \frac{1}{2^4}\xi(2^2x) + \frac{1}{2^6}\xi(2^3x) + \cdots,$$

where $\xi(x)$ is the function of Sect. 5.1 but assumed to be extended outside $(0, 1)$, in a way that it has a period 2 and, is even. $F(x)$ is continuous and of bounded variation in every interval, it is not absolutely continuous in any interval, so that the set of singularities of $F(x)$ is everywhere dense and nevertheless it does not necessarily contain any particular point. Every point is equally likely to be part of this set of measure zero.[18]

Let E_s be the set of singularities of $F(X)$; then for any sequence of set of open intervals enclosing E_s and of measures tending towards zero, we have

$$\lim\left[\sum \Delta P + \sum \Delta N\right] = \lim \sum \Delta V = V_s(b) = P_s(b) + N_s(b);$$

now $\lim \sum \Delta P$ and $\lim \sum \Delta N$ cannot exceed respectively

$$P_s(b) = p_0, \quad N_s(b) = n_0,$$

therefore, we have exactly

$$\lim \sum \Delta P = p_0, \quad \lim \sum \Delta N = n_0;$$

[18] This fact is paradoxical; however, it becomes less surprising when we consider that, to compute $\int f(x)\,dx$, it is necessary to retain some points of interval or set over which the integral is extended, and yet, we can still remove any point from the set.

E_s is also the set of singularities of $P(X)$ and of $N(X)$.

Conversely, the way we have determined the set of singularities of $P(x)$ and $N(x)$, shows that their sum yields the set of singularities of $F(X)$; because it is obvious that the sum of sets of singularities of the terms of a sum is the set of singularities of the sum. E_s being the set of singularities of $F(x)$, for any sequence of open intervals enclosing E_s and of measure tending towards zero, the sum

$$\sum \Delta F = \sum (\Delta P - \Delta N) = \sum \Delta P - \sum \Delta N$$

tends towards

$$p_0 - n_0 = P_s(b) - N_s(b) = F_s(b).$$

In other words, the method which has allowed to associate with each measurable set E, an increment $\mathfrak{A}_F(E)$, when F was absolutely continuous—a method consisting of enclosing E in a sequence of sets of open intervals[19] whose measures tend towards that of E, and taking the limit of the sums $\sum \Delta F$ provided by this sequence of sets of intervals—applies to every function F of bounded variation, when we take its set of singularities as E. But the set of singularities is the only set for which this method still applies, at least when we are dealing with a continuous function.

More precisely, we could show that if we call $F_1(x)$ as a function equal to $F(x)$, except at the points where we have

$$F(x-0) = F(x+0) \neq F(x)$$

for which

$$F_1(x) = F(x-0),$$

—that is, if we call $F_1(x)$ the function which provides the same function of intervals as $F(x)$, but which is free from the unnecessary singularities of $F(x)$—the previously used method of definition for the increment in a set of an absolutely continuous function can still be applied to $F(x)$, but only for the sets of singularities for $F_1(x)$.

By a completely different procedure *we will define the increment of $F(x)$ in an extended class of sets.* For that, let us decompose F into $S + C$. To the function of jumps S, we will attach in E, an increment equal to

$$\sum_E [F(x+0) - F(x-0)],$$

where the summation is taken over those points of discontinuity of F that belong to E.

Let $\mathfrak{V}(x)$ be the total variation of C from a to x, the change of variable

$$\theta = x + \mathfrak{V}(x)$$

[19] When it is a matter of continuous function it does not matter whether the intervals are open or closed.

transforms $C(x)$ in a function of θ, say $\mathcal{C}(\theta)$.[20] $\mathcal{C}$ having in every interval (θ_1, θ_2) a total variation $\mathfrak{V}(x_2) - \mathfrak{V}(x_1)$, less than $\theta_1 - \theta_2$, has, with respect to θ, derivative numbers in absolute value less than 1. $\mathcal{C}(\theta)$, being absolutely continuous, has a definite increment in each measurable set E_x. We agree to define the increment in E_x as follows

$$\mathfrak{A}_C(E_x) = \mathfrak{A}_{\mathcal{C}}(\mathcal{E}_\theta).$$

The sets E_x we will reach at thus, are all measurable; because if we enclose $\mathcal{E}_\theta$ and its complement $\mathfrak{F}_\theta$ in two families of intervals, having common subsets of total measure ε, by the change of variable from θ to x, we deduce two families of intervals enclosing E_x and its complement F_x, and their common subsets have at most measure ε. Indeed, to an interval (x_1, x_2), there corresponds an interval (θ_1, θ_2) of equal or greater length.

However, we cannot guarantee that the increment is defined for all measurable sets E_x. In any case, it is defined for all E_x which reduce to an interval or a point, because for them the $\mathcal{E}_\theta$ are intervals or points. The change of variable transforms the additions and subtractions of sets into additions and subtractions. It follows that the B measurable sets E_x correspond to B measurable sets $\mathcal{E}_\theta$, and consequently, the increment of F is defined, in particular, on every B measurable set E_x.

To state the result, let us note that $V(x)$ will always denote the total variation of $F(x)$ from a to x. We could have reasoned directly on $F(x)$ by setting

$$t = x + V(x),$$

when x is a point of continuity of F, and when x is a point of discontinuity, by including all values of t from

$$t_1 = x + V(x-0) \quad \text{up to} \quad t_2 = x + V(x+0).$$

the function $\mathfrak{F}(t)$ will be such that

$$\mathfrak{F}[x + V(x)] = F(x)$$

if x is a point of continuity and, if x is a point of discontinuity, we will have

$$\mathfrak{F}(t_1) = \mathfrak{F}[x + V(x-0)] = F(x-0),$$
$$\mathfrak{F}(t_2) = \mathfrak{F}[x + V(x+0)] = F(x+0),$$

where $\mathfrak{F}(t)$ is linear between t_1 and t_2.

---

[20] I had used this fact to study the Stieltjes integral; It was M. de la Vallée Poussin who, in the conference of Strasbourg congress, pointed out its current application. Prior to this, M. de la Vallée Poussin had shown how to obtain the set function associated with a non-absolutely continuous function $F(x)$, thanks to a method which exactly generalises the one which led to the measure of the sets. *See* his book already cited and, later, the Chap. 11.

Then, it is sufficient to set

$$\mathfrak{A}_{F(x)}(E_x) = \mathfrak{A}_{\mathfrak{F}(t)}(\mathcal{E}_t)$$

to define the increment of $F(x)$ using a method similar to the previous one and providing the same result, as we immediately see. Therefore: *we can attach to each function $F(x)$ of bounded variation, a completely additive set function, which we call the increment of $F(x)$, and which is defined on a family of measurable sets, that varies with F, but which always includes all the B measurable sets; it is the family of sets which, by the change of variable $t = x + V(x)$ interpreted suitably, provides measurable sets in t.*

We have seen that, using the same procedure that allowed us to construct E_s, we can always ensure that the set of singularities of F is B measurable. Then, for this set, we have two different definitions of increment of F. It is necessary to verify that they are consistent with each other.

Now, let E_x be a B measurable set that corresponds to a set $\mathcal{E}_t$; let us enclose $\mathcal{E}_t$ in open intervals $\mathfrak{J}_t$[21] and let their measures approach that of $\mathcal{E}_t$. If we take care to choose the $\mathfrak{J}_t$ in such a way that they do not have any extremity in the interior of intervals (t_1, t_2) corresponding to the singular points of F, which is possible, then $\mathfrak{J}_t$ correspond to a set of open intervals I_x[22] enclosing E_x and of measure tending towards that of E_x. The sum $\sum_{I_x} \Delta F$, then, tends towards $\mathfrak{A}_F(E_x)$, because it is equal to $\sum_{\mathfrak{J}_t} \Delta\mathfrak{F}$, which tends towards $\mathfrak{A}_{\mathfrak{F}}(\mathcal{E}_t)$.

If, therefore, E_x is the set of singularities of F, the only new aspect is that we are no longer bound to use the particular set I_x deduced from $\mathfrak{J}_t$ for calculating $\mathfrak{A}_{F(x)}(E_x)$. We can use any other set i_x of open intervals enclosing E_x and of variable measure tending towards zero.

If we restrict ourselves[23] to the consideration of the sets E_x to which correspond the measurable sets $\mathcal{E}_t$, we can say that the set of singularities are those for which we have both

$$m_(E_x) = 0, \quad \mathfrak{A}_{V(x)}(E_x) = V_s(b).$$

Hence, we deduce that *if E^1 and E^2 are two sets of singularities for F, of this category, B measurable say, the set E^{12} of points common to E^1 and E^2, is also the set of singularities of F.*

Indeed, we have

$$\mathfrak{A}_{V(x)}(E^1) = V_s(b); \quad \mathfrak{A}_{V(x)}(E^1 + E^2) = V_s(b);$$
$$\mathfrak{A}_{V(x)}(E^1 + E^2) = \mathfrak{A}_{V(x)}(E^1) + \mathfrak{A}_{V(x)}(E^1 - E^{12}).$$

---

[21] The intervals of $\mathfrak{J}_t$ which would have an extremity at a or V would be, however, taken closed.

[22] In I_x there could however be a closed interval at a or at b.

[23] I do not know if this restriction is real, that is, whether there are sets E_1, to which corresponds the non-measurable set $\mathcal{E}_t$.

Hence

$$\mathfrak{A}_{V(x)}(E^1 - E^{12}) = 0,$$
$$\mathfrak{A}_{V(x)}(E^{12}) = \mathfrak{A}_{V(x)}(E^2) - \mathfrak{A}_{V(x)}(E^2 - E^{12}) = \mathfrak{A}_{V(x)}(E^2) = V_s(b);$$

the theorem is proved.

Let us study the set function $\mathfrak{A}_{F(x)}(E)$ which has just been defined. For that, let us note that if, in the beginning of this theory of set functions, we were naturally led to consider functions defined on *all* measurable sets, this condition was by no means essential to our reasoning; the conclusion of these arguments remain valid for the set functions defined only on, or known only on, the B measurable sets. It will be convenient for us to talk about the functions known for all the B measurable sets, because the sets E^p, E^n of Sect. 9.1, which are the set E_s of Sect. 9.3, are B measurable.

The function $\mathfrak{A}_{F(x)}(E)$, or simply $\mathfrak{A}(E)$, is defined for all B measurable sets and is completely additive in the family of these sets. It is therefore of bounded variations and has variations $\mathfrak{P}(E), \mathfrak{N}(E), \mathfrak{V}(E)$, which are also completely additive; how can we calculate these functions knowing F?

We do not always have

$$\mathfrak{V}(E) = \mathfrak{A}_{V(x)}(E);$$

the example of F zero everywhere in (0, 1), except for $x = \frac{1}{2}$ shows it right away. $\mathfrak{A}(E)$ is then identically zero, therefore, so is $\mathfrak{V}(E)$ and for E reduced to point $x = \frac{1}{2}$, we have

$$\mathfrak{A}_{V(x)}(E) = 2\left|F\left(\frac{1}{2}\right)\right|.$$

To rule out such singularities, let us modify $F(x)$ at its points of discontinuity, interior to (a, b), in such a way that the new function $F_1(x)$ does not have two jumps of opposite signs, at any point.

To fix ideas, let us take $F_1(x)$ continuous from the right in the interior of (a, b). Let $V_1(x)$, $P_1(x)$, $N_1(x)$ be the three total variations of $F_1(x)$; we *will prove that we have*

$$\mathfrak{V}(E) = \mathfrak{A}_{V_1(x)}(E); \quad \mathfrak{P}(E) = \mathfrak{A}_{P_1(x)}(E); \quad \mathfrak{N}(E) = \mathfrak{A}_{N_1(x)}(E);$$

$\mathfrak{P}(a \le x \le b)$ is the upper limit of the values of $\mathfrak{A}(E)$; but each value of $\mathfrak{A}(E)$ can be calculated by using a system of intervals I_x, therefore, $\mathfrak{P}(a \le x \le b)$ is the upper limit of numbers $\mathfrak{A}(I_x)$ for the system I_x of open intervals, except, possibly, at a and b. On the other hand, $P_1(b)$ is the upper limit of numbers $\mathfrak{A}(J_x)$ for the system J_x of closed intervals. As we exclude from a system J_x of intervals all those which result in negative ΔF_1, we increase $\mathfrak{A}(J_x)$, and since we can combine two intervals that yield non-negative ΔF_1 and have a common extremity, we can assume that the intervals J_x have no common extremities. Let (l, m) be one of the intervals, $F_1(x)$ being continuous from the right, the interval $l \le x \le m$

yields the same increment as $l < m \leq x$ and almost the same as $l < x < m + h$, for h very small.

Therefore, $\mathfrak{A}(J_x)$, for J_x formed of closed intervals, differ as small as we want from $\mathfrak{A}(I_x)$ for I_x suitably formed of open intervals, except, possibly, at a and b. And we have

$$\mathfrak{P}(a \leq x \leq b) = \mathfrak{A}_{P_1(x)}(a \leq x \leq b).$$

Similarly, we would have had the same

$$\mathfrak{P}(a \leq x \leq X) = \mathfrak{A}_{P_1(x)}(a \leq x \leq X),$$

hence it will follow that, in any open interval I, closed or semi closed, we have

$$\mathfrak{P}(I) = \mathfrak{A}_{P_1(x)}(I);$$

then we will deduce the equality $\mathfrak{P}(E) = \mathfrak{A}_{P_1(x)}(E)$ for all the B measurable sets.

The similar equalities related to $\mathfrak{N}$ and $\mathfrak{V}$ will follow.

Let F_{1s}, V_{1s}, P_{1s}, N_{1s} be the functions of singularities of F_1, V_1, P_1, N_1; V_{1s}, P_{1s}, N_{1s} are the three variations of F_{1s}. To these functions, which are continuous from the right in the interior of (a, b), correspond the increments $\mathfrak{A}_s(E)$, $\mathfrak{V}_s(E)$, $\mathfrak{P}_s(E)$, $\mathfrak{N}_s(E)$ linked together as $\mathfrak{A}$, $\mathfrak{V}$, $\mathfrak{P}$, $\mathfrak{N}$. These functions $\mathfrak{A}_s$, $\mathfrak{V}_s$, $\mathfrak{P}_s$, $\mathfrak{N}_s$ are called the functions of singularities of $\mathfrak{A}$, $\mathfrak{V}$, $\mathfrak{P}$, $\mathfrak{N}$. Since $F_{1s}(x)$ is the function of smallest total variation such that $F_1(x) - F_{1s}(x)$ is absolutely continuous, *the function of singularities of a function $\mathfrak{A}(E)$, completely additive on B measurable sets, is, among all the corrective functions $\mathfrak{A}_s(E)$ making $\mathfrak{A}(E) - \mathfrak{A}_s(E)$ absolutely continuous, the one with the smallest total variation.*

The set of singularities of F_1, taken as B measurable, is called the set of singularities of $\mathfrak{A}(E)$. Thus, *by set of singularities of an additive function $\mathfrak{A}(E)$ for B measurable sets, we mean any B measurable set E, which is of measure zero and for which the function $\mathfrak{V}(E)$, the total variation of $\mathfrak{A}(E)$, attains the largest value that it can attain on the sets of measure zero.*

The set of singularities of $\mathfrak{P}(E)$, is consequently a set $E_{s,\mathfrak{P}}$ at which

$$\mathfrak{P}[E_{s,\mathfrak{P}}] = P_{1s}(b) = \mathfrak{P}_s[E_{s,\mathfrak{P}}];$$

for every set E, not having any point in common with $E_{s,\mathfrak{P}}$, $\mathfrak{P}_s(E)$ will be consequently zero, since $\mathfrak{P}_s$ cannot be either negative or greater than $P_{1,s}(b)$. The set $E_{s,\mathfrak{P}}$, is the set on which $\mathfrak{F}_s(E)$, or $\mathfrak{P}_s(E)$, attains its upper limit. The set $\mathcal{E}_{s,\mathfrak{N}}$ of singularities of $\mathfrak{N}(E)$ is that on which $\mathfrak{F}_s(E)$ or $-\mathfrak{N}(E)$, attains its lower limit. The set $E_{s,\mathfrak{P}} + E_{s,\mathfrak{N}}$ is the set of singularities of $\mathfrak{A}(E)$.

The set of singularities of a completely additive function $\mathfrak{A}$ for B measurable sets decomposes itself in two disjoint sets, which are respectively the sets of singularities of the positive variation of $\mathfrak{A}$ and its negative variation; we have seen, indeed (Sect. 9.1), that the sets E^p

and E^n, on which a function [here $\mathfrak{F}_s(E)$] attains its upper limit and lower limit, can be taken without common points.

Thus, the sets of singularities of $P_1(x)$ and $N_1(x)$ can be chosen to be disjoint, and we can assume that any singular point of $F(x)$ does not belong to these sets, since these singular points form only finite or countable set. But to obtain the set of singularities of $F(x)$ it is sufficient to add to the set of singularities of $P_1(x)$ the points at which $F(x)$ has at least one positive jump. Therefore, *the sets of singularities of positive and negative variations, $P(x)$ and $N(x)$, of a function of bounded variation, $F(x)$, can be taken without any other common points than those at which $F(x)$ has a right jump and a left jump different from zero and of opposite signs.*

From all this analysis it is necessary to remember that *a completely additive function of B measurable set is absolutely continuous if, and only if, it has a value zero on any B measurable set of measure zero.* The function can then be extended to any measurable set.[24]

[24] We can compare this statement with that of Sect. 9.1.

The Search for Primitive Functions. The Existence of Derivatives

10

10.1 The Search for Primitive Functions

Let $\mathfrak{F}(x)$ be a function having a derivative $f(x)$, we know that $f(x)$ is measurable because it is a function of first class (Sect. 7.2). Let us assume that $f(x)$ is bounded, then $r[\mathfrak{F}(x), x, x+h]$ is also bounded, for any x and h. Since $f(x)$ is the limit of $r[\mathfrak{F}(x), x, x+h]$ for $h = 0$ we can write, according to a theorem stated in Sect. 8.4

$$\int_a^x f(x)\,dx = \lim_{h\to 0} \frac{\int_a^x [\mathfrak{F}(x+h) - \mathfrak{F}(x)]\,dx}{h} = \mathfrak{F}(x) - \mathfrak{F}(a),$$

because $\mathfrak{F}(x)$ is a continuous function.

Therefore *the indefinite integral functions of one variable of a derivative are its primitive functions.* We have solved the fundamental problem of integral calculus for bounded functions. Moreover, we have a regular method of calculation that allows us to determine whether a bounded function is a derivative or not.[1]

To go further, let us show that the derivative numbers of a continuous function are measurable and even B measurable. For this purpose, let us consider a sequence of functions $u_1, u_2, \ldots$, and the functions $\overline{u}$ and $\underline{u}$ that are, for each value of x, equal to the largest and the smallest limits of u_n, taken for x constant and n increasing indefinitely. These are *envelopes of indetermination* of limits of u. Here is how we can obtain the upper envelope $\overline{u}$; v_i is the function which, for each value of x, is equal to the largest of the functions $u_1, u_2, \ldots, u_i$; $\mathfrak{w}_i$ is the limit of the increasing sequence $v_1, v_2, \ldots$; $\mathfrak{w}_i$ is defined from $u_i, u_{i+1}, \ldots$, in the same way as $\mathfrak{w}_1$ is defined from $u_1, u_2, \ldots$; $\overline{u}$ is the limit of the decreasing sequence $\mathfrak{w}_1, \mathfrak{w}_2, \ldots$. If the u_i are the continuous functions, the same is true of v_i, so the $\mathfrak{w}_i$ are at most of first class and $\overline{u}$ is at most of second class. A similar argument applies to $\underline{u}$. If we

[1] Compare with Sect. 6.4.

R. Jain, *Lebesgue's Theory of Integration*,
https://doi.org/10.1007/978-981-96-1169-0_10

only assume that the u_i are measurable, we see that $\overline{u}$ and $\underline{u}$ are also measurable, and that does not require $\overline{u_i}$, $\overline{u}$ and $\underline{u}$ to be finite everywhere.

The definition of envelopes of indetermination could have been given for a function $g(x, h)$, where h is a parameter replacing the index of the function u_i. One of the derivative numbers of $f(x)$ is one of the envelopes of indetermination of $r[f(x), x, x+h]$, when we let h tend towards zero, by the values of definite sign. But since $r[f(x), x, x+h]$ is continuous in (x, h) for $h \neq 0$, we can, as we will see, replace, for the determination of these envelopes, the uncountably infinite values of h with a sequence of values of h tending towards zero and chosen suitably. Therefore, derivative numbers are, when they are finite, at most of the second class and in any case B measurable.[2]

But it is necessary to prove that we can, as it has been stated, replace the consideration of the continuous value of h, by that of a sequence of values of h. If we are dealing with the right-hand derivative numbers, that is, positive values of h, we will take a sequence containing the numbers $1, \frac{1}{2}, \frac{1}{4}, \frac{1}{8}, \ldots$ and, furthermore, containing numbers between $\frac{1}{2^i}$ and $\frac{1}{2^{i+1}}$ that divide this interval in sufficient equal parts so that, when h varies in one of these parts, the oscillation of $r[f(x), x, x+h]$ remains less than $\frac{1}{i}$, with x remaining constant and having any value. It is indeed clear that this sequence provides, for the largest and the smallest limit of r, the two right-hand derivative numbers of $f(x)$. *Therefore, derivative numbers are measurable* and we can hope, in the extended case, that integration of summable functions will allow us to find a function when one of the derivative numbers is known.

In Chap. 5, we showed that a continuous function having its right-hand superior derivative number for example, identically zero, was constant by calculating in an approximate way the increment undergone by this function, with the help of the considerations of a chain of intervals. It is the same method of calculation that would be applied to a continuous function $f(x)$, whose right-hand superior derivative number $\Lambda f(x)$ is assumed to be known in an interval (a, b).

Starting from a, we will cover (a, b) with a chain of intervals satisfying the condition that we will indicate[3]:

1. *Each interval has a length at most equal to a positive number* λ, that we will subsequently allow to go to zero. Then, if we take the sums p and $-n$, formed respectively by the positive and negative increments provided by the chain, we know that when λ tends

[2] This distinction is necessary because we have applied the classification of M. Baire only to functions that are finite everywhere. However, we can extend this classification to all functions.

[3] To satisfy the logical requirements, mentioned in the note in Sect. 5.2, it would be appropriate, and easy, to specify these conditions so that the construction of the chain no longer involves choices; we will not dwell on this.

The condition 1 is, in reality, unnecessary; conditions 2 and 3 are sufficient.

We can note that what follows does not assume for a moment that $\Lambda f(x)$ is the right-hand superior derivative number of $f(x)$, but only that $\Lambda f(x)$ is measurable and lies at all points between the two derivative numbers to the right, of $f(x)$; $\lambda_d \leq \Lambda f \leq \Lambda_d$. From our considerations, it follows that f is determined by Λf, when Λf is summable. This is the answer to a question posed by M. Denjoy.

towards zero, p and $+n$ tend uniformly, in some sense, respectively towards the total positive and negative variations P and N, as it has been said in Sect. 5.1.

We propose only to calculate P. We know that we obtain an approximate value of P by summing the positive increments of $f(x)$, provided by the intervals of chain. Now, the interval $(x_0, x_0 + h)$ yields an increment equal to $hr[f(x), x_0, x_0 + h]$, and this number r is an approximate value of $\Lambda f(x_0)$. It is necessary to specify this approximation; to do so we will partition (a, b) into the sets

$$E[l\varepsilon \leq \Lambda f(x) < (l+1)\varepsilon] = E_l$$

relative to the various values of integers l, positive, negative and zero, where ε is a very small positive number, and in the sets

$$E[\Lambda f(x) = +\infty] = E^{ip}, \quad E[\Lambda f(x) = -\infty] = E^{in}.$$

Then we will set the condition:

2. *If x_0 is origin of an interval $(x_0, x_0 + h)$ of the chain, we will have*

$$\begin{aligned} (l-1)\varepsilon \leq & r[f(x), x_0, (x_0 + h)] \leq (l+1)\varepsilon, && \text{if } x_0 \text{ is a point of } E_l, \\ M \leq & r[f(x), x_0, (x_0 + h)], && \text{if } x_0 \text{ is a point of } E^{ip}, \\ & r[f(x), x_0, (x_0 + h)] \leq -M, && \text{if } x_0 \text{ is a point of } E^{in}; \end{aligned}$$

where M is an arbitrarily large positive number.

The intervals of the chain with points of E^{ip} as origin, points of E_l with positive index and some points of E_0 are the only ones to be considered in the calculation of P.

The intervals with points of E^{ip} as origin, have a contribution of the form $M'\Lambda^{ip}$, where M' is at least equal to M and Λ^{ip} is the total length of intervals originating from points of E^{ip}. Those intervals, which originate from the points of E_l, $l > 0$, have a contribution equal to their total length Λ_l multiplied by a number lying between $(l-1)\varepsilon$ and $(l+1)\varepsilon$. In other words, this contribution is $\Lambda_l l\varepsilon$ with an error of $\Lambda_l\varepsilon$ at most. This result is also true for the intervals originating from the points of E_0, because the contribution of these intervals, in the approximate value of P, is at most $\Lambda_0\varepsilon$. Therefore, we can take for the approximate value of P

$$p = \sum_{0}^{+\infty} \Lambda_l l\varepsilon + M'\Lambda^{ip},$$

since we deviate from the value given by the chain only by $\varepsilon \sum_{0}^{\infty} \Lambda_l$ at most, which is at most $\varepsilon(b-a)$.

Now, it is necessary to be able to evaluate the Λ_l and Λ^{ip}. To do this, we will suppose that we have enclosed each of the sets, that we have considered, in a set of intervals. We will thus have the sets A_l, A^{ip}, A^{in}, each formed of non-intersecting intervals. Each of them exceeds in measure the set of the same index by a quantity ε_l, ε^{ip}, ε^{in} that we can choose as small as we want.

Therefore, let us suppose that the series $\varepsilon^{ip} + \varepsilon^{in} + \sum_{-\infty}^{+\infty} \varepsilon_l$ and $\sum_{-\infty}^{+\infty} |l|\varepsilon_l$ are convergent and have arbitrarily small sums η and ζ. And let us set the condition:

3. *Each interval of the chain is completely contained in set A which contains the set to which its origin belongs.*

Then Λ_l is at most equal to $m(E_l) + \varepsilon_l$, and at least equal to the largest of the two numbers 0 or $m(E_l) - \eta$; for Λ^{ip} we have similar limits. Hence, the lower and upper limits for p are deduced; the lower limit is

$$\sum_{0}^{\infty}{}^{p} [m(E_l) - \eta] l\varepsilon + M[m(E^{ip}) - \eta],$$

only the positive terms are retained in $\sum^p$.

Therefore, this sum contains only a finite number of terms, the number varying with η. These terms are less than those of

$$\sum_{0}^{\infty} m(E_l) l\varepsilon,$$

but they tend respectively towards these for η tending towards zero. Now, we can let η tend towards zero, hence the limit

$$\sum_{0}^{\infty} m(E_l) l\varepsilon + Mm(E^{ip});$$

then we can take ε arbitrarily small, M arbitrarily large, therefore we have, the integral having a finite or infinite value,

$$P \geq \int_{E[0<\Lambda f<+\infty]} \Lambda f \, dx + Mm(E^{ip})$$

for any M. Therefore, P will be finite only for

$$m(E^{ip}) = 0 \quad \text{and} \quad \int_{E[0<\Lambda f<+\infty]} \Lambda f \, dx$$

finite. Let us conclude by recalling that we could have taken for $\Lambda f(x)$, any one of the four derivative numbers, and there exists a similar inequality for N, as found for P.

A continuous function of bounded variation has finite derivative numbers ALMOST EVERYWHERE[4]*; its derivative numbers are summable over the set of points where they are finite; its three total variations P, N, V satisfy the relations*

$$P \geq \int_{E[0<\Lambda f]} \Lambda f \, dx = \int_b^a \frac{1}{2}[\Lambda f + |\Lambda f|] \, dx,$$

$$N \geq \int_{E[\Lambda f<0]} -\Lambda f \, dx = \int_b^a \frac{1}{2}[|\Lambda f| - \Lambda f] \, dx,$$

$$V \geq \int_b^a |\Lambda f| \, dx.$$

where Λf *is any one of the four derivative numbers. The points where* Λf *is infinite are excluded from the sets or the intervals of integration.*

Let us now assume that the right-hand superior derivative number Λf is summable over the set $E[0 < \Lambda f < +\infty]$, and assume the set E^{ip} is of measure zero; let us compute the upper limit of P

$$\sum_0^\infty [m(E_l) + \varepsilon_l] l\varepsilon + M'[m(E^{ip}) + \varepsilon^{ip}].$$

The first term is less than $\sum\limits_0^\infty m(E_l) l\varepsilon + \zeta\varepsilon$; for ζ and ε tending towards zero, therefore, it tends towards $\int_{E[0<\Lambda f<+\infty]} \Lambda f \, dx$.

Since $m(E^{ip})$ is zero, the second term reduces to the product $M'\varepsilon^{ip}$ of an arbitrarily large number by an arbitrarily small number, therefore it tells us nothing.

---

[4] We will agree to say that a property holds *almost everywhere* in an interval (a, b), or on a set $\mathcal{E}$, if the points of (a, b) or of $\mathcal{E}$ where it does not hold either do not exist, or form a set of measure zero.

This expression, introduced in the first edition of this book, has been generally adopted. If we recall that M. Denjoy did not find it sufficiently precise and he rejected the expression: *the point P is a point of set E*, we would not be surprised that he considered the expression *almost everywhere* unacceptable. This is because, in his opinion, it had two meanings: one qualitative or descriptive, the other quantitative or metric. I believe it's reasonable to say that we could have agreed to give *almost everywhere* the following meaning: *exceptions made of the points forming a non-dense set everywhere.* Certainly, this is one possible interpretation; however, M. Denjoy says that a property holds over a *full thickness* when I say that it holds *almost everywhere.* Could full thickness not have received a different meaning than the one M. Denjoy chose to give it.

Almost everywhere would be unacceptable if, in ordinary language, this expression had a precise meaning—but it does not. Consequently, the reader, when encountering a statement such as the one above, cannot assign it any specific meaning without referring to its defined usage. Therefore, any error is not possible.

Compelling the reader to refer to a definition had its own drawbacks. I would gladly concede this to anyone except M. Denjoy, who has used a tremendous number of new words in his Memoirs. Moreover, altering or even improving a vocabulary that is already gaining traction hardly mitigates this inconvenience.

Let us enclose E^{ip} in a set I of non-intersecting intervals, let us suppose that the chosen A^{ip} are enclosed in I. Then, if $P(I)$ denotes the sum of the positive variation of f in the intervals I, the second term $M'\varepsilon^{ip}$ is at most $P(I)$. Therefore, if we denote by $P(E^{ip})$, one of the limits of $P(I)$, when we vary I in such a way that $m(I)$ tends towards zero, we will have

$$P \leq \int_a^b \frac{1}{2}[\Lambda f + |\Lambda f|]\,dx + P(E^{ip}),$$

when points of $E^i = E^{ip} + E^{in}$ are excluded from the interval of integration.

Let I_1 be a set formed of a finite number of intervals of I, chosen in such a way that, when we let $m(I)$ tend towards zero, $P(I_1)$ tends, as $P(I)$, towards $P(E^{ip})$. Let J_1 be the complement of I_1 with respect to (a, b); J_1 is formed of a finite number of intervals, we can, therefore, apply a previous formula and write

$$\begin{aligned} P(J_1) &\geq \int_{J_1} \frac{1}{2}[\Lambda f + |\Lambda f|]\,dx \\ &= \int_a^b \frac{1}{2}[\Lambda f + |\Lambda f|]\,dx - \int_{I_1} \frac{1}{2}[\Lambda f + |\Lambda f|]\,dx, \end{aligned}$$

hence, by adding $P(I_1)$ to two sides,

$$P \geq \int_a^b \frac{1}{2}[\Lambda f + |\Lambda f|]\,dx + P(I_1) - \int_{I_1} \frac{1}{2}[\Lambda f + |\Lambda f|]\,dx.$$

And since, as $m(I)$ tends towards zero, $m(I_1)$ also tends towards zero, so does the last integral of the right-hand side, while $P(I_1)$ tends towards $P(E^{ip})$, we have

$$P \geq \int_a^b \frac{1}{2}[\Lambda f + |\Lambda f|]\,dx + P(E^{ip}).$$

By combining the two inequalities obtained in the opposite directions, we finally conclude

$$P = \int_a^b \frac{1}{2}[\Lambda f + |\Lambda f|]\,dx + P(E^{ip}).$$

Therefore $P(E^{ip})$, which denotes one of the limits of $P(I)$, has a definite value; we can state this result as follows:

If Λf is one of the derivative numbers of a continuous function $f(x)$, for $f(x)$ to be of bounded variation, it is necessary and sufficient that Λf be summable over the set of points, where it is finite and positive, that the set E^{ip} of points where Λf is infinitely positive is of measure zero and it may be enclosed in the intervals I providing a sum of total positive variations $P(I)$ that is bounded.

P(I) then tends towards a finite and definite limit, when we vary I in such a way that its measure tends towards zero. This limit is the difference between the total positive variation *of $f(x)$ in the considered interval and the integral of Λf over the set of points where it is positive.*

We have, of course, a similar statement by changing positive to negative, E^{ip} to E^{in}, P to N, which expresses the equality

$$N = \int_a^b \frac{1}{2}[|\Lambda f| - \Lambda f]\,dx + N(E^{in}).$$

From P and N, by addition and subtraction, we deduce the total variation V, and the increment $f(b) - f(a)$ of $f(x)$ in (a, b). The set $E^i = E^{ip} + E^{in}$, of points where Λf is infinite, is of measure zero. Any set I of non-intersecting intervals enclosing E^i, and whose measure tends towards zero, can, therefore, be used simultaneously to calculate $P(E^{ip})$ and $N(E^{in})$, which can consequently be denoted by $P(E^i)$, $N(E^i)$; the limit of the sum $V(I)$ of total variations in the intervals of I would be $P(E^i) + N(E^i) = V(E^i)$; because, in an interval δ, $P(\delta) + N(\delta) = V(\delta)$. But, in $\delta = (\alpha, \beta)$, we have, for the increment $A(\delta)$,

$$A(\delta) = f(\beta) - f(\alpha) = P(\delta) - N(\delta);$$

therefore, we have for the sum of increments of $f(x)$ in the intervals of I,

$$A(I) = P(I) - N(I),$$

hence, for E^i by passing to the limit,

$$P(E^i) - N(E^i) = P(E^{ip}) - N(E^{in}) = A(E^i).$$

Thus, *if Λf is one of the derivative numbers of a continuous function $f(x)$ of bounded variation in an interval (a, b), its three variations and its increment in (a, b) are given by formulae*

$$P = \int_a^b \frac{1}{2}[|\Lambda f| + \Lambda f]\,dx + P(E^i),$$
$$N = \int_a^b \frac{1}{2}[|\Lambda f| - \Lambda f]\,dx + N(E^i),$$
$$V = \int_a^b |\Lambda f|\,dx + V(E^i),$$
$$A = f(b) - f(a) = \int_a^b \Lambda f\,dx + A(E^i).$$

$P(E^i)$, $N(E^i)$, $V(E^i)$, $A(E^i)$ are the finite and well-determined limits, towards which tend the sums of the various total variations and increments of $f(x)$ in the non-intersecting intervals of I, which enclose the set E^i of points where Λf is infinite, when we vary the system I of intervals, in such a way that $m(I)$ tends towards zero.

Let us leave this general statement aside for a moment and focus on the case where E^i does not exist[5]; the numbers $P(E^i)$, $N(E^i)$, $V(E^i)$, $A(E^i)$, are then zero. If, then, we apply the previous theorem to the interval (a, x), we will have:

The necessary and sufficient condition for the derivative number Λf, which is finite everywhere, of a continuous function $f(x)$ to be summable, is that $f(x)$ is of bounded variation.

The primitive function $f(x)$ of Λf is, then, the indefinite integral, function of one variable, of Λf.

Therefore, we can say that we know how to solve the problems A', B', C', (Sect. 6.2), and as a result, the problems A, B, C, for all summable functions.

There is another case in which $P(E^i)$, $N(E^i)$, $V(E^i)$, $A(E^i)$, are all zero, and that is when $f(x)$ is absolutely continuous; because then $V(I)$ tends towards zero with $m(I)$. Therefore: *an absolutely continuous function is the indefinite integral of each of its four derivative numbers*[6] *its total variation is the indefinite integral of the absolute value of any of its derivative numbers.*

Now that we have this statement at our disposition, we could replace some of the previous results by simple sufficient conditions of absolute continuity:

A continuous function f, which has summable and everywhere finite derivative numbers Δf in (a, b), is absolutely continuous.

A continuous function $f(x)$ of bounded variation, which has its derivative numbers Λf finite everywhere in (a, b), is absolutely continuous.

Let us return to the general case in which E^i exists, and where the numbers $P(E^i)$, $N(E^i)$, $V(E^i)$, $A(E^i)$ are not necessarily zero.

Then, if we consider a set J of non-intersecting intervals, we have, by appealing to e^{ip} the subset of E^{ip} in J

$$P(J) = \int_J \frac{1}{2}[\Lambda f + |\Lambda f|]\, dx + P(e^{ip});$$

---

[5] The order of the text has been adopted to avoid redundancy, but it is worthwhile to note how much the previous considerations simplify when we limit ourselves to the case of an always finite derivative number Λf; in particular, it should be noted that in this case, the notion of a set function no longer comes into play.

The historical order is reverse of the one in the text; the theorems related to the case of always finite Λf were included in the first edition of this book; the idea of evaluating the difference $f(b) - f(a) - \int_a^b \Lambda f\, dx$ as limit of $A(I)$ is due to M. Vallée Poussin, who first obtained the general theorem, initially for monotonic functions (*Cours d'Analyse infinitesimale*, 2nd *edition, t.I, p.* 269).

We will note that the idea of M. de la Vallée Poussin already inherently contains the notions of function and the set of singularities. These notions, along with all those related to set functions, were examined in my Memoire: *Sur l'integration des fonctions discontinues (Ann. sc. de l'Ec. Norm,* 1910*).*

[6] It is understood that, in the definition of this indefinite integral, we neglect the set E^i of measure zero, formed by points at which the considered derivative number Λf is infinite. Or, equivalently, we take integral of the function that is zero on the points of E and equal to Λf elsewhere.

but $P(e^{ip})$ is at most equal to $P(E^{ip})$ and the integral of the right-hand side tends towards zero with $m(J)$. Therefore, when $m(J)$ tends towards zero, the largest of the limits of $P(J)$ is $P(E^{ip})$, and this largest limit is attained when we take J enclosing E^{ip}. In other words: *E^{ip} is the set of singularities of the function $P(x)$ representing total positive variation from a to x.*

The function of singularities, $P_s(x)$, of the function $P(x)$ is

$$P_s(x) = P(x) - \int_a^x \frac{1}{2}[\Lambda f + |\Lambda f|]\,dx.$$

Similarly, we have, with the notations whose meaning is clear,

$$N_s(x) = N(x) - \int_a^x \frac{1}{2}[|\Lambda f| - \Lambda f]\,dx; \quad \text{hence}$$

$$V_s(x) = V(x) - \int_a^x |\Lambda f|\,dx$$

$$f_s(x) = f(x) - f(a) - \int_a^x \Lambda f\,dx;$$

formulae which reveal the functions of singularities of $P(x)$, $N(x)$, $V(x)$, $A(x)$. As for the set of singularities, it is E^{ip} for $P(x)$, E^{in} for $N(x)$, it is E^i for $P(x)$, $N(x)$, $V(x)$, $f(x)$. We have, in particular

$$P_s(b) = P(E^i), \quad N_s(b) = N(E^i), \quad V_s(b) = V(E^i), \quad f_s(b) = A(E^i),$$

therefore, these numbers are all zero only in the previously examined case, where f is absolutely continuous.[7] Earlier, we decomposed (Sect. 9.3) a continuous function of bounded variation into its function of singularities and an absolutely continuous kernel, let AC be the kernel of $f(x)$; we just proved that we have

$$AC(x) = f(a) + \int_a^b \Lambda f\,dx,$$

and similar formulae for the kernels of $P(x)$, $N(x)$, $V(x)$.

10.2 The Differentiation of Functions of Bounded Variation

We will now take into account that each of our results has four interpretations depending on whether, by Λf, we have denoted the right-hand superior derivative number of f, or the right-hand inferior derivative number, etc.

[7] For f non-absolutely continuous, one of the numbers $P(E^i)$, $N(E^i)$, $A(E^i)$ can be zero, but only one.

We first saw that E^{ip} plays a very special role, as we have just shown by proving that E^{ip} is the set of singularities of $P(x)$, when $P(x)$ is not absolutely continuous. There are four sets E^{ip}, let $\mathcal{E}^{ip}$ be their common subset; $\mathcal{E}^{ip}$ is the set of points at which $f(x)$ has a derivative equal to $+\infty$, but the common subset to several sets of singularities, if they are B measurable, is also a set of singularities (Sect. 9.3), therefore, $\mathcal{E}^{ip}$ is a set of singularities of $P(x)$.

Thus, *in all that precedes, we can replace* E^{ip}, E^{in}, E^{i}, *respectively with the sets* $\mathcal{E}^{ip}$, $\mathcal{E}^{in}$, $\mathcal{E}^{i} = \mathcal{E}^{ip} + \mathcal{E}^{in}$ *formed by the points at which* $f(x)$ *has a definite derivative, in magnitude and sign, equal to* $+\infty$, $-\infty$, $+\infty$, *or* $-\infty$, *respectively. These three sets,* $\mathcal{E}^{ip}$, $\mathcal{E}^{in}$, *and* $\mathcal{E}^{i}$ *are sets of singularities of* $P(x)$, $N(x)$, $f(x)$ *respectively, when these functions are not absolutely continuous.*

In particular, let us mention that *a continuous function of bounded variation, which, at any point does not have a well defined derivative, in magnitude and sign, equal to* $+\infty$ *or* $-\infty$, *is absolutely continuous.*

We then saw that the kernel $AC(x)$ of $f(x)$ was given by the formula

$$AC(x) = f(a) + \int_a^x \Lambda f \, dx,$$

therefore, the integral of right-hand side is same for each of the four derivative numbers; thus (Sect. 9.2) these four derivative numbers are equal, except possibly at points of a set of measure zero. *A continuous function*[8] *of bounded variation* $f(x)$ *has a derivative almost everywhere and the kernel of* $f(x)$ *is, therefore, given by the formula*

$$AC(x) = f(a) + \int_a^x f'(x) \, dx.$$

Let us apply this result to $AC(x)$, which is its own kernel:

$$AC(x) = AC(a) + \int_a^x AC'(x) \, dx;$$

$f'(x)$ and $AC'(x)$ are, therefore, equal almost everywhere as they have the same indefinite integral, and consequently, the derivative of $f(x) - AC(x)$ is zero almost everywhere. In other words: *the derivative of the function of singularities* $f_s(x)$ *of a continuous function*[9] *of bounded variation* $f(x)$ *is zero almost everywhere.* The converse is true; in a more precise way: *a continuous function*[10] *of bounded variation and whose derivative is zero almost everywhere, is its own function of singularities.* Indeed, from the previous formula, its kernel is identically zero.

---

[8] We will see in a moment that the word ≪ continuous ≫ can be omitted.

[9] We will see in a moment that the word ≪ continuous ≫ can be omitted.

[10] We will see in a moment that the word ≪ continuous ≫ can be omitted.

As it has been mentioned in note, the previous three theorems could be extended to all functions of bounded variation, whether continuous or discontinuous.[11] To do so, it will be sufficient to use the formula of Sect. 9.3,

$$F = S + C = F_s + AC = S + C_s + AC,$$

and to demonstrate that *a function of jumps has a zero derivative almost everywhere.*

To each point of discontinuity x_0 of the considered function S, let us attach two functions $\varphi(x)$; the first equal to

$$S(x_0) - S(x_0 - 0) \quad \text{for} \quad x \geq x_0$$

and zero elsewhere. The second equal to

$$S(x_0 + 0) - S(x_0) \quad \text{for} \quad x > x_0$$

and zero elsewhere. S is the sum of the series $\sum \varphi(x)$. The total variation $VS(x)$ of S is, for the interval (a, x), the sum of the series $\sum |\varphi(x)|$; series which is majorised by the series of constants $\sum |\varphi(b)|$.

Let us modify each function $|\varphi(x)|$ to the left of its points of discontinuity x_0, in an interval $(x_0 - h, x_0)$, in order to obtain a function $\psi(x)$ that is continuous everywhere, equal to $|\varphi(x)|$ outside $(x_0 - h, x_0)$, and linear in $(x_0 - h, x_0)$.

The function $\Psi(x) = \sum \psi(x)$, being defined by a series of the continuous functions, majorised by the series of the constants $\sum |\varphi(b)|$ is continuous. Furthermore, it is non decreasing.

Let E be the set of points not belonging to any of the intervals $(x_0 - h, x_0)$; in E we have

$$\Psi(x) = VS(x);$$

in the complement CE of E, we have

$$\Psi(x) > VS(x).$$

since each function ψ is greater than the corresponding function $|\varphi|$ in the interval where they differ.

The functions Ψ and ψ being non-decreasing and continuous, they simultaneously have derivatives almost everywhere. At a point where all these derivatives exist, we have, for any positive integer p.

$$\Psi = \psi_1 + \psi_2 + \cdots + \psi_p + \rho_p,$$

where ρ_p is a non-decreasing function, therefore,

[11] The differentiation of discontinuous functions was first studied by M. and Mrs. W. H. Young (Quarterly Journal, 1910 and Proceedings of the London math. Soc., 1910).

$$\Psi' = \sum_1^p \psi_i' + \rho_p' = \sum_1^\infty \psi' + \theta,$$

where θ is positive or zero. Hence, according to the theorems of Sects. 10.1 and 8.4,

$$\begin{aligned}\Psi(b) \geq \int_a^b \Psi'(x)\,dx &= \sum_1^\infty \int_a^b \psi'\,dx + \int_a^b \theta\,dx \\ &= \sum_1^\infty \psi(b) + \int_a^b \theta\,dx = \Psi(b) + \int_a^b \theta\,dx,\end{aligned}$$

and, since θ is never negative, θ is zero almost everywhere. However, at points of E, all derivatives ψ' are zero. Therefore, almost everywhere in E, we have

$$\Psi' = \sum \psi' = 0.$$

Now, according to the relations mentioned above,

$$\Psi(x) = VS(x)$$

at points of E, $\Psi(x) > VS(x)$ at points of CE; at any point of E the function $VS(x)$ has derivative numbers to the right at most equal to those of $\Psi(x)$, therefore almost everywhere in E the total variation VS, and *therefore* S, has a right-hand derivative that is equal to zero. But the measure of CE is at most the sum $\sum h$ which we could make arbitrarily small. Therefore, $S(x)$ admits zero as right-hand derivative almost everywhere. A similar conclusion is obviously applicable to the left-hand derivative; the theorem is proved. We are now in a position to answer the question posed previously and to justify some assertions. We have already, Sect. 9.1, alluded to the following property: *for a function of one variable $F(x)$ to be the indefinite integral of an unknown function $f(x)$, it is necessary and sufficient that $F(x)$ be absolutely continuous.*[12] The legitimisation of this statement is now immediate; we have seen, indeed, that any indefinite integral is absolutely continuous, Sect. 9.2, and then in Sect. 10.1, that every absolutely continuous function is an indefinite integral.

From this, it immediately follows that: *for a function of set or interval to be the indefinite integral of an unknown function $f(x)$, it is necessary and sufficient that be completely additive and absolutely continuous;* according to what we know about the transition from such a function to an absolutely continuous function of one variable and on the converse transition.

---

[12] In the first edition of this book, I had indicated this statement, in note of the page 128, quite incidentally and without proof. M. Vitali rediscovered this theorem and published the first proof (*Acc. Reale delle Sc. di Torino,* 1904–1905) of it. It was on the occasion of this theorem that M. Vitali introduced, for the functions of one variable, the concept of absolute continuity and showed the simplicity and clarity that the whole theory acquires when we put this notion into its foundation.

In Sect. 9.2, we formulated this problem: *to find a function when its indefinite integral is known.* Let us take this indefinite integral in the form of a function of one variable $F(X)$; $F(X)$ is absolutely continuous, therefore it has a derivative almost everywhere and is the indefinite integral of this derivative, But two functions which have the same indefinite integral are almost everywhere equal, therefore the function whose integration gave $F(X)$ is almost everywhere equal to $F'(X)$. In other words: *an indefinite integral* $F(X)$ *admits the integrated function as its derivative almost everywhere.*

If the indefinite integral is given as function of set or interval, the problem is no less solved. Let us examine what happens to the differentiation operation in this case. Let h be positive or negative, the ratio $r[F(x), x_0, x_0 + h]$ is the quotient of the function Ψ, evaluated on the interval δ with extremities x_0 and $x_0 + h$, divided by the measure of this interval

$$r[F(x), x_0, x_0 + h] = \frac{\Psi(\delta)}{m(\delta)}.$$

Therefore, it is through the consideration of such ratios that we obtain the derivative; with the ordinary derivative, we would be led to the use the two families of intervals $(x_0 - h, x_0)$, $(x_0, x_0 + h)$ having the studied point as the extremity or origin. However, it is better to note that the derivative can just as well be defined using the intervals $(x_0 - h, x_0 + k)$, because we have

$$\begin{aligned} r[F(x), x_0 - h, x_0 + k] = {} & \frac{h}{h+k} r[F(x), x_0 - h, x_0] \\ & + \frac{k}{h+k} r[F(x), x_0, x_0 + k], \end{aligned}$$

which shows that the value of r in the left-hand side is included between those which appear in the right-hand side. We are thus led to say: *to differentiate at a point* x_0*, the function of intervals* $\Phi(\delta)$*, we seek the limit of the ratio* $\Phi(\delta)$ *when* $m(\delta)$ *tends to zero,* δ *denotes a variable interval containing* x_0.[13]

This definition is satisfactory for a function of an interval, but for a function of a set, we would like to define the limit of $\frac{\Psi(E)}{m(E)}$ as $m(E)$ tends towards zero and the set E tends towards x_0; that is, E is contained in an interval Δ containing x_0 and whose measure tends towards zero. If E were subject only to these conditions, we could take it, in particular, reduced to a point x_0 and an interval δ not containing x_0. Therefore, it would be necessary that the derivative appeared as the limit of

$$\frac{\Psi(\delta)}{m(\delta)} = r[F(x), x_0 + h, x_0 + h + k],$$

[13] We will compare this definition, which is merely the translation of the definition of ordinary derivative, with the one given by M. Volterra for the derivative of a function of line (*Leçons sur les équations intégrales et les équations intégro-différentielles, p.* 12). The definition of the text can obviously be applied to set functions of points in a plane, in space, etc. If we apply it to a set of points in a plane, the resemblance with the definition of M. Volterra becomes more apparent.

when h and k tend towards zero, even by the values of the same sign. According to the mean value theorem, the right-hand side is included between the extreme values taken in $(x_0 + h, x_0 + h + k)$, by the function f whose indefinite integral is F. Moreover, let $x_0 + h$ be a point in the neighbourhood of x_0 at which F' exists, we could then find an interval $(x_0 + h, x_0 + h + k)$ such that hk is positive and as small as we want, and $r[F(x), x_0 + h, x_0 + h + k]$ is as close to $F'(x_0 + h)$ as we want. Therefore, the methods that we are examining will succeed, with all the intervals δ or all the sets E, only if F' is continuous at x_0 in the set of points where it exist. Furthermore, in this case, the function f whose indefinite integral is F, can be assumed to be continuous at x_0—since it is subjected only to the condition of being equal to F' almost everywhere—and as a result the methods of defining the derivative using all the sets whose all the points tend towards x_0, can be used[14] in this case as well.

Apart from this trivial case, it is necessary, therefore, to specify the family of sets E used to minimise the influence of points at which f differs significantly from $F'(x_0)$. To achieve this, with ε arbitrarily chosen positive, let E_i be the set of points

$$E[i\varepsilon \leq f(x) < (i+1)\varepsilon],$$

let f_i be the function equal to f at points of E_i and zero elsewhere, let finally $g_i = f - f_i$. A set of points of measure zero being excepted, the indefinite integrals $\int f_i\, dx$ and $\int |g_i|\, dx$ have derivatives, in the ordinary sense of word, equal to the functions f_i and $|g_i|$, respectively. Then, if the non exceptional point x_0 belongs to E_l, we have

$$\frac{\int_E f\, dx}{m(E)} = \frac{\int_E f_l\, dx}{m(E)} + \frac{\int_E g_l\, dx}{m(E)},$$

$$\left|\frac{\int f\, dx}{m(E)} - \frac{\int_E f_l\, dx}{m(E)}\right| \leq \frac{\int_E |g_l|\, dx}{m(E)} \leq \frac{\int_\Delta |g_l|\, dx}{m(\Delta)}\frac{m(\Delta)}{m(E)};$$

where Δ denotes the smallest interval containing both E and x_0. The first ratio of the right-hand side tends towards the derivative of $\int |g_l|\, dx$ at x_0, which is zero by hypothesis. Therefore, we will only have to deal with the influence of f_l, that is, with the points of E_l, if the ratio $\frac{m(\Delta)}{m(E)}$ does not increase indefinitely. Hence the definition: *We will call the derivative at x_0 of a set function Ψ the limit, if it exists, the ratio $\frac{\Psi(E)}{m(E)}$, for sets E belonging to a regular family. By this, we mean that for an arbitrarily chosen positive integer k, we will consider a set E only if, Δ being the smallest interval containing E and x_0, we have*

$$m(E) > km(\Delta),$$

and we will vary E in such a way that $m(\Delta)$ tends towards zero.

---

[14] In short, it is necessary and sufficient that f be continuous at x_0 when we neglect the sets of measure zero [Sect. 9.1, in note].

We have just seen that with this choice of set E the two incremental ratios $\frac{\int_E f\,dx}{m(E)}$, $\frac{\int_E f_l\,dx}{m(E)}$ have the same limits almost everywhere. Let us study the second. Let F_l be the set of points common to E and E_l, and let $G_l = E - F_l$; we have

$$\frac{\int_E f_l\,dx}{m(E)} = \frac{\int_{F_l} f_l\,dx}{m(E)} = \frac{\int_{F_l} f_l\,dx}{m(F_l)}\frac{m(F_l)}{m(E)}.$$

The first factor of the last expression is included between $l\varepsilon$ and $(l+1)\varepsilon$; as for the second, it can also be written $1 - \frac{m(G_l)}{m(E)}$. Now, if h_l denotes the zero function in E_l, and equal to 1 outside E_l, we have

$$\frac{m(G_l)}{m(E)} \le \frac{m(G_l)}{km(\Delta)} = \frac{\int_{G_l} h_l\,dx}{km(\Delta)} \le \frac{1}{k}\frac{\int_\Delta h_l\,dx}{m(\Delta)}.$$

Outside an exceptional set of measure zero, all the indefinite integrals $\int h_i\,dx$ have derivatives equal to h_i. Therefore, if x_0 is also taken outside this new exceptional set, the last expression of the previous relations tends towards zero; therefore, the incremental ratio $\frac{\int_E f\,dx}{m(E)}$ has all its limits included between $l\varepsilon$ and $(l+1)\varepsilon$, that is, it differs from $f(x_0)$ by at most ε. But, as ε is arbitrarily small, if we take x_0 outside the sum of the exceptional sets attached to the values $\varepsilon = 1, \frac{1}{2}, \frac{1}{3}, \frac{1}{4}, \ldots$, that is, outside a set of measure zero, the derivative of set function $\int_{\mathcal{E}} f\,dx$ will be $f(x_0)$. Therefore, *an indefinite integral of a set function has as its derivative, the integrated function, almost everywhere.*

Therefore, we have the same statement for three kinds of indefinite integral. However, it is quite necessary to note that, in its last form, it expresses a property which is much more precise than in the earlier two forms, which were equivalent. If the indefinite integral $\Psi(E)$ has a derivative at point x_0, then $F(X)$ also has one, and these two derivatives are equal; but the converse may not hold.

Therefore, it is appropriate to express the results we have just obtained in terms of $F(X)$; further developments related to the calculation of $\Psi(E)$, when $F(X)$ is known, lead immediately to the statement: *$F(X)$ being an absolutely continuous function in (a, b) and x_0 a point of (a, b), we enclose x_0 in an interval Δ and we choose, in Δ, non-intersecting intervals (α_1, β_1) such that we have*

$$\sum(\beta_i - \alpha_i) \ge km(\Delta),$$

where k is a fixed positive number. Then, we have almost everywhere

$$F'(x_0) = \lim_{m(\Delta)\to 0} \frac{\sum[F(\beta_i) - F(\alpha_i)]}{\sum[\beta_i - \alpha_i]}.$$

We can give the following form to this statement: *$f(x)$ being a summable function, the function $\int |f(x) - f(x_0)|\,dx$ admits a zero derivative for $x = x_0$, provided that x_0 is not taken in some exceptional set of measure zero.*

Indeed, let us show that there is an identity between this exceptional set $\mathcal{E}$ and the $\mathcal{E}_1$ of points x_0 at which the set function of indefinite integral of $f(x)$ does not admit a derivative equal to $f(x_0)$.

Indeed, let us suppose that x_0 belongs to this latter set $\mathcal{E}_1$, which means that we can find sets E whose all the points approach x_0 indefinitely, which belong to a regular family of parameter k and for which we have

$$\left|\frac{\int_E f(x)\,dx}{m(E)} - f(x_0)\right| > \alpha,$$

where α is a fixed non-zero number.

Therefore, we have *a fortiori*,

$$\frac{\int_E |f(x) - f(x_0)|\,dx}{m(E)} \geq \left|\frac{\int_E f(x)\,dx - \int_E f(x_0)\,dx}{m(E)}\right| > \alpha,$$

and, as a result, if Δ is the smallest interval containing $E + x_0$,

$$\begin{aligned}\frac{\int_\Delta |f(x) - f(x_0)|\,dx}{m(\Delta)} &\geq \frac{\int_E |f(x) - f(x_0)|\,dx}{m(\Delta)}\\ &= \frac{\int_E |f(x) - f(x_0)|\,dx}{m(E)}\frac{m(E)}{m(\Delta)} > k\alpha;\end{aligned}$$

which shows that x_0 is point of $\mathcal{E}$.

Conversely, let us assume x_0 is a point of $\mathcal{E}$, then we can find a sequence of intervals Δ, enclosing each x_0 and of decreasing length and tending towards zero, such that for each of them we have

$$\frac{\int_\Delta |f(x) - f(x_0)|\,dx}{m(\Delta)} > \alpha,$$

where α is a positive number.

Let us partition the points of Δ in two sets E_p and E_n, in which the difference $f(x) - f(x_0)$ is, respectively, positive and non positive and let us keep only a partial sequence of Δ in such a way that

$$\frac{\int_{E_p}[f(x) - f(x_0)]\,dx}{m(\Delta)}, \frac{\int_{E_n}[f(x) - f(x_0)]\,dx}{m(\Delta)}$$

tends towards two definite limits β and $-\gamma$, and we have

$$\frac{\int_{E_p}[f(x) - f(x_0)]\,dx}{m(\Delta)} > \frac{\beta}{2} \quad \text{or} \quad < \frac{\gamma}{4},$$

according to β is positive or zero, and

$$\frac{\int_{E_n}[f(x) - f(x_0)]\,dx}{m(\Delta)} < -\frac{\gamma}{2} \quad \text{or} \quad > -\frac{\beta}{4}.$$

according to γ is positive or zero.

Furthermore, let us note that $\beta - \gamma$ is at least equal to α, therefore that β and γ are never both zero, and

$$m(E_n) + m(E_p) = m(\Delta),$$

therefore, at least one of the two sets, E_n and E_p, belongs to the regular family of sets of parameter $k = \frac{1}{4}$.

If, E_p is regular and $\beta > 0$, by taking $E \equiv E_p$, we have

$$\frac{\int_E f(x)\,dx}{m(E)} > f(x_0) + \frac{\beta}{2}\frac{m(\Delta)}{m(E)} > f(x_0) + \frac{\beta}{2}.$$

If these two conditions are not satisfied at the same time and if E_n is regular and $\gamma > 0$, we take $E \equiv E_n$,

$$\frac{\int_E f(x)\,dx}{m(E)} < f(x_0) - \frac{\gamma}{2}\frac{m(\Delta)}{m(E)} < f(x_0) - \frac{\gamma}{2}.$$

If E_p is irregular, $\beta > 0$, and if $\gamma = 0$, then E_n is regular; we take $E \equiv \Delta$ and we will have

$$\begin{aligned}\frac{\int_E f(x)\,dx}{m(E)} &= \frac{\int_{E_p} f(x)\,dx}{m(\Delta)} + \frac{\int_{E_n} f(x)\,dx}{m(\Delta)} \\ &> \left[f(x_0)\frac{m(E_p)}{m(\Delta)} + \frac{\beta}{2}\right] + \left[f(x_0)\frac{m(E_n)}{m(\Delta)} - \frac{\beta}{4}\right] = f(x_0) + \frac{\beta}{4}.\end{aligned}$$

Finally, there remains only the case where E_p being regular and E_n irregular, $\beta = 0$; with $E \equiv \Delta$, then we have

$$\frac{\int_E f(x)\,dx}{m(E)} < f(x_0) - \frac{\gamma}{4}.$$

Therefore, if $\lambda > 0$ is less than both $\frac{\beta}{4}$ and $\frac{\gamma}{4}$, we always have

$$\left|\frac{\int_E f(x)\,dx}{m(E)} - f(x_0)\right| > \lambda.$$

The sequence of these sets E, which belong to the regular family for $k = \frac{1}{4}$, allows us to see that x_0 belongs to $\mathcal{E}_1$.[15]

In the course of the previous arguments, we introduced a concept that needs to be explained. Let E be a measurable set, the ratio $\frac{m(e)}{\beta-\alpha}$ of the measure of the subset e of

[15] In a paper concerning Fourier series that I published in *Math. Annalen* (Bd LXI, 1905) we will find another proof of this theorem, from which we can deduce the proposition regarding the differentiability of set functions which are indefinite integrals, in a manner completely different from the one which we have used here.

E located in an interval (α, β) to the measure of that interval, is called *the mean density of E in* (α, β). If the mean density of E in the intervals (α, β) containing a point x_0, tends towards a definite limit when $\beta - \alpha$ tends towards zero, this limit is called *the density at* x_0; x_0 need not be a point of E for this definition to apply.[16]

Let φ be the measurable function which is equal to one at points of E and zero elsewhere. The mean density of E in (α, β) is the incremental ratio $\frac{\int_\alpha^\beta \varphi\, dx}{\beta-\alpha}$, therefore the theorem on differentiation of $\int_\alpha^x \varphi\, dx$ yields: *the density of a measurable set E is equal to one almost everywhere at point of E, and equal to zero outside E, almost everywhere.*

We can consider this property as fundamental geometric theorems on differentiation of definite integrals. To establish these theorems, we can follow a course exactly opposite to the one presented here, namely, to prove first the theorem on density.[17] we can simply deduce from this theorem an interesting proposition concerning any measurable function f, using the notations introduced for the study of the differentiation of the functions of set (Sect. 10.2).

Except at the points of an exceptional set of measure zero, the different sets E_i have density equal to either one or zero, depending on whether it concerns the density at a point of E_i or not. We can suppose this exceptional set is chosen in such a manner that it corresponds to the values $1, \frac{1}{2}, \frac{1}{3}, \ldots$ assigned to ε. After choosing x_0 outside this exceptional set, let $l(\varepsilon)$ be the index of the E_l corresponding to ε which contains x_0. Then, at x_0, for each of the indicated values of ε, $E_{l(\varepsilon)}$ has a density equal to one, while the other E_i have a density zero.

Let us choose an interval Δ_1 with center x_0 and sufficiently small so that, in every interval contained in Δ_1 and containing x_0, the mean density of $E_{l(1)}$ is greater than $1 - \varepsilon_1$, where ε_1 is a number chosen arbitrarily between zero and one. Let $1 - \varepsilon_1 + \eta$ be the lower limit of this mean density in the considered intervals.

Let us choose a number ε_2 smaller than both $\frac{\varepsilon_1}{2}$ and η_1, and let Δ_2 be an interval centered at x_0, of length less than $\frac{\Delta_1}{4}$ and such that in any interval contained in Δ_2 and containing x_0, the mean density of $E_{l(\frac{1}{2})}$ is greater than $1 - \varepsilon_2$. We then define

$$1 - \varepsilon_2 + \eta_2,$$

[16] I introduced these notations, which are almost necessary, given the physical meaning of same expressions, in the cited Memoire in the *Math. Annalen.* M. Denjoy replaces the word *density* with *thickness;* he is bothered by phrases such as: ≪ A set everywhere non-dense can have a density equal to one almost everywhere≫. It is certain that such a phrase may seem like a pun. However, between the two terms—'dense' and 'density'—it is the former which is poorly chosen because it, in my opinion, evokes an idea of measure, a qualitative idea.

Be that as it may, the reader who has followed me so far will not allow himself to be disturbed by these semantic concerns, as he would have agreed to delve into the quest for definition of the definite integral, which may seem absurd, and the definition of the integral that remains indefinite even after being defined, which is, to some extent, a humbling realisation.

[17] *See*, for example, my Memoire of *Ann, Sc. de l'Ec. Norm.,* 1910.

the lower limit of this mean density and, starting from η_2 and ε_2, just as before with η_1 and ε_1, we will define, through the intermediary of $E_{l(\frac{1}{3})}$ a number ε_3 and an interval Δ_3, and so on.

Let E be the set formed by the part of $E_{l(1)}$ which is exterior to Δ_2 and contained in Δ_1, and the part of $E_{l(\frac{1}{2})}$ which is exterior to Δ_3 and contained in $\Delta_2, \ldots$. It is clear that $f(x)$ is continuous at x_0 on E. We will see that E is of density one at point x_0.

To do this, consider a set $\mathcal{E}$ consisting of $E_{l(1)}$ in $\Delta_1 - \Delta_2$ and, in Δ_2, would have a mean density exceeding $1 - \varepsilon_2$ in any interval contained in Δ_2 and containing x_0. For example we could take $\mathcal{E}$ to be identical to $E_{l(\frac{1}{2})}$ in Δ_2.

The intervals contained in Δ_1 and containing x_0 are, either contained in Δ_2, and then we know for them the lower limit $1 - \varepsilon_2$ of their mean density, or not contained in Δ_2. Let L be such an interval, l be its subset located in Δ_2. In L, $E_{l(1)}$ had a measure at least equal to $(1 - \varepsilon_1 - \eta_1)m(L)$; but any l can be a subset of $E_{l(1)}$ and, in l, $\mathcal{E}$ has a measure greater than $(1 - \varepsilon_2)m(l)$. Finally, in L, $\mathcal{E}$ has a measure greater than

$$(1 - \varepsilon_1 + \eta_1)m(L) - m(l) + (1 - \varepsilon_2)m(l),$$

therefore, a mean density greater than

$$1 - \varepsilon_1 + \eta_1 - \varepsilon_2 \frac{m(l)}{m(L)} > 1 - \varepsilon_1 + \eta_1 - \varepsilon_2 > 1 - \varepsilon_1.$$

Now, let $\mathcal{E}_2$ be the set $\mathcal{E}$ identical to $E_{l(\frac{1}{2})}$ in Δ_2. Let $\mathcal{E}_i$ be the set identical to $\mathcal{E}_{i-1}$ to the exterior of Δ_i and identical to $E_{l(\frac{1}{i})}$ in Δ_i. Now, it is clear that the mean densities of $\mathcal{E}_2, \mathcal{E}_3, \ldots,$ are all greater than $1 - \varepsilon_1$ in the intervals containing x_0, contained in Δ_1 and containing Δ_2; that the mean densities of $\mathcal{E}_3, \mathcal{E}_4, \ldots,$ are all greater than $1 - \varepsilon_2$ in the intervals containing x_0, contained in Δ_2 and containing $\Delta_3, \ldots$.

Therefore, the mean density of E is at least equal to $1 - \varepsilon_1, 1 - \varepsilon_2, \ldots,$ respectively, in the various types of considered intervals; the density of E at point x_0 is therefore equal to one and that of the complement CE of E is consequently zero, because the sum of the mean densities, in the same interval, of two complementary sets is quite obviously equal to one. Thus:

A measurable function $f(x)$ is continuous at each point x_0, provided we neglect the points of a set of zero density at x_0,[18] *with the exceptions, however, of the points x_0 belonging to an exceptional set of measure zero.*

Hence, it follows, in particular, that the derivatives of functions $F(x)$ of bounded variation, whose existence has been shown, are continuous almost everywhere, except at the sets of zero density. Everyone can explore the relations between this continuity and possibility

[18] The neglected set, of zero density at x_0, varies with the point x_0; it is the set CE of the text.

of differentiating the indefinite integrals of functions of set. Without pausing further, we now briefly discuss the rectification of the curves, which will bring us back to the methods discussed at the beginning of this section.

10.3 The Rectification of Curves

Let a curve be given by the continuous functions $x(t), y(t), z(t)$, defined on some interval (a, b). By applying to the radical, which represents the length of a chord of this curve, the inequality

$$|l| \leq \sqrt{l^2 + m^2 + n^2} \leq |l| + |m| + |n|,$$

we concluded in Chap. 4, that the curve is rectifiable if and only if, $x(t), y(t), z(t)$ are of bounded variation. When this condition is met, the arc given by the values of t in an interval δ, has a length between the total variation $Vx(\delta)$ of x in δ [or $Vy(\delta)$, or $Vz(\delta)$], and the sum $Vx(\delta) + Vy(\delta) + Vz(\delta)$.

If, $s(t)$ denotes the arc length from $t = a$ to an arbitrary t, we can write:

$$Vx(\delta) \leq Vs(\delta) \leq Vx(\delta) + Vy(\delta) + Vz(\delta).$$

Let us apply this double inequality to all the intervals of a set I of non-intersecting intervals, and add. We obtain a result that can be denoted by

$$Vx(I) \leq Vs(I) \leq Vx(I) + Vy(I) + Vz(I).$$

Let us now vary I in such a way that $m(I)$ tends towards zero, we will see that *$s(t)$ is absolutely continuous if, and only if, $x(t), y(t), z(t)$ are absolutely continuous. The set of singularities of $s(t)$ is the sum of the sets of singularities of $x(t), y(t), z(t)$.*

Therefore, we can consider the set $\mathcal{E}_s$ of values of t where $x(t), y(t), z(t), s(t)$ do not all have finite and definite derivatives as a set of singularities.

This being said, let us enclose $\mathcal{E}_s$ in a set A^i of non-intersecting intervals. Then a positive ε being arbitrarily chosen, let us enclose in another set A_p of non-intersecting intervals, the set E_p of the points at which we have

$$p\varepsilon \leq \sqrt{x'^2 + y'^2 + z'^2} < (p+1)\varepsilon.$$

Now, let us cover (a, b), starting from a, using a chain of intervals chosen among the following:

a. To a point t_0 of $\mathcal{E}_s$ let us attach the interval with its origin at t_0 and its extremity coinciding with the extremity of the interval from the set A^i which contains t_0;
b. To a point t_0 of E_p, let us attach an interval of origin t_0, contained in the interval from the set A_p which contains t_0, and such that the length of the chord provided by this interval

δ differs from

$$m(\delta)\sqrt{x'(t_0)^2 + y'(t_0)^2 + z'(t_0)^2}$$

by less than ε

Let's use this chain, or rather the corresponding inscribed polygon, to calculate an approximate value of $s(b)$. The contribution of the intervals of type a is at most $V_s(A^i)$. Moreover, it will also be as close as we want to this value if we modify the construction of the chain as follows: we will proceed as it indicated only outside the first p intervals, $\delta_1, \ldots, \delta_p$ constituting A^i, and we will subdivide $\delta_1, \ldots, \delta_p$ into extremely small partial intervals. With this modification, we can thus assume, by suitably choosing the A^i, that the contribution of the intervals of type a differ as little as we want from the lower limit of $V_s(A^i)$, when we let $m(A_i)$ tend towards zero, that is, from the value taken by $s_s(b)$ for $t = b$ by the function of singularities $s_s(t)$ of $s(t)$.

On the other hand, the contribution of intervals of kind b lies between

$$\sum[|p| - 2]\varepsilon m(\alpha_p) \quad \text{and} \quad \sum[|p| + 2]\varepsilon m(A_p),$$

where α_p denotes the part of A_p not covered by A^i, and the other A_p intervals. By appropriately choosing $\varepsilon, m(A^i), m(A_p)$ to be sufficiently small, we can ensure that the two previous sums differ arbitrarily little from

$$\int_{\mathcal{E}} \sqrt{x'^2 + y'^2 + z'^2}\, dt,$$

where $\mathcal{E}$ is the set of points of (a, b) which do not belong to $\mathcal{E}_s$. Before concluding, let us note that if we had formed $\mathcal{E}_s$ using the points where one of the numbers x', y', z' was infinite with a definite sign and $\mathcal{E}$ from points where x', y', z' were all three finite and definite, it would not have changed either the lower limit of $V_s(A^i)$, or the integral $\int_{\mathcal{E}} \sqrt{x'^2 + y'^2 + z'^2}\, dt$, therefore:

If we denote by $\mathcal{E}$, the set of values of t for which $x'(t)$, $y'(t)$, $z'(t)$ are finite and definite, and by $\mathcal{E}_s$ the set of values of t for which at least one of the derivatives $x'(t)$, $y'(t)$, $z'(t)$ has an infinite value of definite sign, the length of the rectifiable curve $x(t)$, $y(t)$, $z(t)$ is equal to

$$\int_{\mathcal{E}} \sqrt{x'^2 + y'^2 + z'^2}\, dt + s_s(b);$$

where $s_s(b)$ *is the lower limit of the sum of the arc lengths for the curve, containing all points on this curve given by values of* $\mathcal{E}_s$.[19]

If we correct $s(t)$ by its function of singularities, we have an absolutely continuous function

[19] The proof shows that we can replace in this statement the words ≪ the lower limit of the sum of the arc lengths ≫ by ≪ the lower limit of the limit of the sum of the chord lengths of the arcs ≫.

$$AC_s(t) = s(t) - s_s(t) = \int_0^t \sqrt{x'^2 + y'^2 + z'^2}\, dt,$$

which has $\sqrt{x'^2 + y'^2 + z'^2}$ as its derivative almost everywhere; but $s_s(t)$ has a zero derivative almost everywhere. Therefore, *we have almost everywhere*

$$\overline{s'(t)}^2 = \overline{x'(t)}^2 + \overline{y'(t)}^2 + \overline{z'(t)}^2,$$

for any rectifiable curve.

Let us express the coordinates of points of the curve by function of parameter $s = s(t)$; this formula becomes

$$1 = \overline{x'(s)}^2 + \overline{y'(s)}^2 + \overline{z'(s)}^2;$$

it holds almost everywhere; the expression 'almost everywhere' now being relative to the measure with respect to the variable s. Wherever it holds $x'(s)$, $y'(s)$, $z'(s)$ exist, and are not all three zero, therefore, *if we exclude the points of a rectifiable curve, given by a set of measure zero for the values of the arc length s, at every point of the curve there exists a well-defined tangent, and we have*

$$x_s'^2 + y_s'^2 + z_s'^2 = 1.$$

The Totalisation 11

11.1 The Functions of the First Class

We will now demonstrate the theorem that we have already stated in Sect. 7.2:

For a function to be of class one at most, it is necessary and sufficient that it be point-wise discontinuous for any perfect set.

This theorem, due to M. Baire,[1] beyond the scope of our subject. However, on one hand, we will use the necessary condition it expresses, and on the other hand, and most importantly, the transfinite method that we will use to prove that the stated condition is sufficient, is the very same method that enabled M. Denjoy to completely solve the problem of primitive functions.[2] Also, we will henceforth consistently use transfinite numbers. Readers who are not familiar with the use of these numbers should study the Note located at the end of the volume before reading this chapter.

Let us show that the condition is necessary, meaning that if f is the limit of a convergent sequence of continuous functions $f_1, f_2, \ldots$, and if E is a perfect set, there exist points in E at which the function f, considered only on E, is continuous. Let us denote by $E_{n,p}$ the set of points in E at which we have $|f_n - f_{n,p}| \leq \varepsilon$, where ε is a positive number chosen arbitrarily, and E_n the set of points common to $E_{n,1}, E_{n,2}, E_{n,3}, \ldots$. The set $E_{n,p}$ is closed, therefore, E_n is also closed. E is the sum of E_n.

I assert that we can find an interval I, in which there are points of E, and in which E and E_n are identical for a sufficiently large value of n.[3] Let I_0 be an interval containing points of E. If E_1 is not identical to E in I_0, let us consider a point x_0 of E and of I_0 not belonging

[1] *Annali di Matematica,* 1899. *See* also the book published by M. Baire in this collection.

[2] The Memoires of M. Denjoy published in *Journal de Mathematiques,* 1915, *in Bulletin de la Societe mathematique de France,* 1915 and in *Annales scientifiques de l'Ecole Normale,* 1916, 1917.

[3] By point contained in an interval, we mean a point included between the extremities and different from these extremities.

R. Jain, *Lebesgue's Theory of Integration*,
https://doi.org/10.1007/978-981-96-1169-0_11

to E_1,[4] where I_1 is an interval with midpoint x_0 and is taken to be sufficiently small. I_1 will be interior to I_0, it will contain points of E and no points of E_1.

If, in I_1, E and E_2 are not identical, then in I_1, we can find a point x_1 belonging to E but not belonging to E_2, starting from which we can define an interval I_2 contained in I_1, containing points of E and no points from E_2, etc.

Now the sequence of $I_1, I_2, \ldots$ cannot be infinite because, in the interior of all these intervals, there would be points; these points would belong to E and would not belong to any of E_n, which is impossible.[5] Therefore, we will arrive at one last interval, say I_p, and, in I_p, E and E_{p+1} are identical.

Thus, we could find an interval I and a value of n such that, in I, E and E_n are identical and indeed have the points. In I let us take i, containing points of E, and such that the oscillation of f_n, in i, is smaller than ε. Then, for x_1 and x_0 taken in i and in E, we have, for any p positive or zero,

$$|f_{n+p}(x_1) - f_n(x_1)| \leq \varepsilon, \quad |f_n(x_0) - f_n(x_1)| \leq \varepsilon,$$

from where

$$|f_{n+p}(x_1) - f_n(x_0)| \leq 2\varepsilon,$$

and as a result

$$|f(x_1) - f(x_0)| \leq 2\varepsilon.$$

In other words, *in i, f and f_{n+p} (p positive or zero) are each constant to within* 2ε *and, when x and p vary, they are equal to each constant to within* 4ε. If now we take the numbers $\varepsilon_1, \varepsilon_2, \varepsilon_3, \ldots$ tending towards zero, we can find intervals $i_1, i_2, \ldots$ contained within each other, containing points of E, and in which the oscillation of f, over ε[6] would be at most $2\varepsilon_1, 2\varepsilon_2, \ldots$ respectively. In the interior of all these intervals i_k, there will exist at least one point of E; at this point, f is continuous on E.

To show that the condition is sufficient, let us first prove that it suffices that, for any $\varepsilon > 0$, a function f differs by less than ε from a function f_ε of class one at most, for f to be of class one at most.

Indeed, let us assume that f differs by less than $\frac{1}{2^k}$, from a function $f_{\frac{1}{2^k}}$ of class one. Then we can write

$$f = f_{\frac{1}{2}} + \left(f_{\frac{1}{2^2}} - f_{\frac{1}{2}}\right) + \left(f_{\frac{1}{2^3}} - f_{\frac{1}{2^2}}\right) + \cdots = \sum f^k,$$

[4] I will refrain from specifying the rule for the choices of $x_0, x_1, \ldots, I_0, I_1, \ldots$ as it would be required to meet the logical requirements mentioned in Sect. 5.2.

[5] In other words, a perfect set E cannot be the sum of a finite or a countably infinite number of sets that are everywhere non-dense on E.

[6] **Translator's note:** Instead of ε, it is probably, $i_1, i_2, \ldots$, that the author intends to say.

the series is uniformly convergent, since it has its terms majorised by those of the series $\sum\left(\frac{1}{2^k}+\frac{1}{2^{k-1}}\right)$, and all its terms are of class one at most.

Therefore, f^k is the limit of a sequence of continuous functions φ_p^k, and as $|f^k|$ does not exceed $\frac{1}{2^k}+\frac{1}{2^{k-1}}$, if we set, for $k>1$,

$$\psi_p^k=\varphi_p^k, \qquad \text{when we have } |\varphi_p^k|\le\frac{1}{2^k}+\frac{1}{2^{k-1}};$$

$$\psi_p^k=\frac{1}{2^k}+\frac{1}{2^{k-1}}, \qquad \text{when we have } \varphi_p^k>\frac{1}{2^k}+\frac{1}{2^{k-1}};$$

$$\psi_p^k=-\left(\frac{1}{2^k}+\frac{1}{2^{k-1}}\right), \qquad \text{when we have } \varphi_p^k<-\left(\frac{1}{2^k}+\frac{1}{2^{k-1}}\right);$$

for fixed k and p increasing indefinitely, the sequence of ψ_p^k has the same limit as that of φ_p^k.

Let us finally set

$$\Psi_p=\varphi_p^1+\psi_p^2+\psi_p^3+\cdots$$

Ψ_p is a continuous function because the series of the second member has its terms majorised by, starting from the second term, those of the series

$$\sum\left(\frac{1}{2^k}+\frac{1}{2^{k-1}}\right).$$

Now, it is clear that we have

$$f=\lim\Psi_p,$$

which proves that f is of class one at most.

Thanks to this lemma, to prove that a function $f(x)$, point-wise discontinuous on any perfect set is of class one at most, it suffices to show that, for any $\eta>0$, we can construct a function $\varphi(x)$ of class one at most, that does not differ from $f(x)$ by more than η. This construction is based on the properties of the function $f(x)$ considered on the closed set F.

When, in Chap. 2, we defined the maximum, minimum, and oscillation of a function $f(x)$ at a point, we made a remark that these definitions, and as a result those of continuity and discontinuity, do not require that $f(x)$ be defined on the whole interval. If we apply these definitions to a closed set, they lead us to say that $f(x)$ is, continuous at each isolated point of closed set F or, equivalently, the oscillation of $f(x)$ on F is zero at any isolated point of closed set F.

Hence, it follows that a function $f(x)$ that is point-wise discontinuous on any perfect set is also point-wise discontinuous on any closed set because, either this closed set is perfect, or it contains isolated point at which f must be regarded as continuous.

The reasoning in Sect. 3.1 leads, for functions defined on a closed set, to the following statement: *if, at all points of a closed set F, the oscillation of a function $f(x)$ on F is at* *most equal to ω, the oscillation of $f(x)$ is less than $\omega + \varepsilon$, on a subset of F contained in an interval of length λ, as long as λ is small enough; where ε is any positive number.*

With these remarks in mind, given a function $f(x)$ that is point-wise discontinuous on any perfect set in an interval (a, b), we can obtain a function $\varphi(x)$ that differs from $f(x)$ by at most η through the repetition of the following operation: let us suppose that $\varphi(x)$ is already constructed except at the points of a closed set F, we consider the set Φ of points of F at which the oscillation of $f(x)$ on F is at least equal to $\frac{\eta}{2}$. Φ is a closed set non-dense everywhere on F, since $f(x)$ is point-wise discontinuous on F.

(α, β) being any of the intervals contiguous to Φ, let us subdivide it in two equal intervals (α, z_0), (z_0, β); let us subdivide each of them into two equal interval, giving us, in particular, the two extreme intervals (α, z_{-1}), (z_1, β) which we further subdivide in two equal intervals by the points z_{-2} and z_2, respectively. Let us continue similarly by subdividing the extreme interval (α, z_{-2}), (z_2, β) in two equal intervals, etc. The considered interval (α, β) is, thus, divided into a sequence, infinite in both directions, of intervals (z_i, z_{i+1}). In (z_i, z_{i+1}) there are no points at which the oscillation of $f(x)$ on F exceeds $\frac{\eta}{2}$, therefore, by subdividing (z_i, z_{i+1}) in sufficient number of equal subsets, the oscillations of $f(x)$ on the subsets of F contained in each of these partial intervals would not exceed η.

Let us suppose, for fixing our ideas, that we always take the smallest number of the subsets which lead to this result; in this way, we will have subdivided (α, β) using the points

$$\cdots < x_{-2} < x_{-1} < x_0 = \frac{\alpha + \beta}{2} < x_1 < x_2 < \cdots$$

We agree that, at points of F where we have

$$x_i \leq x < x_{i+1},$$

to take

$$\varphi(x) = \frac{l_i + L_i}{2}$$

where l_i and L_i are the lower and upper limits of the values taken by $f(x)$ on subset of F located in (x_i, x_{i+1}).

If a belongs to F and not to Φ we will set $\varphi(a) = f(a)$, and similarly, if b belongs to F without belonging to Φ, we will set $\varphi(b) = f(b)$.

It is clear that $\varphi(x)$, now defined for every point not belonging to Φ, differs from $f(x)$ by η at most outside of Φ.

We will explain how the transfinite repetition of this process allows determining $\varphi(x)$ throughout (a, b), then we will prove that $\varphi(x)$ is at most of class one outside Φ.

Let us apply our method to the case where F is the entire interval (a, b); this operation O_1 provides us with φ except at the points of a closed set Φ, which we denote by e_1. Let us take this set e_1 as the new F. The subsequent operation, the operation O_2 will provide us with φ except at the points of a set Φ that we denote by e_2. Then we will take e_2 as the new F, leading to an operation O_3, etc. If it happens that e_n does not exist, then φ will be entirely determined by the operation O_n. This can occur for an arbitrary value of n, for example, for $n = 1$. However, it can also happen that we exhaust the sequence of entire finite indices n without determining $\varphi(x)$ in whole (a, b). All the sets $e_1, e_2, \ldots$ exist; they are closed, each of them contains the next. Therefore there are points common to all these sets and these points constitute a closed set which we will denote by e_ω.

The operation O_ω, which is of a different nature from the previous ones, will simply involve constructing e_ω and recognising that, after the preceding operations prior to O_ω, the function φ is known except at the points of e_ω.

The operation $O_{\omega+1}$ will be the one in which we take e_ω as the known set F. In general, the operation $O_{\alpha+1}$, following immediately the operation O_α (which provides $\varphi(x)$ except at the points of e_α, will be the one in which we take e_α as the new set F.

Whenever we have exhausted the finite and transfinite indices less than a transfinite number of second kind α, without arriving at the definition of φ over the (a, b), it means that the sets e_β exist for all $\beta < \alpha$. In this case, there are points common to all the e_β sets and which constitute a closed set e_α. The operation O_α then reduces to constructing e_α and recognising that, after the operations $O_\beta(\beta < \alpha)$, we know $\varphi(x)$ except at the points of e_α.

The family of operations O is thus defined and provides a well-ordered sequence of closed sets $e_1, e_2, \ldots$ such that each one contains all the subsequent ones, and each one is non-dense everywhere in those which precede it, since f is point-wise discontinuous on any closed set F. Therefore, two sets e_i, e_j of different indices cannot be identical. Also (*see* the note at the end of this Volume) the sequence of e_i is at most countable. In other words, after a finite or countably infinite number of operations, we arrive at an operation O_μ for which there is no exceptional set e_μ, meaning that it reveals $\varphi(x)$ over the entire (a, b).

Let us now show that $\varphi(x)$ is of class one at most in the interval (a, b) which we denote by e_0.

Considering an arbitrary point X from e_0, X belongs to the sets $e_0, e_1, \ldots$. However, since e_μ does not exist, there is a first index α, at most equal to μ, beyond which X no longer belongs to e_α. Moreover, α is of first kind; a point, belonging to all the e_β sets of indices less than a second kind number γ, indeed belongs, to e_γ by very definition of e_γ. Therefore, the function $\varphi(x)$ has been defined at point X during the operation O_α and through the intermediary of an interval (x_i, x_{i+1}) containing X obtained from the subdivision of an interval contiguous to e_α. If X is at a distance of at least $\frac{1}{p}$ from e_α, and if we have

$$x_i \leq X \leq \frac{px_{i+1} + x_i}{p+1}$$

we set

$$\varphi_p(X) = \varphi(X).$$

We set

$$\varphi_p(a) = \varphi(a) = f(a) \quad \text{and} \quad \varphi_p(b) = \varphi(b) = f(b).$$

The points at which $\varphi_p(x)$ is thus defined, excluding the points a and b, naturally form closed sets; E_α will be the set of those points of X belonging to all the e_β sets for which β is less than α and not belonging to e_α to which our definition of $\varphi_p(x)$ applies.

The different E_α sets are at least at a distance of $\frac{1}{p}$ from each other, therefore there are at most $p(b-a)$ of them that actually exist. Each of them decompose through the consideration of intervals $\left(x_i, \frac{px_{i+1}+x_i}{p+1}\right)$ into a finite number of closed sets and, on each of these, $\varphi(x)$, and therefore $\varphi_p(x)$, are constant.

Finally, $\varphi_p(x)$ is defined by the condition of being constant on a finite number of closed sets separated from each other; therefore, we can complete the definition of $\varphi_p(x)$ in (a, b) in such a way that $\varphi_p(x)$ is continuous in entire (a, b).

It is clear that $\varphi(x)$ is the limit of functions $\varphi_p(x)$ when p increases indefinitely; at any point X we indeed have $\varphi(x) = \varphi_p(x)$ from a certain value of p, which we determine as follows: let α be the index from which X no longer belongs to e_α, and let l be the distance of X from e_α. Consider (x_1, x_{i+1}), the interval obtained from the subdivision of intervals contiguous to e_α and such that we have

$$x_i \leq X \leq x_{i+1},$$

p is smallest integer at least equal to both $\frac{1}{l}$ and $\frac{X-x_i}{x_{i+1}-X}$.

11.2 The Primitive Functions of Everywhere Finite Derivatives

Let $f(x)$ be a derivative function which is finite everywhere in (a, b). From what we have learned so far, we know how to find the primitive function of $f(x)$ when $f(x)$ is summable. It is clear that the processes of Cauchy and Dirichlet allow us to attain, from there, the primitive function of $f(x)$ when the points of non summability[7] of $f(x)$ form a reducible set. Let us limit ourselves to the case where the two extremities of the considered intervals (a, b) are the only points of non summability of $f(x)$; then we can obtain the primitive function $F(x)$ through a passage to the limit, and in particular, we have

$$F(b) - F(a) = \lim[F(\beta) - F(\alpha)].$$

When we let α and β tend towards b, in such a way that we have

$$a < \alpha < \beta < b.$$

From this particular case we will immediately deduce a very extensive result. For this purpose, let us note that the points of non summability of a function $f(x)$ necessarily form a closed set E, since, if x_0 is a point of summability, meaning that if x_0 is *interior* to an

[7] Points in the neighbourhood of which $f(x)$ is not summable.

interval in which $f(x)$ is summable, all the points sufficiently close to x_0, in order to be interior to the same interval, are also points of summability.

To say that a point belongs to E means that $f(x)$ is not *summable over any interval* containing this point. However, it does not necessarily imply that $f(x)$ is not *summable over* E around that point. We will examine precisely this case where $f(x)$ is summable over E.

We already know $F(x)$, up to an additive constant in any interval contiguous to E. To complete the determination of $F(x)$, it would be sufficient to be able to construct the continuous function $F_1(x)$, equal to $F(x)$ at the points of E, as well as at a and at b, and is linear in the intervals where $F(x)$ is already known.[8]

Now, we know at every point the right-hand and left-hand derivatives of $F_1(x)$. Indeed, $F_1(x)$ and $F(x)$ have the same derivative $f(x)$ at points of E which are neither origin, nor the extremities of the intervals contiguous to E. If (α, β) is an interval contiguous to E, the derivative of $F_1(x)$ at any interior points of (α, β) is the known quantity $r[F(x), \alpha, \beta]$. This quantity is also the right-hand derivative of $F_1(x)$ at α and the left-hand derivative at β. At α, the left-hand derivative of $F_1(x)$ is $f(\alpha)$ except if α is an isolated point of E, in which case α is also an extremity of an interval contiguous to E, and we know left-hand derivative of $F_1(x)$ at α. Similarly, we know the right-hand derivative of $F_1(x)$ at β. Therefore, if $F_1(x)$ is of bounded variation, meaning that if its right-hand derivative, for example, is summable, we can compute $F_1(x)$. Now, the right-hand derivative of $F_1(x)$ is summable only if it is summable, on one hand over E, that is, if $f(x)$ is summable over E, and on the other hand, over the set of intervals contiguous to E, that is, if the series $\sum[F(\beta) - F(\alpha)]$ extended to the intervals contiguous to E is absolutely convergent. Therefore, when the previous conditions are fulfilled, we have

$$F(b) - F(a) = \int_E f(x)\,dx + \sum[F(\beta) - F(\alpha)].$$

To give this result its complete scope, let us note that we have only used the fact that E is closed, therefore:

If we know the primitive function $F(x)$ of a function $f(x)$, given on an interval (a, b), in every interval (α, β) contiguous to a closed set E:

If the series $\sum[F(\beta) - F(\alpha)]$ is absolutely convergent; if $f(x)$ is summable over E, then we have

$$F(b) - F(a) = \int_E f(x)\,dx + \sum[F(\beta) - F(\alpha)].$$

[8] We introduce here the extremities a and b of the considered interval, because we have agreed to consider (a, x_1) and (x_ω, b) as two intervals contiguous to E_1, where x_1 and x_ω are, respectively, the points of smallest and greatest abscissa of E_1.

Now, we will see that, as soon as the first of the three conditions of the previous statement[9] is met, there exists a partial interval i, in the interior of which E has points, and in which the two other conditions of the statement are also satisfied.

Let us suppose, indeed, that $F(x)$ is known in every interval contiguous to a closed set E. Then:

Either E is not perfect; let us take an interval i containing in its interior only one point of E, which is possible since E has isolated points; in i, the three conditions of the statement are fulfilled;

Or E is perfect. We have learned, Sect. 10.1, to choose a sequence of values h_n tending towards zero and such that the ratio $r[F(x), x, x+h]$ has an oscillation equal to $\frac{1}{n}$ at most, for h included between h_{n+1} and h_n. Let us consider $f(x)$ as the limit of continuous functions $r[F(x), x, x+h] = f_k$. We know that we can determine an interval i, containing points of E, and in which f and f_{n+p} are, on E, equal and within a distance of 4ε, for all positive values of p, with n chosen suitably, Sect. 11.1. I assert that this interval i answers the question. Indeed, for x located in i and on E, $r[F(x), x, x+h]$ is, for $h < h_n$, differs by $\frac{1}{n}$ at most from one of the functions f_i. Therefore, the value of $r[F(x), x, x+h]$ is within $4\varepsilon + \frac{1}{n}$. Thus, except may be for the intervals contiguous to E, which have a length greater than h_n, and there are finite number of them, every interval (α, β) contiguous to E and contained in i, provide for $r[F(x), \alpha, \beta]$ a value within $4\varepsilon + \frac{1}{n}$. Therefore, we have, for *any* interval contiguous to E and included in i,

$$|r[F(x), \alpha, \beta]| < M,$$

where M is a finite number. It follows that

$$|F(\beta) - F(\alpha)| < M(\beta - \alpha)$$

for all the contiguous intervals contained in i. Therefore, it is quite clear that, for the subsets of E located in i, the series $\sum[F(\beta) - F(\alpha)]$ is absolutely convergent. However, on the other hand, $f(x)$ is constant to within 2ε on E, is bounded, and therefore summable, and all the requisite conditions for the application of our theorem are satisfied in i.

Since then, in any interval containing points of E, we can find another one, containing points of E, and in which we know how to determine the primitive function of $f(x)$. The points of E which are not interior to such intervals, therefore form a set, which is obviously closed and non-dense everywhere on E. Let H be this set; if (l, m) is an interval contiguous to H and if we take (λ, μ) such that

$$l < \lambda < \mu < m,$$

[9] I mentioned this statement in my Thesis in note in Sect. 4.1. It represents the extreme point that I had reached in the research of primitive functions.

we know how to calculate the primitive function of $f(x)$ in (λ, μ), and therefore a passage to the limit will provide us this function in (l, m).

Thus: *if we know how to determine, up to an additive constant, the primitive function of a derivative function $f(x)$, in any interval contiguous to a closed set E, we know through this very fact how to determine it in any interval contiguous to a closed set H, formed of points of E and non-dense everywhere on E.*

This proposition will allow us to proceed by transfinite induction. First, let us consider the interval (a, b) itself, for E. Then, the set H is, the set E_1 of points of non summability of $f(x)$, and we will call O_1 the operation which reveals $F(x)$ in the intervals contiguous to E_1. Then, let us take E_1 for the set E. We will denote by E_2 the set H given by E_1, and O_2 will denote the operation which provided $F(x)$ in the intervals contiguous to E_2. And so on. If we exhaust all the finite indices without arriving at $F(x)$ in entire interval (a, b), it means that all the sets $E_1, E_2, E_3, \ldots$ exist. As they are closed and each one contains all the following ones, there are, then, common points to all these sets. These points form a closed set E_ω, contained in E_n and non-dense on any of them.The operation O_ω will consist of deducing $F(x)$ in the intervals contiguous to E_ω from the knowledge of $F(x)$ in the intervals contiguous to E_n, by passing to the limit.

In a more general way, if the operations of indices less than α do not provide $F(x)$ in entire (a, b), the operation O_α is defined as follows:

If α is finite or transfinite of first kind, the operation O_α is the one which consists of calculating $F(x)$ in intervals contiguous to a closed set E_α, which is obtained as set H, when one assigns the role of E to $E_{\alpha+1}$.

If α is transfinite of second kind, then all the E_β sets, for $\beta < \alpha$ exist; there are points common to all these E_β sets. These points form a closed set E_α. The operation O_α consists of determining $F(x)$ in intervals contiguous to E_α, by passing to the limit.

This finite or transfinite sequence of operations, which constitutes the *totalisation*, was conceived by M.A. Denjoy. It is clear that the sequence of closed sets $E_1, E_2, E_3, \ldots$, all different from those which precedes them and contained within them, can contain only a finite or countably infinite number of terms. Therefore, *the totalisation permits, in all cases, the determination of the primitive function of a known derivative function, in the entire interval where this derivative is given.*

We will return to the operation of totalisation and investigation of the primitive functions of derivative numbers later. For the time being, we will modify our operational procedure, while preserving the essential transfinite reference, and in doing so, we will find the primitive functions of the derivatives without using the concept of integral of the summable function.[10] This procedure generalises that of Sect. 7.1.

---

[10] The possibility of dispelling this notion is certain since we can always replace an integral by one of the sums which serve its definition chosen in such a way as to commit only a negligible error, then passing to the limit. However, this always involves the use of the integral, albeit in a concealed manner.

Let η be an arbitrarily chosen positive number. We will construct a function $\Phi(x)$, which differs from $F(x)$ only by η at most, from the point of view of differential; meaning that, which is such that, in any positive interval (α, β), we have

$$|[\Phi(\beta) - \Phi(\alpha)] - [F(\beta) - F(\alpha)]| < \eta(\beta - \alpha).$$

It is clear that, if we know how to construct this function $\Phi(x)$, for any η, we will deduce $F(x)$ by passing to the limit.[11] The construction of this function $\Phi(x)$ is based on the following remarks:

If we know a function $\Phi_1(x)$ for the positive interval (x_1, x_2), and a function $\Phi_2(x)$ for the positive interval (x_2, x_3), then the function $\Psi(x)$ equal to $\Phi_1(x)$ in (x_1, x_2) and given by

$$\Psi(x) = \Phi_2(x) - \Phi_2(x_2) + \Phi_1(x_2)$$

in (x_2, x_3), is a function $\Psi(x)$ for (x_1, x_3). Indeed, if we take an interval (α, β) located in (x_1, x_2) or (x_2, x_3), it is clear that we have

$$|[\Psi(\beta) - \Psi(\alpha)] - [F(\beta) - F(\alpha)]| < \eta(\beta - \alpha);$$

if we have

$$x_1 \leq \alpha < x_2 < \beta \leq x_3,$$

we have

$$\begin{aligned} |[\Psi(\beta) - \Psi(\alpha)] - [F(\beta) - F(\alpha)]| &< |[\Psi(\beta) - \Psi(x_2)] - [F(\beta) - F(x_2)]| \\ &+ |[\Psi(x_2) - \Psi(\alpha)] - [F(x_2) - F(\alpha)]| \\ &\leq \eta(\beta - x_2) + \eta(x_2 - \alpha). \end{aligned}$$

If we have an increasing (or decreasing) sequence of numbers x_i tending towards a limit X, then the repeated application of previous procedure yields, for every (x, X), a function Φ derived from functions $\Phi_i(x)$ relative to the intervals (x_i, x_{i+1}).

It is obviously sufficient to prove that the function Ψ resulting from the construction in the statement is continuous at point X. Now we have, by assuming, for example, an increasing sequence with $x_n < x < X$,

$$|[\Psi(x) - \Psi(x_n)] - [F(x) - F(x_n)]| < \eta(x - x_n) \leq \eta(X - x_n);$$

The method which is about to be presented, and that I introduced in an article of *Acta Mathematica (t.* 49*)*, deviates more significantly from the totalisation as just presented and, on the contrary, approaches the construction of the proof of the sufficient condition of theorem of M. Baire.

[11] The right-hand superior derivative number $\Lambda_d\Phi(x)$, for example, tends towards $f(x)$ when η tends towards zero; we will compare the construction of $\Lambda_d\Phi(x)$ and the function $\varphi(x)$ defined in Sect. 11.1.

therefore, in (x_n, X), the oscillation of $\Psi(x)$ is, at most, the oscillation of $F(x)$, augmented by $\eta(X - x_n)$. Therefore, the oscillation of $\Psi(x)$ in (x_n, X) tend towards zero with the length of (x_n, X).

In particular, it follows from this that *when we know how to determine a function* $\Psi(x)$, *in any interval* (α, β), *entirely interior to an interval* (a, b),

$$a < \alpha < \beta < b,$$

then, we know how to determine one for (a, b).

Finally, let us note that, *if we know a function* $\Phi(x)$ *for any interval that is contiguous to a closed set* $\mathcal{E}$,

if the corresponding series $\sum[\Phi(\beta) - \Phi(\alpha)]$ *is absolutely convergent,*

if the function $f(x)$ *is constant within* η *on* $\mathcal{E}$,

we obtain, at any point x *of* (a, b), *a function* $\Phi(x)$ *by using the expression*

$$\Phi(x) = \sum_a^x [\Phi(\beta) - \Phi(\alpha)] + f_0 m(\mathcal{E}_a^x);$$

where f_0 *is one of the values taken by* $f(x)$ *on* $\mathcal{E}$; *the indices* a *and* x *indicate that we will deal only with the subsets of* $\mathcal{E}$ *and with the intervals contiguous to* $\mathcal{E}$ *located in* (a, x).

It will be sufficient for us to show this proposition for $x = b$. To do this, let us cover (a, b) starting from a with a chain of intervals, some of which will be intervals contiguous to $\mathcal{E}$, and the other intervals of length λ at most, having as origin and extremity points $x, x + h$ from $\mathcal{E}$, and such that we have

$$m \leq r[F(x), x, x + h] \leq M,$$

where m and M are two numbers at a distance η at most, and all the values taken by $f(x)$ on $\mathcal{E}$ are contained between them. We will evaluate $F(b) - F(a)$ using this chain, but first, it must be noted that the series $\sum[F(\beta) - F(\alpha)]$, extended to the intervals (α, β), contiguous to $\mathcal{E}$, is absolutely convergent, because $F(\beta) - F(\alpha)$ differs from $\Phi(\beta) - \Phi(\alpha)$ by $\eta(\beta - \alpha)$ at most.

This being said, the intervals of the chain will give us as contribution to $F(b) - F(a)$ as follows:

On one hand, a part of the sum $\sum[F(\beta) - F(\alpha)]$, containing in particular all the terms arising from the intervals (α, β) with a length greater than λ, therefore tending towards $\sum[F(\beta) - F(\alpha)]$ when λ tends towards zero; and on the other hand, the measure of the lengths of intervals of the second kind, that is, a measure tending towards $m(\mathcal{E})$ when λ tends towards zero, multiplied by a number included between m and M.

The sum of these two contributions is

$$\sum[\Phi(\beta) - \Phi(\alpha)] + f_0 m(\mathcal{E})$$

within

$$\eta \sum (\beta - \alpha) + \eta m(\mathcal{E}) = \eta(b - a).$$

The statement is legitimate.

We have seen that, a closed set $\mathcal{E}$ being given, it is possible to determine an interval i containing points of $\mathcal{E}$ in its interior, in which $f(x)$ is within η on $\mathcal{E}$, and for which the sum $\sum_i [F(\beta) - F(\alpha)]$, extended to subsets of the intervals contiguous to $\mathcal{E}$, which are situated in i, is absolutely convergent. Then, if we know the functions $\Phi(x)$ for each interval contiguous to $\mathcal{E}$, the sum $\sum_i [\Phi(\beta) - \Phi(\alpha)$ is also absolutely convergent, and we are within the conditions of application of the previous theorem.

In other words, as soon as the first of the conditions in the previous statement is met, the points of $\mathcal{E}$ which are not interior to the intervals i, in which the three conditions of that statement are satisfied, necessarily form a closed set, which is non-dense everywhere on $\mathcal{E}$. As a result, *if we are able to determine functions Φ for all intervals contiguous to a closed set $\mathcal{E}$, we can determine functions Φ for all intervals contiguous to a closed set $\mathcal{H}$, formed from points of $\mathcal{E}$ and non-dense everywhere on $\mathcal{E}$.*

It is then clear that this statement allows us the construction of $\Phi(x)$ through transfinite induction:

The operation O_1 will be the one in which we take (a, b) as the set $\mathcal{E}$. The set $\mathcal{H}$ will be a set $\mathcal{E}_1$, which contains all the points at which the oscillation of $f(x)$ is greater than η, and some of those at which the oscillation is equal to η. O_1 will reveal $\Phi(x)$ in every interval not containing points of $\mathcal{E}_1$ in its interior.

If α is finite or transfinite of first kind, the operation $O_{\alpha-1}$ would have revealed Φ in any interval not containing points of a closed set $\mathcal{E}_{\alpha-1}$ in its interior. The operation O_α will be the one in which $\mathcal{E}_{\alpha-1}$ plays the role of $\mathcal{E}$, it will lead as set $\mathcal{H}$ to a set $\mathcal{E}_\alpha$ and reveal $\Phi(x)$ in any interval not containing any point of $\mathcal{E}_\alpha$ in its interior.

If α is of second kind, the points common to all $\mathcal{E}_\beta$ sets for $\beta < \alpha$, form a set $\mathcal{E}_\alpha$. The operations O_β have formed $\Phi(x)$ in every interval not containing any point of $\mathcal{E}_\alpha$, neither in its interior nor as origin or extremity.

The operation O_α will provide $\Phi(x)$ in every interval not containing the points of $\mathcal{E}_\alpha$ in its interior.

The function $\Phi(x)$ will be provided in entire interval (a, b) through this transfinite induction. To determine this function, which, in the preceding discussion, depends on arbitrary choices, it would be sufficient to fix these choices by laws. This would be easy, but it is completely futile to insist on it.

Then, since $\Phi(x)$ being defined for each number η, by letting η tend towards zero, we would have $F(x)$ as the limit of $\Phi(x)$.

Therefore, we have two sightly different transfinite procedures, both of which allow us to obtain the primitive function $F(x)$ from a given derivative $f(x)$; let us show by examples that all the planned steps of these transfinite processes are necessary.[12]

[12] A new operation O_α may be required due to the fact that one or another of the various conditions stated in our propositions are not satisfied in the entire (a, b). We may aim to show, through examples,

Let us denote by $\varphi(x)$, a function defined on (0, 1), which is continuous and differentiable, such that we have

$$|\varphi(x)| \leq x^2, \quad |\varphi(x)| \leq (1-x)^2,$$

which is of bounded variation in $(h, 1-h)$, and of unbounded variation in $(0, h)$, $(1-h, 1)$, however small and positive h may be, and whose derivative is continuous except for $x = 0$ and $x = 1$.

We could take for example,

$$\varphi(x) = x^2(1-x^2)\sin\frac{1}{x^2(1-x^2)};$$

the set of roots of $\varphi'(x) = 0$ then forms a set whose derivative reduces to 0 and 1. But we can also choose $\varphi(x)$ in such a way that, among the roots of $\varphi'(x) = 0$, all the points of any perfect or closed set can be found.

Let λ be any finite or transfinite number. We will define the functions $F_1(x), F_2(x), \ldots, F_\lambda(x)$ whose determination from their derivatives will respectively require the operations O_1; O_1 and O_2; O_1, O_2 and O_3; ...; $O_1, O_2, \ldots, O_\lambda$. And this applies to both of our two transfinite processes, whether one or the other.

For that let us arrange in a simply infinite sequence S, the finite and transfinite numbers up to λ, say $\lambda_1, \lambda_2, \ldots$. If β is a transfinite number of second kind, at most equal to λ, we will call the sequence determining β, the one obtained by crossing out, in S, first every number greater than or equal to β, and then any number in the resulting sequence that is preceded by a larger number.

Let us denote by E a closed set non-dense everywhere, chosen once and for all in (0, 1). $F_1(x)$ would be the function $\varphi(x)$.

$F_\alpha(x)$ would be, for an index α, finite or transfinite of first kind, the function zero on E and equal to $(m-l)F_{\alpha-1}\left(\frac{x-l}{m-l}\right)$, in the interval (l, m) contiguous to E. $F_\beta(x)$ would be, for transfinite β of second kind and determined by sequence $\beta_1, \beta_2, \ldots$, the function equal to

$$\frac{1}{2^p F_{\beta_p}}\left[2^p\left(x-\frac{1}{2^p}\right)\right]$$

for

$$\frac{1}{2^p} \leq x \leq \frac{1}{2^{p-1}}.$$

It is clear that the functions thus constructed are continuous and have derivatives which are formed from φ' just as the F are formed from φ. We immediately see that the determination of the function F_γ, whatever be its index γ, from its derivative, requires the operations

---

that each of these conditions, by itself, requires all the steps of transfinite. This is what M. Denjoy has done. The reader may refer to his work. Here, we will only show that the set of conditions in our propositions necessitates the use of transfinite in all its generality.

$O_1, O_2, \ldots$, up to O_γ, whether we are using one of the two procedures of investigation we have described.[13]

In reality, some of these operations are very simple; for example, for $F_2(x)$, the operation O_2 of totalisation is reduced to choosing the primitive functions in intervals contiguous to E, which are zero at the origin of these contiguous intervals. There is no need to integrate over E, since $F_2'(x)$ is zero on E. But it would be sufficient to add to each function $F_\alpha(x)$ a function $u_\alpha(x)$ of continuous derivative, in order to obtain a function $\mathfrak{F}_\alpha(x)$, whose determination from its derivative would require all the operations O_1, O_2, up to O_α, these operations comprising integrations over sets-integration that are exact in the first method, that is, the totalisation method, and approximate in the second.

11.3 The Primitive Functions of Everywhere Finite Derivative Numbers

When attempting to extend the methods from the previous section to the determination of the primitive function of a given derivative number, which is everywhere finite, we encounter difficulties from the start. These methods are indeed based on the fact that a derivative is a function of class one at most and, as a result, is point-wise discontinuous on any perfect set. However, we only know, Sect. 10.1, that the derivative numbers are of the second class at most, and from this we have only been able to deduce that they are B measurable.

If we examine the two methods from previous section a little more closely, we note that while the second one indeed makes use of the fact that the derivative is point-wise discontinuous on any perfect set, the first one, in reality, relies on a proposition that we can formulate thus: *An everywhere finite derivative function is point-wise unbounded on any perfect or closed set.*[14]

However, M. Denjoy obtained a proposition concerning any derivative number, which precisely replaces the previous one in the search for the primitive function, and it can be stated as follows:

When the right-hand superior derivative number $\Lambda_d F(x)$ of a continuous function $F(x)$ is everywhere finite or at least never equal to $+\infty$, it is point-wise unbounded from above on any closed set.

Or, in a more precise way: *when the right-hand superior derivative number $\Lambda_d F(x)$, of a continuous function $F(x)$ is not equal to $+\infty$ at any point of a closed set E, then there exists a positive number M and an interval I, containing the points of E in its interior, such*

[13] Indeed, it happens that, on the closed sets that the transfinite sequence of operations lead us to consider, there is an identity between the points of discontinuity of the derivative, its points of non summability and the points around which the derivative is unbounded.

[14] It is appropriate to make, in context of this statement and the following ones, an observation similar to the one formulated in note, Sect. 7.2; therefore, we must understand that any derivative is either bounded or point-wise unbounded on any closed set.

that for any interval (α, β) *whose origin is the point of* E *and of* I, *we have*

$$r[F(x), \alpha, \beta] < M.$$

It is clear that the second statement implies the first[15] one. It is also clear that, when we have demonstrated them for any perfect set, they will be proved by that very fact for any closed set E, since any isolated point of E is the point at which Λ_d is not equal to $+\infty$, and as a result, can be enclosed in an interval I satisfying the conditions of the second statement.

Proof of the second statement.[16]

Let us denote by $E_{n,p}$ the set of points x_0 of a perfect set E for which we have

$$r[F(x), x_0, x_0 + h] \leq n,$$

as soon as we have $h \geq \frac{1}{p}$, $E_{n,p}$ is a closed set, since $r[F(x), x_0, x_0 + h]$ is, for $h \geq \frac{1}{p}$, a continuous set function of two variables x_0 and h, on which this ratio depends.

The set E_n, consisting of the points common to all $E_{n,p}$, of the same index n, is therefore also a closed set. E is the sum of E_n since $\Lambda_d F(x)$ is assumed to be finite at any point of E or at least not equal to $+\infty$. Therefore, by reasoning as in Sect. 11.1, we see that there exists an interval I in which E is identical to one of the E_n, say E_m; then, for any point belonging simultaneously to E and I, we have

$$r[F(x), \alpha, \beta] \leq M,$$

for any $\beta > \alpha$; which proves the theorem.

We will also use the following property[17]:

If the ratio $r[F(x), x_0, x_0 + h]$ *relative to a continuous function* $F(x)$ *is uniformly bounded from above for all the points* x_0 *belonging to a closed set* $E, h > 0$,

the series $\sum[F(\beta) - F(\alpha)]$, *extended to the intervals contiguous to E is then convergent,*

the right-hand superior derivative number $\Lambda_d F(x)$ *is, on* E, *equal to* $-\infty$ *only at points of a set of measure zero,* E^{in}.

$\Lambda_d F(x)$ *has, in the set* $E - E^{in}$, *a definite finite integral, and we have*

$$F(b) - F(a) \leq \int_{E-E^{in}} \Lambda_d F(x)\, dx + \sum[F(\beta) - F(\alpha)].$$

When E^{in} *does not exist, then the equal sign is appropriate.*

Let us denote by E_l the set of points of E at which we have

[15] This second statement refines the first one, just as the one in Sect. 11.1 refines the necessary condition for a function to be of class one.

[16] The two previous statements replace the proposition that M. Denjoy calls the first fundamental theorem (descriptive) concerning derivative numbers.

[17] This property will replace here the second fundamental theorem (metric) concerning the derivative numbers, of M. Denjoy.

$$l\varepsilon \leq \Lambda_d F(x) < (l+1)\varepsilon,$$

ε being an arbitrarily chosen positive number.

Let $G(x)$ be the continuous function equal to $F(x)$ at points of E and linear in intervals contiguous to E. For x_0 to be the origin or the interior of such an interval (α, β), we have

$$\Lambda_d G(x) = r[F(x), \alpha, \beta] < k,$$

if k is the upper limit mentioned in the statement. At the points of E, which are not origins of intervals contiguous to E, we have moreover

$$\Lambda_d G(x) \leq \Lambda_d F(x) \leq k.$$

where $\Lambda_d G(x)$ is bounded from above in any (a, b), $G(x)$ is of bounded variation and we have, and where we denote (a, b) as H_0, and the set of points where $\Lambda_d G(x) = -\infty$ as E_G^{in}

$$\begin{aligned} G(b) - G(a) &= \int_{H_0 - E_G^{in}} \Lambda_d G(x)\, dx - N(E_G^{in}) \\ &= \int_{H_0 - E_G^{in}} \Lambda_d G(x)\, dx + \sum [G(\beta) - G(\alpha)] - N(E_G^{in}); \end{aligned}$$

the symbol $N(E_G^{in})$ has the meaning indicated in Sect. 10.1. At any point of E, we have $\Lambda_d F(x) \geq \Lambda_d G(x)$; inequality of which the second member is bounded except at the points of E_G^{in}. Therefore, the set E^{in} is contained in E_G^{in}, and as a result, is of measure zero.

Furthermore, the series $\sum l\varepsilon m(E_l)$ cannot be less than

$$\int_{E - E_G^{in}} [\Lambda_d G(x) - \varepsilon]\, dx;$$

since, almost everywhere on E, $\Lambda_d F(x)$ is at least equal to $\Lambda_d G(x)$. To be more precise, we can say that negative values of l provide sets E_l which contribute at least as much to $\sum l\varepsilon m(E_l)$ as they do to the previous integral, while the positive values of l give $l\varepsilon$ at most equal to λ, therefore $\int_{E-E^{in}} \Lambda_d F(x)\, dx$ exists and is at least equal to $\int_{E-E_G^{in}} \Lambda_d G(x)\, dx$. It is true that the two integrals are extended to the different intervals, $E - E^{in}$ and $E - E_G^{in}$, but they differ only by a set of measure zero.

Finally, if we note that

$$G(b) - G(a) = F(b) - F(a).$$
$$G(\beta) - G(\alpha) = F(\beta) - F(\alpha).$$

We have the inequality of the text

$$F(b) - F(a) \leq \int_{E - E^{in}} \Lambda_d F(x)\, dx + \sum [F(\beta) - F(\alpha)].$$

In the case where E^{in} does not exist, let us cover the entire interval (a, b) from a, using a chain of intervals chosen as it follows.[18]

Let us define, as it was done in Chap. 9, Sect. 10.1 and following, sets E_l and A_l by considering numbers ε, η, ζ. However, this time we are constructing E_l using only points of E. The intervals in the chain with origins at points in E are chosen to satisfy the three conditions indicated in Chap. 9. The interval with origin at a point x_0 in an interval (α, β) contiguous to E is the interval (x_0, β).

The intervals of the first kind yield a contribution to expression $F(b) - F(a)$, which tends towards $\int_E \Lambda_d F\, dx$ when ε, η, ζ tend towards zero.

An interval (x_0, β) of the second kind gives a contribution

$$F(\beta) - F(x_0) = F(\beta) - F(\alpha) - [F(x_0) - F(\alpha)];$$

$F(x_0) - F(\alpha)$ is at most $k(x_0 - \alpha)$, therefore the sum of the terms $F(x_0) - F(\alpha)$ is at most $k\eta$ (*see*, in Chap. 9 the significance of η) and as a result, have zero or negative limits when ε, η, ζ tend towards zero.

As a result we have

$$F(b) - F(a) \geq \int_E \Lambda_d F\, dx + \sum [F(\beta) - F(\alpha)].$$

Therefore, according to the statement, we indeed have

$$F(b) - F(a) = \int_E \Lambda_d F\, dx + \sum [F(\beta) - F(\alpha)].$$

From these two theorems it follows, in particular, that: *If E is a closed set, at the points of which $\Lambda_d F(x)$ is finite, there exists an intervals i containing points of E in its interior, in which $\Lambda_d F(x)$ is summable over E, and for which the series $\sum[F(\beta) - F(\alpha)]$ of the increments of F in intervals contiguous to E is absolutely convergent.*

Let us show, without assuming this time $\Lambda_d F(x)$ to be bounded from above in i, that the *increment of F in i is the sum of the integral of $\Lambda_d F(x)$ on a subset of E located in i, and of the series $\sum[F(\beta) - F(\alpha)]$ relative to E and i, which we would denote*

$$\mathfrak{A}_{F(x)}(i) = \int_{i,E} \Lambda_d F(x)\, dx + \sum_i^E [F(\beta) - F(\alpha)].$$

Indeed, if it was not so, the points of i which are not interior to intervals j in which we would have

[18] It is clear that we can replace in the above, the use of results borrowed from Chap. 9, with their demonstration using chains of intervals; thus, at the cost of some length, we will have a more homogeneous presentation.

$$\mathfrak{A}_{F(x)}(j) = \int_{j,E} \Lambda_d F(x)\,dx + \sum_{j}^{E}[F(\beta) - F(\alpha)],$$

would form a closed set H. An interval contiguous to H is the sum of a countably infinite number of j intervals without common interior points. For each of these j intervals we have the previous equality. The sum of all these equalities may be carried out since, by hypothesis, $\int_{i,E} |\Lambda_d F(x)|\,dx$ and $\sum_i^E |F(\beta) - F(\alpha)|$ exist. And this proves that any interval contiguous to H is itself a j interval.

Now, from the theorems of this section, it follows that we can find an interval λ containing points of H in its interior, for which $r[F(x), x_0, x_) + h]$ is bounded when x_0 is a point of H, so that we have

$$\begin{aligned}\mathfrak{A}_{F(x)}(\lambda) &= \int_{\lambda,H} \Lambda_d F(x)\,dx + \sum_{\lambda}^{H}[F(m_n) - F(l_r)]\\ &= \int_{\lambda,H} \Lambda_d F(x)\,dx + \sum_{\lambda} \mathfrak{A}_{F(x)}(\delta_n),\end{aligned}$$

$\delta_r = (m_r, l_r)$ denotes the intervals contiguous to H and located in λ. Each δ_r being a j interval, we have

$$\mathfrak{A}_{F(x)}(\delta_r) = \int_{\delta_r,E} \Lambda_d F(x)\,dx + \sum_{\delta_r}^{E}[F(\beta) - F(\alpha)].$$

Hence, it follows

$$\mathfrak{A}_{F(x)}(\lambda) = \int_{\lambda,E} \Lambda_d F(x)\,dx + \sum_{\lambda}^{E}[F(\beta) - F(\alpha)],$$

thus λ would be a j interval, which is contrary to the definition of H.

It is clear that our two new statements are entirely similar to those upon which we based the construction of a function from its derivative. *Therefore, we can similarly obtain the primitive function of a bounded derivative number by transfinite induction.*

Let us extend this result to the following case: *We know the finite value of $f(x)$, right-hand superior derivative number to the right $\Lambda F(x)$, except at for the points of a countable set D. We do not know if, at the points of D, $\Lambda F(x)$ is finite or infinite,* in which case $F(x)$ is still determined up to an additive constant, as shown in Sect. 6.3.

The theorem of Sect. 11.3 will be replaced by the following: *If, at points of a perfect set E, $F(x)$ is equal to $+\infty$ only at countably infinite number of points, say points P_i, then there exists a positive number M and an interval I containing in its interior, points of E, such that, for every interval (α, β), whose origin is a point from E and I, we have*

$$r[F(x), \alpha, \beta] < M.$$

Indeed, E is still the sum of a countably infinite number of closed sets, namely the E_n, and various points P_i, each considered as forming its own closed set. These sets cannot all be non-dense everywhere on E, Sect. 11.1; however, all the P_i are non-dense on E. Therefore, one of E_n is dense in E within an interval, meaning that it is identical to E within that interval, and the proof is complete as previously described.

Then, agreeing as always that we will leave aside the points P_i at which $\Lambda_d F(x)$ is infinite in the study of the summability and for the calculation of the integral, *given a closed set E, there exists an interval i:*

a. *Containing the points of E in its interior;*
b. *Such that* $\int_{i,E} \Lambda_d F(x)\,dx$ *and* $\sum\limits_{i}^{E}[F(\beta) - F(\alpha)]$ *exist;*
c. *And for which we have*

$$\mathfrak{A}_{F(x)}(i) = \int_{i,E} \Lambda_d F(x)\,dx + \sum_{i}^{E}[F(\beta) - F(\alpha)].$$

Indeed, either E contains an isolated point, in which case an interval i containing only that single point of E in its interior answers the question; or E is perfect, and nothing changes in the previous reasoning.

From this, it can be further deduced that whenever the conditions a. and b. are met, c. follows. The only difference is that the interval λ containing points of H, will be defined by the condition that $r[F(x), x_0, x_0 + h]$ is bounded for x_0, a point of i and H, only if H is perfect. If H is not perfect, an interval containing a single point of H will be taken for λ.

Thus, totalisation still allows the calculation of $F(x)$ when we know the finite value of one of its derivative numbers, except at the points of a countable set.

There is a distinction to be made between the results in this section and the preceding one. The fundamental operation of the totalisation consists of, given a closed set E and the function F to be obtained already known up to an additive constant in the intervals contiguous to E, determining an interval i containing points from E for which $\int_{i,E} f(x)\,dx$, $\sum_i^E [F(\beta) - F(\alpha)]$ exists. However, in the case where f is a derivative, we were able to restrict i *further* to ensure that f is bounded on the subset E_i of E located in i. If f is a superior derivative number, we were able to restrict i *further* to ensure that f is bounded above on E_i. Thus, there are methods of *special totalisation* alongside *the general totalisation*, which requires only the summability of f on E_i and the convergence—therefore, the absolute convergence—of $\sum_i^E [F(\beta) - F(\alpha)]$. We will study this general totalisation.

11.4 The Totalisation

Let us begin by clarifying the definition of a finite or transfinite sequence of operations that constitutes *totalisation*. This sequence of operations is performed based on a given function, denoted by $f(x)$, assumed to be finite everywhere[19] in (a, b). When the indicated operations are feasible, $f(x)$ is considered *totalisable*. Totalisation then associates a continuous function $F(x)$ with $f(x)$ in the entire interval (a, b), which is the indefinite total of $f(x)$. $F(x)$ is determined only up to an additive constant. If (l, m) is any interval contained in (a, b), the increment $F(m) - F(l)$ of $F(x)$ in (l, m) is the *definite total* of $f(x)$ in (l, m). Therefore, totalisation can be considered for two different purposes; for obtaining a function, which is *indefinite totalisation;* or obtaining a number, which is *definite totalisation.*[20]

The operations of the totalisation are constructed based on the following two:

A. *We assume the indefinite totals $F_k(x)$ are known in intervals (a_k, b_k) such that we have*

$$l < \cdots < a_2 < a_1 < b_1 < b_2 < \cdots < m,$$

the a_k tend towards l and the b_k tend towards m. Then, we form the continuous function $F(x)$, equal to $F_1(x)$ in (a_1, b_1), equal to

$$[F_k(x) - F_k(a_{k-1})] + [F_{k-1}(a_{k-1}) - F_{k-1}(a_{k-2})] + \cdots + F_1(a_1)$$

in (a_k, a_{k-1}) and equal to

$$[F_k(x) - F_k(b_{k-1})] + [F_{k-1}(b_{k-1}) - F_{k-1}(b_{k-2})] + \cdots + F_1(b_1)$$

in (b_{k-1}, b_k), that we take for indefinite total of $F(x)$ in (l, m).

B. *We have a closed set E contained in an interval (l, m); we assume that the totals of $f(x)$ are known in various intervals (α, β) contiguous to E, with respect to (l, m), and we suppose that the series $\sum[F(\beta) - F(\alpha)]$ provided by these totals is convergent and $f(x)$ is summable over E. Then we form the quantity*

$$\int_E f\,dx + \sum[F(\beta) - F(\alpha)],$$

that we take for the definite total of $f(x)$ in (l, m).

---

[19] Previously we had neglected the points where Λf was infinite, and therefore, took $\Lambda f(x) = 0$ at these points.

[20] This very coherent language introduced by M. Denjoy, clearly shows a complete parallelism between totalisation and integration. However, it should be noted that, in the case of integration, it is the concept of definite integral which is primordial, while, in the case of totalisation, it is the notion of indefinite total which is most important. Totalisation is more directly related to the integration of differential equations than to the calculations of quadratures: the definite total is equivalent to the definite integral of Duhamel (Chap. 6).

The conditions for $f(x)$ to be totalisable are the following:

1. *The operation A must lead to a function $F(x)$ which is continuous at l and m;*
2. *For any closed set $\mathcal{E}$, there must exist an interval (l, m) enclosing points of $\mathcal{E}$ such that, on the subset E of $\mathcal{E}$ located in (l, m), $f(x)$ must be summable, and the series $\sum[F(\beta) - F(\alpha)]$, extended to the intervals contiguous to E, is convergent.*[21]

These conditions being satisfied; taking $\mathcal{E}$ as the interval (a, b) itself, we see that the points of (a, b) where $f(x)$ is not summable, form a set E_1, non-dense everywhere in (a, b). Operations B, followed by operations A, reveal $F(x)$ in every interval contiguous to E_1; this set of operations constitutes the first operation O_1 of the totalisation.

If α is a finite or a transfinite number and if the operations with index lower than α have not revealed $F(x)$ in the entire (a, b), they have revealed $F(x)$ in every interval that does not contain any points of a certain closed set H. If α is not of second kind, this closed set H is called $E_{\alpha-1}$, and it has been provided by operation $O_{\alpha-1}$, which revealed $F(x)$ in every interval contiguous to $E_{\alpha-1}$. Then, by taking $E_{\alpha-1}$ as the set $\mathcal{E}$, it follows from the second condition satisfied by $f(x)$ that points of $E_{\alpha-1}$, which are not interior to intervals in which we can carry out the operation B, form a closed set E_α non-dense everywhere on $E_{\alpha-1}$. Operations B, followed by operations A, reveal $F(x)$ in every interval contiguous to E_α. The set of these operations constitute the operation O_α of the totalisation.

If α is of second kind, the set H consists of the points common to all the sets E_i with indices smaller than α. This set H is then denoted by E_α, and the operation O_α is reduced to the operations A necessary to construct $F(x)$ in intervals contiguous to E_α from the known functions $F(x)$ in intervals contiguous to E_i with indices lower than α.

The sets $E_1, E_2, \ldots$ are closed, and since each contain all that follow, we know they are finite or countably infinite in number. Therefore, the totalisation clearly determines $F(x)$ in the entire (a, b) after a finite or countably infinite number of operations O_i.

Thus, the stated conditions are sufficient for the operations of totalisation to be possible, but it is not so obvious whether they are necessary. The first condition is certainly necessary since it is indispensable for the continuity of $F(x)$ at l and m. However, it is sufficient for the second condition to be satisfied when taking $\mathcal{E}$ as the sets E_α to which the very operations of totalisation lead, along with the subsets of E_α located in various interval (l, m) contained in (a, b), for the operations of the totalisation to be legitimate. Now, we will see that as soon as the second condition is met for these special sets—meaning when totalisation is possible—the second conditions is fulfilled for any closed set $\mathcal{E}_0$.

Let O_N be last operation of totalisation. Therefore, the set E_n does not exist, while all E_i with indices lower than N do[22] exist.

[21] We could associate with this statement the following properties: the sum of several totalisable functions is totalisable; the total is the sum of the totals.

[22] The index N of this last operation is, therefore, never a transfinite number of second kind.

Let α be the smallest index such that $\mathcal{E}_0$ is not entirely in E_α. The number α exists and is less than N; moreover, α is not of second kind, as otherwise E_α would contain all the points belonging to E with indices lower than α, and would contain $\mathcal{E}_0$. Therefore there is a set $E_{\alpha-1}$ which contains $\mathcal{E}_0$. Since E_α does not contain all of $\mathcal{E}_0$, we can find an interval (l, m) *entirely interior* to an interval contiguous to E_α and in which there are points of $\mathcal{E}_0$, and therefore of $E_{\alpha-1}$.

Let $e_{\alpha-1}$ be the subset of $E_{\alpha-1}$ located in (l, m); also, let e^0 be the subset of $\mathcal{E}_0$ located in (l, m). Let us denote by (α_i, β_i) the various intervals contained in (l, m) and contiguous to e^0, and let e^i be the subset of $e_{\alpha-1}$ entirely interior to (α_i, β_i). We have

$$e_{\alpha-1} = e^0 + \sum e^i,$$

where the sets of the right-hand side are pairwise disjoint. Now, since in (l, m) there are no points of E_α, the operation O_α reveals $F(x)$ in (l, m), and therefore in any portion of (l, m). Therefore, f is summable over $e_{\alpha-1}$, and hence, also on e^0, and on the e^i, and we have

$$\int_{e_{\alpha-1}} f\,dx = \int_{e^0} f\,dx - \sum \int_{e^i} f\,dx.$$

The series $\sum\limits_{(l,m)}^{e_{\alpha-1}} [F(\beta) - F(\alpha)]$ of increments of $F(x)$, in intervals contiguous to $e_{\alpha-1}$ and contained in (l, m) is therefore convergent, and hence, so is the series $\sum\limits_{(\alpha_i,\beta_i)}^{e^i} [F(\beta) - F(\alpha)]$.

And we have

$$F(m) - F(l) = \int_{e_{\alpha-1}} f\,dx + \sum_{(l,m)}^{e_{\alpha-1}} [F(\beta) - F(\alpha)];$$

which can be written, since convergence of the series necessitates its absolute convergence,

$$\begin{aligned} F(m) - F(l) &= \int_{e^0} f\,dx + \sum \int_{e^i} f\,dx + \sum \left\{ \sum_{(\alpha_i,\beta_i)}^{e^i} [F(\beta) - F(\alpha)] \right\} \\ &= \int_{e^0} f\,dx + \sum \left\{ \int_{e^i} f\,dx + \sum_{(\alpha_i,\beta_i)}^{e^i} [F(\beta) - F(\alpha)] \right\}. \end{aligned}$$

Now, if $F(\beta_i) - F(\alpha_i)$ was provided by an operation prior to O_α, then e^i does not exist and the summation sign $\sum$ reduces to $F(\beta_i) - F(\alpha_i)$. If $F(\beta_i) - F(\alpha_i)$ was provided by O_α, then we have

$$F(\beta_i) - F(\alpha_i) = \int_{e^i} f\,dx + \sum_{(\alpha_i,\beta_i)}^{e^i} [F(\beta) - F(\alpha)].$$

Therefore, in all cases,

$$F(m) - F(l) = \int_{e^0} f\,dx + \sum_{(l,m)}^{e^0} [F(\beta_i) - F(\alpha_i)].$$

In other words, the operation B applies to e^0. Therefore, the second condition stated in Sect. 11.4 is satisfied by $\mathcal{E}_0$.[23]

Therefore, We have characterised the totalisable functions. Let us try to characterise the functions given by totalisation: the indefinite totals.[24]

A function $F(x)$ is an indefinite total if, and only if:

1. *It is continuous;*
2. *E being a closed set, the continuous function $G(x)$, equal to $F(x)$ at the points of E and linear in any interval contiguous to E, is absolutely continuous in an interval containing points of E in its interior.*

These conditions are necessary; the two properties, stated in Sect. 11.4, that a totalisable function $f(x)$ possesses, immediately imply the previous properties for the indefinite total $F(x)$ of $f(x)$.

These conditions are sufficient; because once they are met, we can, by reference to transfinite, construct a function $f(x)$ whose $F(x)$ is the indefinite total, operating as follows:

---

[23] The reader could also use this mode of reasoning to prove that if we have managed to attach a definite total to $f(x)$ taken in (a, b) using operations A and B, but subjecting the closed sets E appearing in the statement of B to the conditions indicated in this statement and, *in addition*, to the supplementary conditions (Sect. 11.3), the number obtained is the one that the general totalisation would have attached to $f(x)$ taken in (a, b).

In other words, there is never a disagreement between the numbers or functions provided by special totalisation and the general totalisation.

There is no need to elaborate on this reasoning here because in Sects. 11.2 and 11.3 of this chapter, we have justified the use of the general totalisation for the search for the primitive functions. However, we have noted that certain special totalisations are sufficient to obtain the result.

[24] In *Comptes rendus* in 1912, M. Denjoy solved the problem of primitive functions for derivatives using a special totalisation which he latter referred to as complete totalisation. In *Comptes rendus* in 1915, he implicitly introduced the general totalisation for solving the problem of the primitive functions for derivative numbers. It was not until his Memoirs, published from 1916 onwards, that he embarked on the study of all questions relating to totalisation. Before the publication of these Memoirs, various authors, building on the results already published by M. Denjoy, had also delved into the study of these questions.

It is worth mentioning that M. Lusin was the first to characterise the indefinite totals (*Comptes rendus*, in 1912,) and to study the differentiation of the indefinite totals (*Moscow thesis,* 1915). However, he did so only for the indefinite totals provided by complete totalisation.

M. Khintchine (*Comptes rendus* 1916) introduced a new method of differentiation to study the differentiation of the indefinite totals provided by general totalisation, which he called the asymptotic derivative and which we will refer to as the approximate derivative, following M. Denjoy.

Let us take the set E as the interval $H_0 = (a, b)$ itself, where $G(x)$ is identical to $F(x)$. Therefore, the set of points of non-absolute continuity of $F(x)$ is non-dense everywhere in (a, b). Let H_1 be this set. Outside of H_1, we take for $f(x)$ the derivative of $F(x)$ where it exists and, for example, zero where it does not exist.

Let us, then, take H_1 for set E. The corresponding function $G(x)$ will be called $G_1(x)$. The points where $G_1(x)$ is not absolutely continuous, form a closed set H_2 that is non-dense everywhere on H_1. At the points of $H_1 - H_2$, we take for $f(x)$ the derivative of $G_1(x)$ where it exists and zero at the other points.

In a general way, continuing like this, when we have defined $f(x)$, except at the points of a closed set H_α, we take H_α for the set E. The corresponding function $G(x)$, that we will call G_α, is absolutely continuous, except at the points of a closed set $H_{\alpha+1}$. At the points of $H_\alpha - H_{\alpha+1}$, we take for $f(x)$ the derivative of $G_\alpha(x)$ at the points where it exists and zero elsewhere.

To complete the definition of $f(x)$, it is sufficient to state that by H_β, where β is a transfinite number of second kind, we mean the set of points common to all the H with indices lower than β.

Our statement is thus legitimate. We will now show an equivalent statement due to M. Denjoy.[25]

When a continuous function $F(x)$ satisfies that the series

$$\sum [F(\beta) - F(\alpha)],$$

extended to the intervals contiguous to a closed set E, is convergent, we say that the increment[26] of $F(x)$ over E is *defined* and equal to

$$F(b) - F(a) - \sum [F(\beta) - F(\alpha)].$$

[25] The main interest of M. Denjoy's statement, lies in the fact that it is the culmination of a fine analysis of certain notions, such as that of a function of bounded variation, for example. The reader may refer to the Memoirs of M. Denjoy.

It is necessary to state that the sufficient condition of M. Denjoy appears to demand less, but potentially more immediately applicable than the previous one. However, in what follows, we have not needed the statement of M. Denjoy.

To better illustrate the difference between the two statements, let us note that the first one is equivalent to the following: *For a continuous function $F(x)$ to be an indefinite total, it is necessary and sufficient that, for any closed set E, there exists an interval (l, m) containing points of E in its interior, such that if we take a set of non-intersecting intervals I whose extremities belong to E and (l, m), the sum of the increments of $F(x)$ in the intervals in I, tends towards zero with the measure of I.*

If we compare this statement to the one given later in the text, we see that the latter requires, as a sufficient condition, that a certain property holds for every closed set of measure zero, whereas the statement of this note requires that the same property holds in a more uniform manner.

[26] M. Denjoy used the word variation in place of the increment.

When this is the case, the increment of $F(x)$ over E is the limit of the increment of $F(x)$ in a family of intervals I, consisting of a finite number of non-intersecting intervals whose origins and extremities are the points of E, which enclose E, and whose measure $m(I)$ tends towards that of E. Indeed, the complement of I is formed of intervals contiguous to E, and any interval contiguous to E ends up being a subset of this complement when $m(I)$ tends towards $m(E)$.[27] In particular, when the function $G(x)$ associated with E is absolutely continuous, the increment $\mathfrak{A}_{F(x)}(E)$ of $F(x)$ over E is defined and equal to the number $\mathfrak{A}_{G(x)}(E)$ which follows from the previous definitions, applicable only to the absolutely continuous functions, Sect. 9.2.[28]

A continuous function $F(x)$ is said to be *resolvable* if, for any closed set of measure zero E, there exists an interval (l, m) containing points of E such that the increment of $F(x)$ over the subset of E located in (l, m) is defined and equal to zero.

Now, here is M. Denjoy's statement: *For a function to have an indefinite total, it is necessary and sufficient that it be resolvable.*

This condition is necessary. Indeed, if $F(x)$ is an indefinite total and if E is a perfect set of measure zero, the function $G(x)$ constructed using $F(x)$ and E is absolutely continuous in an interval (l, m) containing points of E. Consequently, for the subset e of E located in (l, m), we have

$$\mathfrak{A}_{F(x)}(e) = \mathfrak{A}_{G(x)}(e) = 0.$$

This condition is sufficient. According to the first statement, if $F(x)$ is not an indefinite total, there exist a closed set E such that the corresponding function $G(x)$ is not absolutely continuous in any interval containing points of E in its interior. We will show that we can even replace this set E with another of measure zero. For that, we will distinguish three cases.

a. Let us suppose that for any interval (l, m) containing in its interior points of E, the series $\sum_{(l,m)}^{E} [F(\beta) - F(\alpha)]$, extended to the subsets of the intervals contiguous to E located in (l, m), contains an infinite number of positive terms with an infinite sum.

   Let us choose a finite number of terms of $\sum_{(l,m)}^{E} [F(\beta) - F(\alpha)]$ such that their sum exceeds 1; let $(\alpha_1, \beta_1), \ldots, (\alpha_i, \beta_i)$ be the corresponding intervals. Let us take the intervals $(a_1, \alpha_1), (\beta_1, b_1), \ldots, (a_i, \alpha_i), (\beta_i, b_i)$, that are not included in the chosen intervals (α, β), have their origins and extremities at point of E, and form a set I_1 of measure ε at most. This is possible because E is perfect since, in an interval containing only an isolated point of E, the function G would necessarily be absolutely continuous. Thus a point like α_1, which is a point of E and is the origin of the interval (α_1, β_1) contiguous

[27] The increment of $F(x)$ on E is therefore defined by one, and determined by one, of the methods that can be adopted when we take into account the B measurable nature of E. *See Sect.* 9.1.

[28] The same conclusion is true of all the continuous functions of bounded variation $F(x)$, if for them, the increment on a closed set is defined as described in Chap. 8.

to E, must necessarily have points of E to its left within as small a neighbourhood as desired. We can obviously ensure that the points of E of smallest and greatest abscissa are the origin and extremity points of intervals of I_1.

Similarly, let us choose a set I_2 formed of a finite number of intervals contained in I_1, bounded by points of E among which are found all the origins and extremities of I_1, and such that, among the intervals constituting the complement of I_2 are found the intervals (α, β) such that we have the inequalities

$$\sum_{(l,m)}^{E} [F(\beta) - F(\alpha)] > 0,$$

for every interval (l,m) of I_1. Finally, I_2 must have the measure smaller than $\frac{\varepsilon}{2}$.

It is clear that by continuing in this manner, we would construct the sets $I_1, I_2, \ldots$ of measures tending towards zero, each of which contains the following ones and, consequently, the points common simultaneously to all of these sets form a set $\mathcal{E}$ of measure zero, closed, and even perfect. Among the intervals contiguous to $\mathcal{E}$, there are, in particular, all the intervals (α_p, β_p) that are used to construct I_1, all those used for constructing $I_2, \ldots$. Therefore the series $\sum_{(\lambda,\mu)}^{E} [F(\beta) - F(\alpha)]$ extended to the intervals contiguous to $\mathcal{E}$ and located in an interval (λ, μ) containing in its interior points of $\mathcal{E}$, is divergent. The increment of $F(x)$ over the subset of $\mathcal{E}$ located in (λ, μ) is not defined; $F(x)$ is non resolvable.

b. When the series $\sum_{(l,m)}^{E} [F(\beta) - F(\alpha)]$ does not always contain an infinite number of positive terms with a sum $+\infty$, nor does it always have an infinite number of negative terms with a sum $-\infty$, for any interval (l, m) containing points of E, it means that there exists such an interval (l, m) for which the series $\sum_{(l,m)}^{E} [F(\beta) - F(\alpha)]$ is convergent. If we remove the points of E outside of (l, m), we can say that $\sum_{(a,b)}^{E} [F(\beta) - F(\alpha)]$ is convergent.

Let us suppose that $G(x)$ is not of bounded variation in any interval (l, m) containing points of E. Then, let (l, m) be an interval which contains points of E in its interior. In (l, m), $G(x)$ is of unbounded variation; therefore, we can find a finite number of intervals in (l, m) whose set I provides a value $\mathfrak{A}_{G(x)}(I)$ as large as we want. Moreover, we can remove any number of intervals from I not containing the points of E in their interior, as this modifies $\mathfrak{A}_{G(x)}(I)$ only by

$$\sum_{(l,m)}^{E} |F(\beta) - F(\alpha)|$$

at most. As a result, we can choose the intervals I with origins at the points of E, which are not origins of intervals contiguous to E, and with extremities at points of E, which are not extremities of intervals contiguous to E, assuming I to be of measure as small as we want. In such a set I, we have $\mathfrak{A}_{F(x)}(I) = \mathfrak{A}_{G(x)}(I)$.

With this in mind, let us choose in (a, b) a sequence of sets of intervals $I_1, I_2, \ldots$, satisfying the following conditions: each of them contains the following ones, the measures of I_p tend towards zero, each origin of an interval must have points of E to its right as close as we want, and each extremity must have points of E to its left, of which it is the limit point. The origins and extremities of the intervals of I_{k-1} are origin and extremities of intervals of I_k. Finally, the subset i_k of I_k contained in any of the intervals of I_{k-1}, must yield a value greater than k for $\mathfrak{A}_{G(x)}(i_k)$. It is clear that the complement J_k of I_k provides an increment $\mathfrak{A}_{F(x)}(J_k)$ which tends towards $-\infty$ since we have

$$F(b) - F(a) = \mathfrak{A}_{F(x)}(I_k) + \mathfrak{A}_{F(x)}(J_k);$$

and the same holds if we envisage only the subset of J_k located in an interval (l, m) containing points of E. In this case, the function $F(x)$ is not resolvable; because the increment of $F(x)$ is not, in fact, defined on any subset of the set $\mathcal{E}$, which is perfect and of measure zero, formed of the points common to all the I_k.

c. If none of the previously examined cases apply, then there exists an interval containing points of E in which $G(x)$ is of bounded variation. By removing points of E exterior to this interval, we can assume that $G(x)$ is of bounded variation in (a, b) and it admits E as the set of points of non-absolute continuity. We know that if we decompose $G(x)$ in its absolutely continuous kernel and the positive and negative variations of its function of singularities according to the formula

$$G(x) = AC(x) + P_s(x) - N_s(x),$$

at least one of the two numbers $P_s(b)$, $N_s(b)$ is non-zero; let us assume that $P_s(b)$ is positive.

We can find a set I_1 consisting of a finite number of intervals whose origins and extremities are points at which $P_s(x)$ is increasing respectively to the right and left, and therefore are points of E. Additionally, we can assume that the measure of I_1 is less than ε, and that we have the two inequalities

$$\mathfrak{A}_{P_s(x)}(I_1) > P_s(b)(1 - \varepsilon_1)$$
$$\mathfrak{A}_{N_s(x)}(I_1) < P_s(b)\varepsilon_1.$$

In I_k we can find a set I_{k+1} of measure less than $\frac{\varepsilon}{2^k}$, formed of a finite number of intervals whose extremities and origins satisfy the same conditions as above. The family of these extremities and origins for I_{k+1} includes those related to I_k, and for any interval (λ, μ) in I_k, the subset i_{k+1} of I_{k+1} contained in it satisfies the inequalities

$$\mathfrak{A}_{P_s(x)}(i_{k+1}) > [P_s(\mu) - P_s(\lambda)](1 - \varepsilon_{k+1}),$$
$$\mathfrak{A}_{N_s(x)}(i_{k+1}) < [P_s(\mu) - P_s(\lambda)]\varepsilon_{k+1},$$

where ε_k are the numbers such that the product $\Pi(1 - 2\varepsilon_{k+1})$ is convergent and of value $\frac{1}{2}$.

It is clear that the set $\mathcal{E}$ formed from points common to all these I_k is perfect and of measure zero, the increment in $F(x)$, that is, $G(x)$, on $\mathcal{E}$ is defined and its value is the limits of $\mathfrak{A}_{G(x)}(I_k)$. Now, we have

$$\mathfrak{A}_{G(x)}(I_k) = \mathfrak{A}_{AC(x)}(I_k) + \mathfrak{A}_{P_s(x)}(I_k) - \mathfrak{A}_{N_s(x)}(I_k),$$

and as the first of these numbers tend towards zero with the measure of I_k, we have

$$\mathfrak{A}_{F(x)}(\mathcal{E}) = \mathfrak{A}_{G(x)}(\mathcal{E}) = \lim \mathfrak{A}_{G(x)}(I_k) \geq P_s(b)\Pi(1 - 2\varepsilon_k) = \frac{P_s(b)}{2} > 0.$$

Thus, the increment in $F(x)$ on $\mathcal{E}$ is not zero, and a similar conclusion holds for any subset of $\mathcal{E}$ contained in an interval containing points of $\mathcal{E}$ in its interior. $F(x)$ is non-resolvable, and M. Denjoy's criterion is entirely legitimate.

In our first criterion, we saw how, given an indefinite total $F(x)$, we could determine a function $f(x)$ of which $F(x)$ is the indefinite total. We did this using some sets $H_1, H_2, \ldots,$ defined by considering points of non-absolute continuity of $F(x)$ and certain functions $G(x)$. Let us show that these sets H can be defined just as well from any function $\varphi(x)$ that admits $F(x)$ as its indefinite total.

Indeed, the first operation of the totalisation of $\varphi(x)$ reveals the exceptional set E_1, formed of points at which $\varphi(x)$ is not summable. We claim that E is identical to H_1. First of all, since the indefinite total of $\varphi(x)$ is absolutely continuous at every point not belonging to E_1, H_1 is contained in E_1. If it was not identical to it, there would exist an interval (α, β) contiguous to H_1 and containing several points of E_1, therefore also an interval (α_1, β_1) contiguous to E_1. In any interval *entirely interior to* (α_1, β_1), $\varphi(x)$ is summable, by hypothesis, and we have almost everywhere

$$\varphi(x) = F'(x).$$

Therefore, almost everywhere in *entire* (α_1, β_1) we have $\varphi(x) = F'(x)$, and $\varphi(x)$ is summable over the *entire* interval (α_1, β_1)[29] since $F(x)$ is absolutely continuous in the interior of (α, β) therefore in (α_1, β_1).

However, there exists an interval (l, m), entirely interior to (α, β), containing points of E_1 and no point of the exceptional set E_2, formed by the second operation of the totalisation of φ. Let us even suppose that (l, m) contains no points of E_2, either as origins or extremities. Then $\varphi(x)$ is summable over the subset e_1 of E_1, located in (l, m). However, (l, m) can be

[29] It is sometimes necessary to reduce (α_1, β_1), on the side of α_1 if α_1 was at α, and on the side of β_1 if β_1 was at β, so that this remains true in one or the other of these hypothesis.

expressed as the sum of e_1 and of intervals, or subsets of the intervals contiguous to E_1; let us write

$$(l, m) = e_1 + i_1 + i_2 + \cdots$$

We know that, in i_1, φ is summable and we have, almost everywhere $\varphi(x) = F'(x)$. Therefore, the series

$$\int_{e_1} \varphi \, dx + \int_{i_1} \varphi \, dx + \int_{i_2} \varphi \, dx + \cdots$$

is convergent, since we have

$$\int_{i_1} |\varphi| + \int_{i_2} |\varphi| \, dx + \cdots \leq \int_l^m |F'(x)| \, dx.$$

And this shows that $\varphi(x)$ will be summable over entire (l, m), which implies a contradiction.

Thus E_1 and H_1 are identical.

But the exceptional set E_2 provided by the second operation of totalisation, which yields $F(x)$ from $\varphi(x)$ is the first exceptional set one would encounter when searching, by totalisation, of the function $G(x)$ constructed from E_1, Meanwhile, the set H_2 related to $F(x)$ serves as counterpart of H_1 for $G(x)$, meaning it is the set of points of non-absolute continuity of $G(x)$. Therefore, It is clear that E_2 and H_2 are identical; the same holds for E_3 and H_3, E_4 and H_4, and so on. With E_ω being the set of points common to all E_n of indices lower than ω, while H_ω is the set of points common to all H_n of indices lower than ω, E_ω and H_ω are also identical. Continuing this, we see that there is an identity between the E and H of the same index.

At the same time it can be observed that we have: $\varphi(x) = F'(x)$ almost everywhere at any point exterior to H_1; $\varphi(x) = G_1'(x)$ almost everywhere at any point of $H_1 - H_2$, where $G_1(x)$ is the function G constructed using H_1; $\varphi(x) = G_2'(x)$ almost everywhere at any point of $H_2 - H_3$, where $G_2(x)$ is constructed using H_2, and so on. Now, $H_0 = (a, b)$ can be expressed as the sum of the sets $H_{\alpha-1} - H_\alpha$ in a finite or countably infinite number; therefore, finally, $\varphi(x)$ *is determined almost everywhere by its indefinite total* $F(x)$.[30]

And this determination is always achieved through derivations. Let us examine these derivations more closely. To differentiate $G_1(x)$ at a point x_0 in E_1 amounts to differentiating $F(x)$ at this point, but by taking into account only points of E_1; in other words, we examine the ratio $r[F(x), x_0, x_0 + h]$ in which $x_0 + h$ is also a point of E_1. This is what we call the derivative of $G_1(x)$ over E_1. Therefore, *for any closed set E, there exists an interval (l, m) containing points of E, and within (l, m), $f(x)$ is almost everywhere on E, the derivative (taken over E) of its indefinite total.*

30 We will reach this result more quickly by proving that a function, non-zero almost everywhere has an indefinite total, not identically equal to zero. Throughout this chapter, I have preferred, despite the consequent lengths, the analytic examination of the transfinite operational process of totalisation over the synthetic reasoning, which is though a more rapid reasoning, but in my opinion, would lead to a less profound understanding.

In reality, this statement is proved by the preceding text only if E is one of the sets $H_1, H_2, \ldots$. But if α is the smallest index such that E is not completely in H_α, it means there is an interval (λ, μ) containing a subset e of E which is contained in $H_{\alpha-1} - H_\alpha$. Consequently, almost everywhere on e, $f(x)$ is the derivative of $G_\alpha(x)$, and therefore, the derivative over E of $F(x)$.

This statement can be advantageously replaced by the following, which is due to M. Khintchine, and based on the concept of approximate derivative.

A continuous function $F(x)$ is said to possess, at x_0, an *approximate derivative* equal to $f(x_0)$, if $f(x_0)$ is the derivative of $F(x)$ over a set of density 1 at point x_0.

It is clear that $F(x)$ cannot have, at x_0, two different approximative derivatives $f(x_0)$ and $g(x_0)$, because they would be the derivatives of $F(x)$ over the two sets $E(f)$ and $E(g)$, both having density of 1 at x_0. As a result, $E(f)$ and $E(g)$ would have common points in as small a neighbourhood of x_0 as we want, and as a result, $f(x_0)$ and $g(x_0)$ would be equal.

From this, it follows, in particular, that the approximate derivative coincides with the ordinary derivative when the later exists. Moreover, if $F'_a(x_0)$ is the approximate derivative of $F(x)$ at x_0, and if λ_g, Λ_g, λ_d, Λ_d are the four derivative numbers of $F(x)$ at x_0, we have

$$\lambda_g < F'_a(x_0) < \Lambda_g, \quad \lambda_d < F'_a(x_0) < \Lambda_d,$$

because $F'_a(x_0)$ is the limit of the ratios $r[F(x), x_0, x_0 + h]$ for which h is positive, and the limit of ratios for which h is negative.

A totalisable function is almost everywhere the approximate derivative of its indefinite total.

Indeed, almost everywhere at the points of $H_{\alpha-1} - H_\alpha$, the totalisable function $f(x)$ is the derivative, taken over $H_{\alpha-1} - H_\alpha$, of its total $F(x)$. However, almost everywhere on $H_{\alpha-1} - H_\alpha$, the density of $H_{\alpha-1} - H_\alpha$ is equal to 1. Therefore, almost everywhere on $H_{\alpha-1} - H_\alpha$, $f(x)$ is the approximate derivative of $F(x)$, and the theorem is proved.

This theorem generalises the one related to summable functions—stating that a summable function is almost everywhere the derivative of its indefinite integral—, but using a generalisation of the concept of an ordinary derivative. For understand the concept itself, let us compare a totalised function with the derivative numbers of its total, which the reasonings in Sect. 11.3 and the following would allow us to do.

First, let us note that to show that a property holds, at most, only at the points of a set of non-zero measure, it is sufficient to show that it never takes place at all points of a *closed* set of measure non-zero.[31] Indeed, if it were to take place at all the points of a set $\mathcal{E}$ of non-zero measure,[32] it would be sufficient to enclose the complement of $\mathcal{E}$ in the interior of intervals whose measure is less than that of the entire interval (a, b) considered, so that the set E,

---

[31] This remark is constantly used by M. Denjoy.

[32] Therefore, here I suppose that we are dealing with properties for which we know that the set $\mathcal{E}$ of points where the property holds is necessarily measurable.

formed of the points not interior to these intervals, becomes a closed set of measure non zero in which the property holds at all points.

With this in mind, let us refer to two statements at the beginning of the previous Sect. 11.3.

We see that *the right-hand superior derivative number* $\Lambda_d F(x)$ *of a continuous function* $F(x)$*, is equal to* $-\infty$ *only, at most, at the points of a set of measure zero.* If it was not so, we could find a closed set E of non-zero measure at all points of which $\Lambda_d F(x) = -\infty$; according to the second statement of Sect. 11.3. We could even suppose that $r[f(x), x_0, x_0 + h]$ is bounded from above for all points x_0 of E, in which case the theorem of Sect. 11.3 would apply. However, it asserts that on E, it asserts that the set of points where $\Lambda_d F(x) = -\infty$ is of measure zero.

More generally, *the set of points where one of the* Λ *is equal to* $-\infty$*, or one of the* λ *is* $+\infty$ *is of measure zero.*

Therefore, in what follows, we can argue on the closed sets of non-zero measure at points of which we have neither $\Lambda = -\infty$ nor $\lambda = +\infty$.

In the set of points where a continuous function $F(x)$ *has a finite right-hand superior derivative number,* $F(x)$ *almost everywhere has an approximate derivative equal to this number.*

Let us suppose, indeed, that there exist a closed set of non-zero measure E at the points of which $\Lambda_d F(x)$ is finite without being the approximate derivative of $F(x)$, and let us choose, Sect. 11.3, an interval where the conditions for application of the theorem in Sect. 11.3 are met. Then, according to this theorem, the function $G(x)$ constructed from E has a derivative equal to $\Lambda_d F(x)$ almost everywhere on E. That is to say, almost everywhere on E, $F(x)$ has a derivative taken over E, equal to $\Lambda_d F(x)$ and consequently $F(x)$ has, almost everywhere on E, an approximate derivative equal to $\Lambda_d F(x)$. Hence it follows, in particular, that a *totalisable function* $f(x)$ *is almost everywhere equal to the right-hand superior derivative number of its indefinite total, in the set of points where this derivative number is finite.*

Naturally, we have the same statements for the four derivative numbers.

However, as we will see, there exist the indefinite totals whose derivative numbers are infinite in sets of non-zero measure.

Let, indeed, P be a closed set, non-dense everywhere and of non-zero measure; let us enumerate the intervals contiguous to P, in any manner.

Successively place these intervals $u_1, u_2, \ldots$ on (a, b). Each interval u_i will be thus placed in an interval u^i, which is one of those obtained by subtracting $u_1, u_2, \ldots, u_{i-1}$ from (a, b). Let ρ_i be the length of u^i, and the number ρ_i will be infinitesimally small with $\frac{1}{i}$.

Let us take $F(x)$ zero at the points of P and equal in each u_i to a continuous function whose derivative is continuous, vanishes at the extremities of u_i and is of absolute maximum value equal to $\sqrt{\rho_i}$.

It is clear that $F(x)$ is continuous since $\sqrt{\rho_i}$ tends towards zero with $\frac{1}{i}$; $F(x)$ is the indefinite total of the zero function over P and is equal to $F'(x)$ outside of P.

Let ξ be a point of P which is neither origin nor the extremity of an interval contiguous to P. ξ is found in a sequence of intervals $u^{i_1}, u^{i_2}, \ldots$. As a point of u^{i_j}, we can affirm that, for suitably chosen x in u_{i_j}, we have

$$|r[f(x), \xi, x]| = \frac{\sqrt{\rho_{i_j}}}{|\xi - x|} > \frac{1}{\sqrt{\rho_{i_j}}}.$$

Therefore, at ξ at least one of the four derivative numbers of $F(x)$ is infinite. Since the set of ξ has the same measure as P, which is of non-zero measure, it follows that one of the four derivative numbers of $F(x)$ is infinite on a set of points of non-zero measure. However, $F(x)$ has zero derivative with respect to E at every point of E, and thus its approximate derivative is zero almost everywhere on E.

Although the preceding discussion is sufficient for our conclusion, let us elaborate on this example for further implications.[33] Consider the following two hypothesis: within u_i, we take a positive or zero function and with a maximum of $\sqrt{\rho_i}$, this would give us $F_1(x)$; we take a function in u_i, with a maximum of $\sqrt{\rho_i}$ and minimum of $-\sqrt{\rho_i}$, this would give us $F_2(x)$.

Let us recall the previously mentioned ratio $r[f(x), \xi, x]$ for ξ in u^{i_j} and x in u_{i_j}. Clearly, for some values of j, the difference $\xi - x$ is negative; otherwise, none of the u_i would lie between the origin of u^{i_1} and ξ, which is impossible. Therefore, we can conclude that we have

$$r[F(x), \xi, x] < -\frac{1}{\sqrt{\rho^{i_j}}}$$

for some values of j, whether it concerns $F_1(x)$ or $F_2(x)$ and as a result, we have

$$\lambda_g F_1(\xi) = \lambda_g F_2(\xi) = -\infty.$$

Similarly, we see that we have

$$\Lambda_g F_2(\xi) = +\infty,$$

then

$$\Lambda_d F_1(\xi) = \Lambda_d F_2(\xi) = +\infty,$$

Additionally, we have

$$\lambda_d F_2(\xi) = -\infty,$$
$$\Lambda_g F_1(x) = \lambda_d F_1(x) = 0.$$

Furthermore, let us assume that in each u_i, $F(x)$ is chosen in a way that, in u_i, in the neighbourhood of the extremities of u_i, the curve $Y = F(x)$ is similar to the one which represents, in the neighbourhood of $x = 0$ the function which, in $\left[\frac{1}{(k+1)\pi}, \frac{1}{k\pi}\right]$ is equal to $\frac{1}{k}\left(\sin\frac{1}{x}\right)^{2k}$ if it is $F_1(x)$, or equal to $\frac{1}{k}\left(\sin\frac{1}{x}\right)^{k}$ if it is $F_2(x)$. Then, the above equalities

[33] This example, in its various forms, is, with minor modifications, extracted from the Memoirs of M. Denjoy.

determine the values of four derivative numbers of $F(x)$ not only at the points ξ, but at every point of P. In addition, if P is perfect and the density of P is equal to 1 at every point ξ of P, then $F(x)$ has an approximate derivative equal to zero at every point of P.

Therefore, two indefinite totals have been constructed which have a finite approximative derivative at every point. However, at every point of the perfect set of non-zero measure P, we have

$$\Lambda_g F_1(x) = \lambda_d F_1(x) = 0, \qquad \lambda_g F_1(x) = -\infty, \qquad \Lambda_d F_1(x) = +\infty,$$
$$\Lambda_g F_2(x) = \Lambda_d F_2(x) = +\infty, \qquad \lambda_g F_2(x) = \lambda_d F_2(x) = -\infty.$$

Thus, we cannot establish the relationships between a totalised function and its indefinite total by solely employing the ordinary differentiation taken over an interval; it is essential to resort to a generalisation of the derivative, namely, differentiation over a set or an approximate derivative. Consequently, the generalisation of this statement: every absolutely continuous function has a derivative almost everywhere, which is stated as: *any resolvable function has an approximate derivative almost everywhere* cannot be replaced by a proposition concerning the existence of the ordinary derivative.[34]

Nevertheless, we can obtain information about the ordinary derivative by simultaneously applying the obtained theorems to several derivative numbers, just as we did in Chap. 9 with the theorems related to integration.

Two derivative numbers of the same function, $\Lambda_d F(x)$ and $\lambda_g F(x)$ for example, are equal almost everywhere in the set of points where they are both simultaneously finite. Indeed, we know, on one hand that $F(x)$ admits almost everywhere in this set, an approximative derivative equal to $\Lambda_d F(x)$, and, on the other hand, this approximate derivative is almost everywhere equal to $\lambda_g F(x)$.

In particular, *a continuous function has a derivative almost everywhere in the set of points where its four derivative numbers are finite.*[35]

A continuous function $F(x)$ has its derivative number $\lambda_g F(x)$ finite almost everywhere in the set of points where $\Lambda_d F(x)$ is finite.

Otherwise, there would exist a closed set E of non-zero measure, where $\lambda_g F(x)$ is equal to $-\infty$, and yet, for points x_0 in this set, the ratio $r[F(x), x_0, x_0 + h]$ would always be less than a fixed number k. Let us then cover (a, b) from b, with a chain of intervals chosen as follows: if η is a point of E, we associate with it an interval (ξ, η) in which we have

$$r[F(x), \xi, \eta] < -M,$$

where M is a very large positive number, chosen arbitrarily.

[34] There exists a mode of totalisation called by M. Denjoy, the complete totalisation, which has been studied by M. Denjoy and M. Lusin. With this mode, on the contrary, all theorems considered in the text can be extended to indefinite totals without requiring any differentiation beyond the ordinary one.

[35] P. MONTEL, *Comptes Rendus,* 1912.

If η is a point interior to an interval (α, β) which is contiguous to E, we would associate to η, the interval (α, η).

Let us use this chain to evaluate $F(b) - F(a)$; the contribution of intervals of the first kind would be less than $-Mm(E)$; while the contribution of intervals of the second kind would be less than their length multiplied by k, that is, less than $k(b-a)$.

This being said

$$F(b) - F(a) < k(b-a) - Mm(E).$$

However, this is impossible since the right-hand side is as small as we want; the proposition is therefore proved. We have, of course, a similar statement relative to Λ_g and Λ_d.

By combining the various results obtained, we have the remarkable theorem due to M. Denjoy:

Except at most at the points of a set of measure zero, the four derivative numbers of a continuous function,[36] *$F(x)$, exhibits one of the following arrangements:*

1. *The four derivative numbers are equal, meaning there is an ordinary derivative;*
2. $$\begin{aligned}\Lambda_g &= +\infty,\\ \lambda_g &= \Lambda_d = \text{finite value},\\ \lambda_d &= -\infty\end{aligned}$$

   or

   $$\begin{aligned}\Lambda_g &= -\infty,\\ \Lambda_g &= \lambda_d = \text{finite value},\\ \Lambda_d &= +\infty.\end{aligned}$$
3. $$\Lambda_g = \Lambda_d = +\infty, \quad \lambda_g = \lambda_d = -\infty.$$

The two functions $F_1(x)$ and $F_2(x)$, constructed earlier, also show that cases 2 and 3 indeed occur explicitly in sets of points of non-zero measure.

From this theorem, M. Denjoy deduced that *a continuous function which does not have, at any point, all four of its derivative numbers infinite simultaneously is an indefinite total.*

Let us show that any continuous function $F(x)$ that is not an indefinite total has all four of its derivative numbers infinite at some points. For such a function, there exists a perfect set E

[36] Ms Chisholm Young extended this result to any finite measurable function defined on a measurable set (*Comptes rendus,* 1916).

such that the corresponding function $G(x)$ is not absolutely continuous in any interval (l, m) containing points of E in its interior. This means that, in (l, m), the limit of the absolute value of the increment $\mathfrak{A}_{G(x)}(I)$ of $G(x)$ over a set I of non-intersecting intervals does not tend towards zero when $m(I)$ tends towards zero. Since $G(x)$ is linear in the intervals contiguous to E, we may assume that the origins and extremities of the intervals constituting I are points of E, without affecting the greatest limit of $|\mathfrak{A}_{G(x)}(I)|$.

Two cases must be distinguished:

a. In every interval (l, m), as $m(I)$ tends towards zero, the greatest limit of $A_{G(x)}(I)$ is positive and the smallest is negative;
b. Or else, there exists an interval (l,m) for which one of these limits is zero.

a. In this case, in every (l, m) we can find intervals bounded by points x_0 and $x_0 + h$, $h > 0$, of E, for which $r[F(x), x_0, x_0 + h]$ is as large or as small as we want. Because, for example, if this ratio could not exceed k, then $A_{G(x)}(I)$ would be at most $km(I)$.
   $r[F(x), x_0, x_0 + h]$ is therefore, unbounded from above everywhere on E, therefore, there are points on E where $\Lambda_d F(x) = +\infty$.
   $r[F(x), x_0, x_0 + h]$ is also everywhere unbounded from above, there are points on E where $\Lambda_g F(x) = +\infty$, and since these ratios are also unbounded from below, there are points where $\lambda_d = -\infty$ and points where $\lambda_g = -\infty$.
   Let us specify this result by referring to proof of our first theorem on the derivative numbers (Sect. 11.3). We had noted then that the set E_n of points x_0 of E where

$$r[F(x), x_0, x_0 + h] \leq n,$$

   for any positive h, was a closed set. In the current assumption, this is a closed set which is non-dense on E, since in every interval containing points of E, at some of these points, r exceeds n for some positive values of h.
   The set $\mathcal{E}_n$ of points where we have one of the inequalities

$$r[F(x), x_0, x_0 + h] \leq n, \quad r[F(x), x_0, x_0 + h] \geq -n,$$

   either for any positive h, or for any negative h, the sum of the four sets similar to E_n, is closed and non-dense on E.
   If we note that the sum of $\mathcal{E}_n$ is the set of points of E where one of the four derivative numbers is finite, the proof is easily completed. E cannot be the sum of $\mathcal{E}_n$ nowhere dense on E (Sect. 11.1), therefore there are points of E that do not belong to any $\mathcal{E}_n$; at these points, four derivative numbers of $F(x)$ are infinite.
b. Let us admit that, in an interval, say the interval (a, b) itself, the smallest limit of $A_{G(x)}(I)$, for $m(I)$ tending towards zero, is zero. It is clear that the total negative variation of $G(x)$ is bounded, therefore $G(x)$ is of bounded variation. Moreover, if we decompose $G(x)$ in its absolutely continuous kernel and its function of singularities, the later reduces to

its own positive variation

$$G(x) = AC(x) + P_s(x);$$

we have

$$P_s'(x) = +\infty$$

at points of a set E^{ip}, of measure zero, which is the set of singularities of $P_s(x)$ and of $G(x)$.

The change of variable (Sect. 9.3)

$$t = x + P_s(x),$$

transforms $P_s(x)$, $AC(x)$, $G(x)$ into the absolutely continuous functions $p_s(t)$, $ac(t)$, $g(t)$ and E^{ip} into e^{ip}, a set I of intervals enclosing E^{ip} into a set i of intervals enclosing e^{ip}, the total variations of $P_s(x)$ and $AC(x)$, in I, into the total variations of $p_s(t)$ and $ac(t)$, in i.

However, for $P_s(x)$, the total variation in I is equal to $P_s(b)$, that is, the total variation in entire (a, b) since E^{ip} is the set of singularities of $P_s(x)$. Therefore, the total variation of $p_s(t)$ in e^{ip} is equal to its total variation in (a, b) and, consequently, is non-zero. This shows that e^{ip} is of non-zero measure.

For $AC(x)$, the total variation in I tends towards zero with $m(I)$ since $AC(x)$ is absolutely continuous. Therefore, the total variation of $ac(t)$ is zero in e^{ip} and consequently, we have, almost everywhere on e^{ip},

$$ac'(t) = 0.$$

Now, at every point of E^{ip}, and therefore of e^{ip}, we have

$$P_s'(x) = +\infty;$$

from where

$$t'(x) = 1 + P_s'(x) = +\infty, \quad x'(t) = 0;$$

then

$$1 = x'(t) + p_s'(t), \quad p_s'(t) = 1.$$

Therefore, the derivative of $g(t) = ac(t) + p_s(t)$ exists and is equal to 1 at all points of a set e_0 contained in e^{ip} and having the same measure as e^{ip}.

By restricting e_0 without changing its measure, we can even assume that at the points of e_0, conditions are satisfied that are fulfilled almost everywhere in the transformed set e of E, and thus almost everywhere in e^{ip}. We thus admit that the points of e_0 are neither origin, nor extremities of the intervals contiguous to e and at these points, the four derivative numbers of function $f(t)$ which is transformed from $F(x)$, $f(t) \equiv F(x)$, exhibit one of the four associations indicated by the previous theorem.

From the first of these assumptions, as we have $f(t) = g(t)$ at the points of e, it follows that to any value t_0 belonging to e, we can associate two sequences of values of t tending towards t_0, one by values greater than t_0, and the other by values smaller than t_0, such that $r[f(t), t_0, t]$ has a limit equal to 1. It is sufficient, in fact, to take these sequences of values of t belonging to e; then r tends towards the derivative of $f(t)$, taken over e, which is $g'(t) = 1$.

From the second assumption, then follows:

α. that if we are in the case where the four derivative numbers are equal, we have $f'(t) = 1$;
β. That if we are in the case where $\Lambda_g = +\infty, \quad \lambda_d = -\infty, \quad \lambda_g = \Lambda_d =$ finite value, this finite value is equal to 1, since 1 must be common to two intervals (λ_g, Λ_g), (λ_d, Λ_d);
γ. That if we are in the case where $\lambda_g = -\infty, \quad \Lambda_d = +\infty, \quad \Lambda_g = \lambda_d =$ finite value, this finite value is 1;
δ. Finally, we can have $\Lambda_d = \Lambda_g = +\infty, \quad \lambda_g = \lambda_d = -\infty$.

Now, the formula

$$\frac{\Delta F(x)}{\Delta x} = \frac{\Delta f(x)}{\Delta t}\frac{\Delta t}{\Delta x},$$

which relates the increments, and in which the last ratio tends towards $t_x = +\infty$ at the points in the homologue E_0 of e_0, shows that at a point of E_0 we have, the following cases,

$$\Lambda_d = \Lambda_g = \lambda_d = \lambda_g = +\infty; \tag{A}$$
$$\Lambda_d = \Lambda_g = \lambda_g = +\infty, \quad \lambda_d = -\infty; \tag{B}$$
$$\Lambda_d = \Lambda_g = \lambda_d = +\infty, \quad \lambda_g = -\infty; \tag{C}$$
$$\Lambda_d = \Lambda_g = +\infty, \quad \lambda_d = \lambda_g = -\infty. \tag{D}$$

The theorem of M. Denjoy is proved. It follows that *if we know a finite function $f(x)$, and if we know that it is, at every point, equal to one of the four derivative numbers of a continuous function $F(x)$—equal to Λ_d at some points, to λ_d at others, etc.—then the function $F(x)$ is determined, and obtained by totalisation of $f(x)$*. Indeed, $F(x)$ is an indefinite total, and the totalisable function which when differentiated, is almost everywhere equal to $f(x)$ in the set of points where $f(x) = \Lambda_d F(x)$, in the set of points where $f(x) = \lambda_d F(x)$, etc. These sets are unknown, but it does not matter since it follows that the function to be totalised is almost everywhere equal to $f(x)$ in whole (a, b).

Thus, the totalisation allows not only solving the problems A, B, C posed in the Chap. 5 and their extensions A', B', C'; but it also allows dealing with even broader problems like the one mentioned earlier.

In reality, this one might appear a little strange; It is hard to understand how one could have known that $f(x)$ is everywhere equal to one of the derivative numbers of $F(x)$ without knowing at least which one it is at each point. Therefore, the reader will perhaps wonder if it would be enough to know that the finite function $f(x)$ is at every point, one of the limits of $r[F(x), x, x+h]$, as the values of h tend towards zero, for the knowledge of $f(x)$ to determine $F(x)$ up to a constant? The answer is negative. Let us note, indeed, as M. Denjoy did, that the function $F_2(x)$, Sect. 11.4, has a definite derivative $F_2'(x)$ at every exterior point of the perfect set P and has as its derivative numbers $\Lambda_d = \Lambda_g = +\infty$, $\lambda_d = \lambda_g = -\infty$ at the points of P. The function $F_2(x) + m(x)$, where $m(x)$ denotes the measure of the subset of P located in (a, x), always has the same derivative numbers as $F_2(x)$. Therefore, for each of these functions, one of the limits of r is the function $f(x)$ which is equal $F_2'(x)$ outside of P and zero on P. We can even say that $f(x)$ is one of the limits of r when we let h towards zero by the values of definite sign; by the positive values, for example. Nevertheless, the difference between these functions is not constant.

The Integral of Stieltjes 12

In 1894, Stieltjes, while researching the advancements in the field of continued fractions,[1] defined a new method of integration of continuous functions. It is important to fully understand the originality of Stieltjes' generalisation and how it fundamentally differs from what we have examined so far. In Chap. 1, we recalled what is referred to as integration in the introductory course of infinitesimal calculus. it is a *well-defined* operation, which associates a number to each continuous function $f(x)$. In Chaps. 2, 3, 6, 7, 10 we defined this operation for increasingly larger families of functions $f(x)$. We extended the notion of integration deeper into the realm of functions $f(x)$. Stieltjes, in this case, leaves the family of considered functions $f(x)$ invariable. However, for a given function $f(x)$, he defines as many integrals as we want. Each of them associates a number to $f(x)$. He extends the concept of surface integration into the field of functional operations.

In this chapter, we will present the definition of the Stieltjes integral of a continuous function, which is similar to that in Chap. 1. Then, as done in Chaps. 2, 3, 6, 7, 10, we will extend this concept to increasingly larger class of functions. Finally, as done in Chaps. 4, 5, 8, 9, the concepts and problems associated with the new integration. The execution of this program would assume that research, which has not yet been addressed, has already been conducted. On many points, we will content ourselves with posing the problems.

[1] Annales de la Fac. des Sc. de Toulouse, 1894.

R. Jain, *Lebesgue's Theory of Integration*,
https://doi.org/10.1007/978-981-96-1169-0_12

12.1 The Integral of Stieltjes Defined by the Theory of Summable Functions

Let $\alpha(x)$ be a function of bounded variation in an interval (a, b); we will call it the *determinate function* of the integration to be defined.

Let $f(x)$ *be a continuous function in* (a, b). *We call the Stieltjes integral of* $f(x)$, *taken over* (a, b) *with respect to the determinate function*[2] $\alpha(x)$, *the limit of the sum*

$$S = \sum_{0}^{n} f(\xi_i)[\alpha(x_{i+1}) - \alpha(x_i)],$$

corresponding to a division of (a, b), *using values* x_i, *arranged in order:* $a = x_0, x_1, x_2, \ldots, x_{n+1} = b$, *when the number* n *increases indefinitely and, we choose* x_i *in such a way that the maximum of* $x_{i+1} - x_i$ *tends towards zero.* ξ_i *denotes any value of* x *taken in* (x_i, x_{i+1}).

To justify this notation, it should be proved that the limit exists. Let us consider a sequence of divisions of (a, b), say $D_1, D_2, \ldots$, each obtained by subdivision of the intervals of previous division, and let $S_1, S_2, \ldots$ be the values of S corresponding to these divisions and some choices of ξ_i.

Let $f(\xi_i)[\alpha(x_{i+1}) - \alpha(x_i)]$ be the contribution from one of the intervals of D_k, in S_k. In D_{k+1} the interval (x_i, x_{i+1}) is divided by points y_j and, provide a contribution of form

$$\sum\nolimits_1 f(\eta_j)[\alpha(y_{j+1}) - \alpha(y_j)],$$

where the summation is extended to some values of j. However, with the same values of j, we have

$$f(\xi_i)[\alpha(x_{i+1}) - \alpha(x_i)] = \sum\nolimits_1 f(\xi_i)[\alpha(y_{j+1}) - \alpha(y_j)].$$

The difference between the two contributions from (x_i, x_{i+1}) is, therefore,

$$\sum\nolimits_1 |[f(\eta_j) - f(\xi_i)][\alpha(y_{j+1}) - \alpha(y_j)]| \leq \omega_i \sum\nolimits_1 |\alpha(y_{j+1}) - \alpha(y_j)| \leq \omega_i V_i,$$

where ω_i denotes the oscillation of $f(x)$ in (x_i, x_{i+1}), and V_i is the total variation of $\alpha(x)$ in the same interval.

Hence, by addition,

$$|S_k - S_{k+l}| \leq \Omega_k V,$$

where V is total variation of $\alpha(x)$ in (a, b) and, Ω_k is the maximum oscillation of $f(x)$ in the intervals of division D_k. However, Ω_k tends towards zero as k increases indefinitely. By assumption, therefore, the sequence of S_k is convergent.

[2] **Translator's Note:** The function $\alpha(x)$ is called the *determinate function*, a term used by M. Lebesgue to refer to what is now commonly called the *integrator function* in the Stieltjes integral.

In the particular case of the sequences D_1, $D_2, \ldots$, obtained by successive subdivisions, once this is examined, we then pass to the general case using the reasoning we have applied many times before.

The integral we have just considered is a *definite integral*, denoted by

$$\int_a^b f(x)\,d[\alpha(x)].$$

It obviously possesses the following properties

$$\int_a^b f(x)\,d[\alpha(x)] = -\int_b^a f(x)\,d[\alpha(x)], \tag{a}$$

$$\int_a^b f(x)\,d[\alpha(x)] + \int_b^c f(x)\,d[\alpha(x)] + \int_c^a f(x)\,d[\alpha(x)] = 0, \tag{b}$$

$$\int_a^b [f(x) + g(x)]\,d[\alpha(x)] = \int_a^b f(x)\,d[\alpha(x)] + \int_a^b g(x)\,d[\alpha(x)]. \tag{c}$$

For this integral, *the mean value theorem* states: *If we have*

$$|f(x)| < M,$$

it follows that

$$\left|\int_a^b f(x)\,d[\alpha(x)]\right| < MV,$$

where V *is the total variation of* $\alpha(x)$ *in* (a, b).

All these properties follow immediately from an examination of the sums S.

The function

$$F(x) = \text{const.} + \int_a^x f(x)\,d[\alpha(x)]$$

is called the *indefinite integral function of one variable, in the sense of Stieltjes, of* $f(x)$, *taken with respect to* $\alpha(x)$. This definition is applicable for $a < x \leq b$. We will soon complete the definition for $x = a$.

The indefinite integral is a function of bounded variation. Let us represent by $V(x)$, the total variation of $\alpha(x)$ from a to x. Then, we have

$$|F(l) - F(m)| = \left|\int_m^l f(x)\,d[\alpha(x)]\right| \leq M|V(l) - V(m)|,$$

M denotes the upper bound of the absolute value of $f(x)$ in (a, b). Therefore, if we partition (a, b) into a finite number of intervals (x_i, x_{i+1}) we have

$$\sum |F(x_{i+1}) - F(x_i)| \leq M \sum |V(x_{i+1}) - V(x_i)| = MV.$$

This inequality proves the proposition and provides the upper bound MV for the total variation of the indefinite integral.

When the determinate function is continuous, the indefinite integral is continuous. Indeed, in this case, $V(l) - V(m)$ tends towards zero with length of (m, l); therefore, same holds for $F(l) - F(m)$.

More generally, the indefinite integral is continuous at any point of continuity of $\alpha(x)$. However, *it is discontinuous at* x_0, *if* $f(x_0)$ *is non-zero and if* $\alpha(x)$ *has a discontinuity at* x_0. Indeed, since x_0 is assumed to be different from a, let $\beta(x)$ and $\gamma(x)$ be the two functions defined as follows:

$$\begin{array}{ll} \text{For } x < x_0, & \beta(x) = \alpha(x), \\ & \gamma(x) = 0; \\ \text{At } x_0, & \beta(x_0) = \alpha(x_0 - 0), \\ & \gamma(x_0) = \alpha(x_0) - \alpha(x_0 - 0); \\ \text{For } x > x_0, & \beta(x) = \alpha(x) - \alpha(x_0 + 0) + \alpha(x_0 - 0), \\ & \gamma(x) = \alpha(x_0 + 0) - \alpha(x_0 - 0). \end{array}$$

$\beta(x)$ and $\gamma(x)$ are two functions of bounded variation, whose sum is $\alpha(x)$. The indefinite integral $F(x) = A(x)$, relative to $\alpha(x)$, is therefore the sum of those relative to $\beta(x)$ and $\gamma(x)$; say $B(x)$ and $C(x)$. Now, $B(x)$ is continuous at x_0, because $\beta(x)$ is continuous at x_0, and $C(x)$ is clearly equal to

$$C(x) = \begin{cases} K & \text{if } x < x_0 \\ K + f(x_0)[\alpha(x_0) - \alpha(x_0 + 0)] & \text{if } x = x_0 \\ K + f(x_0)[\alpha(x_0 + 0) - \alpha(x_0 - 0)] & \text{if } x > a \end{cases}$$

Therefore, $C(x)$ is discontinuous at x_0, and consequently, so is $A(x)$, if $f(x_0)$ is non-zero. At the same time, we have proved that: *If we denote by* $s_g(x)$ *and* $s_d(x)$, *respectively, the left and right jumps of* $\alpha(x)$ *at the point* x, *the jump function of the indefinite integral of* $f(x)$, *taken with respect to* $\alpha(x)$, *is*

$$\Phi(x) = \sum_{a \le x_i < x} f(x_i) s_d(x_i) + \sum_{a < x_i \le x} f(x_i) s_g(x_i),$$

where the summation is extended to all values indicated by the inequalities, or, equivalently, all those values which are points of discontinuity of $\alpha(x)$. Nevertheless, by a new convention that complements the definition of the indefinite integral, we include the point a in the first summation.

As a result, the corrected integral of its jump function, $F(x) - \Phi(x)$, is the indefinite integral, in the sense of Stieltjes, with respect to the determinate function obtained by correcting $\alpha(x)$ for its jump function.

$$\alpha(x) - \sum_{a \le x_i < x} s_d(x_i) - \sum_{a < x_i \le x} s_g(x_i).$$

The Stieltjes integral for discontinuous determinate functions is therefore easily calculated using those for continuous determinate functions. In most case, these can be calculated immediately.

Let us suppose, for example, $\alpha(x)$ is continuous and strictly increasing, $\alpha(x_1) > \alpha(x_0)$ for $x_1 > x_0$. The change of variable $\alpha = \alpha(x)$ transforms $f(x)$ into a function $g(\alpha)$ and transforms the definition of $\int f(x)\, d[\alpha(x)]$ into that of an ordinary integral of $g(\alpha)$. Thus

$$\int_a^b f(x)\, d[\alpha(x)] = \int_{\alpha(a)}^{\alpha(b)} g(\alpha)\, d\alpha,$$

and we return to an ordinary integration of a continuous function. In fact, it is the previous formula that is the very origin of the notation of the Stieltjes integral.

The general case, where $\alpha(x)$ is continuous reduces to this one.

Let $p(x)$ and $n(x)$ be the positive and negative total variations of $\alpha(x)$. The two functions

$$\alpha_1(x) = p(x) + x, \quad \alpha_2(x) = n(x) + x$$

are strictly increasing, and we have

$$\alpha(x) = \alpha_1(x) - \alpha_2(x).$$

Therefore, if we set

$$f(x) = g_1(\alpha_1) = g_2(\alpha_2),$$

we deduce

$$\int_a^b f(x)\, d[\alpha(x)] = \int_{\alpha_1(a)}^{\alpha_1(b)} g_1(\alpha_1)\, d\alpha_1 - \int_{\alpha_2(a)}^{\alpha_2(b)} g_2(\alpha_2)\, d\alpha_2.$$

Finally, the most general case is treated similarly. When $\alpha_1(x)$ and $\alpha_2(x)$ are formed using the total variations of $\alpha(x)$, corrected for its jump function, we obtain, using previous notations,

$$\int_a^b f(x)\, d[\alpha(x)] = \sum_{a \le x_i < b} f(x_i) s_d(x_i) + \sum_{a < x_i \le b} f(x_i) s_g(x_i)$$
$$+ \int_{\alpha_1(a)}^{\alpha_1(b)} g_1(\alpha_1)\, d\alpha_1 - \int_{\alpha_2(a)}^{\alpha_2(b)} g_2(\alpha_2)\, d\alpha_2.$$

Therefore, every Stieltjes integral is expressed using the ordinary integrals. Before drawing conclusions from this essential fact, let us provide other equivalent formulae. The advantage of this approach is that it relies only on the basic form of integration: the integral of a continuous function over an interval. However, it requires two integrals, a series summation, and a change of variables.

Let us suppose that $\alpha(x)$ is continuous, strictly increasing, and has a continuous derivative. Then, a change of variable is not necessary because we have

$$\int_a^b f(x)\,d[\alpha(x)] = \int_{\alpha(a)}^{\alpha(b)} g(\alpha)\,d\alpha = \int_a^b f(x)\alpha'(x)\,dx,$$

returning from variable α to x using the classical change of variable formula.[3] In short, we will treat the Stieltjes integral by noting that it reduces to the curvilinear integral $\int_{a,\alpha(a)}^{b,\alpha(b)} f(x)\,d\alpha$, associated with the curve $\alpha = \alpha(x)$.

The previous formula extends to the case where $\alpha(x)$ is only assumed to be absolutely continuous; then, let $'\alpha(x)$ denote the function, determined only on points of sets that are equivalent up to sets of measure zero, for which $\alpha(x)$ is the indefinite integral. This function can be called the almost derivative of $\alpha(x)$. We have

$$S = \sum f(\xi_i)[\alpha(x_{i+1}) - \alpha(x_i)] = \sum f(\xi_i) \int_{x_i}^{x_{i+1}} \alpha(x)\,dx.$$

Therefore

$$S - \int_a^b f(x).'\alpha(x)\,dx = \sum \int_{x_i}^{x_{i+1}} [f(\xi_i) - f(x)].'\alpha(x)\,dx.$$

As a result, with Ω denoting the maximum oscillation of $f(x)$ in the intervals (x_i, x_{i+1}) and V always representing the total variation of $\alpha(x)$ in (a, b),

$$\left| S - \int_a^b f(x).'\alpha(x)\,dx \right| \leq \Omega \sum \int_{x_i}^{x_{i+1}} |'\alpha(x)|\,dx = \Omega V.$$

Hence, by passing to limit,

$$\int_a^b f(x)\,d[\alpha(x)] = \int_a^b f(x).'\alpha(x)\,dx.$$

When the function of bounded variation $\alpha(x)$ is simply assumed continuous, a change of variable is sufficient for bringing us back to the previous case; indeed, it is clear that if we have a uniform change of variable $(t|x)$ in both the directions associating (t_1, t_2) to (a, b), we always have

$$\int_a^b f(x)\,d[\alpha(x)] = \int_{t_1}^{t_2} f[x(t)]\,d\{\alpha[x(t)]\}.$$

---

[3] The extension to various integration methods of classical procedures, for exact or approximate calculation of integrals of continuous functions (integration by parts, by substitution, second theorem of the mean, inequality of Schwarz, etc.) does not find place in our presentation. What follows is actually related to the method of integration by substitution.

It is enough to choose t in such a manner that $\alpha[x(t)]$ is absolutely continuous at t to be able to apply previous formula. This variable t could be the length of curve $\alpha = \alpha(x)$, from a up to x; it is simpler here to take[4]

$$t = x + V(x),$$

where $V(x)$ is, as previously, the total variation of $\alpha(x)$ from a to x. Then we have, by setting $\alpha[x(t)] = A(t)$,

$$\int_a^b f(x)\, d[\alpha(x)] = \int_a^{b+V} f[x(t)].' A(t)\, dt.$$

If $\alpha(x)$ were continuous and variable in entire interval, we could set $\nu = V(x)$ and we would have

$$\int_a^b f(x)\, d[\alpha(x)] = \int_0^V f[x(\nu)].' \alpha[x(\nu)]\, d\nu.$$

We would extend this formula to all the cases: to do this, let us pose the following definitions: *$V(x)$ denotes the total variation, from a to x, of the determinate function $\alpha(x)$, for any value ν_0 contained between 0 and $V = V(b)$ it corresponds to:*

a. *Let one or several values of x be such that we have $\nu_0 = V(x)$, then we choose one of these values x_0 and we will set*

$$x_0 = x(\nu_0), \quad A(\nu_0) = \alpha[x(\nu_0)] = \alpha(x_0),$$

b. *Let a value x_0 be such that, we have*

$$V(x_0 - 0) \leq \nu_0 < V(x_0) \quad \text{or} \quad V(x_0) < \nu_0 \leq V(x_0 + 0);$$

Then, we will set $x_0 = x(\nu_0)$ and in the first case

$$A(\nu_0) = \alpha(x_0 - 0) + \frac{\alpha(x_0) - \alpha(x_0 - 0)}{V(x_0) - V(x_0 - 0)}[\nu_0 - V(x_0 - 0)].$$

in the second

$$A(\nu_0) = \alpha(x_0) + \frac{\alpha(x_0 + 0) - \alpha(x_0)}{V(x_0 + 0) - V(x_0)}[\nu_0 - V(x_0)].$$

With these conventions, we have

$$\int_a^b f(x)\, d[\alpha(x)] = \int_0^V f[x(\nu)].' A(\nu)\, d\nu$$

---

4 *See* Sect. 9.3 and following.

To justify this statement, let us partition (a, b) into partial intervals δ_i, in each of which the oscillation of $f(x)$ is smaller than ε. In each interval $\delta_i = (l_i, m_i)$ let us arbitrarily choose a value $l_i \leq \xi_i \leq m_i$ and set

$$\Delta_i = [V(l_i), V(m_i)], \quad \Delta V_i = V(m_i) - V(l_i).$$

Then, we have

$$\left| \int_a^b f(x)\, d[\alpha(x)] - \sum_i [f(\xi_i)\{\alpha(m_i) - \alpha(l_i)\}] \right|$$
$$= \left| \sum_i \int_{\delta_i} [f(x) - f(\xi_i)]\, d[\alpha(x)] \right|$$
$$\leq \sum_i \left[\varepsilon \int_{\delta_i} dV(x) \right] = \varepsilon \sum_i \Delta_i V = \varepsilon V;$$
$$\left| \int_0^V f[x(\nu)].'A(\nu)\, d\nu - \sum_i [f(\xi_i)\{\alpha(m_i) - \alpha(l_i)\}] \right|$$
$$= \left| \sum_i \int_{\Delta_i} \{f[x(\nu)] - f(\xi_i)\}'A(\nu)\, d\nu \right|$$
$$\leq \sum_i \left[\varepsilon \int_{\Delta_i} |'A(\nu)|\, d\nu \right] = \varepsilon \sum_i \Delta_i V = \varepsilon V.$$

The theorem follows immediately from combining the two inequalities.

The function $A(\nu)$ has a total variation of $\nu_2 - \nu_1$ in (ν_1, ν_2). Therefore, $A(\nu)$ has a derivative almost everywhere equal to ± 1, and as a result *we can assume $'A(\nu)$ to be equal to ± 1 everywhere*. This makes the previous formula particularly simple; but, in what follows, more complex formulae obtained earlier remain valid. It is primarily to clarify the concepts, we will start with a well-defined formula for reducing a Stieltjes integral to ordinary integrals. We will use the previously established formula. It requires the knowledge of theory of summable functions; however, once this theory is understood, the right-hand side of our formula applies to a much broader range of cases than those examined so far, where $f(x)$ was continuous. Thus, *we can take the formula*

$$\int_a^b f(x)\, d[\alpha(x)] = \int_0^V f[x(\nu)].'A(\nu)\, d\nu,$$

as the very definition of the Stieltjes integral of $f(x)$.

This definition will apply, using for now only the integration of summable functions, whenever $f[x(\nu)].'A(\nu)$ is summable. Therefore, for all functions $f[x(\nu)]$ that are summable, since $'A(\nu) = \pm 1$, that is to say, *for all functions $f(x)$ that are summable when* *we take the total variation ν of $\alpha(x)$ from a to x as the variable*. In particular, this definition will apply to all bounded functions that are measurable with respect to ν.

Now, among the functions $f(x)$ that are measurable with respect to ν, we must mention all functions $f(x)$ that are B measurable with respect to x. Indeed, the formula $x = x(\nu)$ associates every interval in x with an interval in ν or a point, a sum of sets in x with a sum of sets in ν, a difference of sets in x, with a difference of sets in ν plus sometimes with some values of ν corresponding to the intervals of constancy of x. Since these values are finite or countable, every B measurable set in x corresponds to a B measurable set in ν. Therefore, if $f(x)$ is B measurable in x, meaning if the set $E[\alpha < f(x) < \beta]$ is B measurable, then the set $E[\alpha < f[x(\nu)] < \beta]$ is also B measurable, and $f[x(\nu)]$ is B measurable with respect to ν.

Therefore, the previous definition is applicable to a class of functions $f(x)$ that vary with the determinate function $\alpha(x)$, while always containing the family of bounded and B measurable functions $f(x)$.

The disadvantage of such a rapid method, which provided us with this result, is that it does not clearly reveal the significance of extending the concept of Stieltjes Integral. A theorem of M. Frédéric Riesz will highlight its significance.[5]

12.2 The Linear Functionals

No one had dealt with the integration of a function with respect to another function since Stieltjes, until in 1909, M.F. Riesz revealed that this concept had, in fact, been the subject of quite a few studies–albeit under a different name, the *linear functional operation.*

A linear functional operation is one that associates with each function $f(x)$, belonging to a certain class of functions, a number $A[f(x)]$ such that:

1. $A[f_1(x) + f_2(x)] = A[f_1(x)] + A[f_2(x)]$;
2. $|A[f(x)]| \leq M \times \max|f(x)|$,

where M is a fixed number. The functional $A[f(x)]$[6] *provided by the linear operation is itself called linear.*

These were primarily questions in mathematical physics that led to the notion of linear functionals. The class of functions which presented itself at that time, varying with the ques-

[5] It is in the context of this theorem of M. Riesz (C.R. *Acad. Sc.*, 1909; *see* also *Annales de l'École Normale superieure,* 1911 *and* 1914) that I introduced (C.R. *Acad. Sc.*, 1909;) the extension of concept of Stieltjes integral by the methods that we just mentioned, sometimes presented in a slightly different form.

[6] Since the term *function of functions* is somewhat ambiguous, M. Volterra had referred to numbers such a $A[f(x)]$ as *functions of lines*. However, the expression *functional* proposed by M. Hadamard has generally prevailed.

tions, always contained the continuous functions but often various types of discontinuous functions as well. Thus the problem of extending the field of application of linear functional operations naturally arose. This extension is achieved through the earlier generalisation of the concept of Stieltjes integral and M. Riesz's theorem.

Among the linear functionals defined on the field of continuous functions, we find those of the form

$$\int_a^b K(x) f(x)\, dx.$$

Expressions of this form were also considered when we tried to construct more general linear functional; M. Hadamard and M.Frechet had obtained quite interesting results in this direction. However, it was left to M. Riesz to completely solve the problem by showing that *any linear functional defined for all continuous functions in* (a, b) *is of the form* $\int_a^b f(x)\, d[\alpha(x)]$; *where* $\alpha(x)$ *is a function of bounded variation that characterises the functional.*[7]

Let $A(f)$ be a linear functional defined on the field C of continuous functions from a to b.[8] The equality

$$A(f_1 + f_2 + \cdots + f_p) = A(f_1) + \cdots + A(f_p)$$

therefore holds in C; there are two important cases where we can even assume that the functions f_i are in infinite number. First, the case where the series of f_i converges uniformly, and second, the case where all f_i, from a certain value of i onwards, are of same sign.

Indeed, let us assume that the series of f_i converges uniformly; if f is its limit and s_p is the sum of its first p terms, we have, for p sufficiently large $|f - s_p| < \varepsilon$, hence

---

[7] I imitate in what follows, the proof given by M. Riesz in his Memoire of Annales de l'École Normale, 1914. M. Riesz honours me by stating that a remark I made regarding the role of monotone sequences of functions guided him. In reality, I had only very imperfectly understood this role; otherwise, I would not have written, in my note of 1909, that it would be very difficult to extend the notion of Stieltjes integral by a different method from the one I was using. Shortly after I made this imprudent statement, M. W.H.Young showed that my method was far from essential and that the Stieltjes integral is defined exactly as the ordinary integral by the method of monotone sequences indicated in Chap. 7, Sect. 8.5 (*Proceed. of the London Math. Society,* 1913).

This work of M.Young was the first of those which eventually clarified what a Stieltjes integral is. We have truly penetrated the depth of this notion thanks to the definition given by M. Radon (*Sitz. d: K. Ak. d. Wiss. in Wien,* 1913) and the work of M. de la Vallée Poussin on the extension of the concept of measure (*See*, in particular, in this collection, the aforementioned book of M. de la Vallée Poussin). However, for these works to be possible, the concepts of set functions (Lebesgue), functions of several variable of bounded variation (Vitali, *Rend. della R. Acc. delle Sc. di Torino,* 1908), Stieltjes integral of a continuous function of several variables (Frechet, *Nouv. Ann. de Math.*, 1905), had to be developed.

Since in this book, I am only concerned with the functions of a single variable, the difficulty and importance of some of these works may not be readily apparent. This is why I want to emphasise that if, with regard to the functions of a single variable, the Memoire of M. Radon has brought us a new definition, which is particularly insightful, of the Stieltjes integral. However, for the case of several variables, this Memoire provides a real extension of the concept of integral.

[8] We could start from a more restricted field, such as that of polynomials, for example.

$$|A(f) - A(s_p)| = |A(f - s_p)| \leq \varepsilon M.$$

The first case is thus examined. Now, the second case reduces to the first, because if f_i are all positive or zero and, *if the sum* f *of the series of* f_i *belongs to the field* C, the set E_p of the points where we have $|f - f_p| \geq \varepsilon$ is a closed set when it exists. Since E_p contains E_{p+1} and there are no common points to all E_p, the set E_p no longer exists as when p is sufficiently large. Therefore, the series of f_i converges uniformly.

Let us now examine the case where we know only that f_i are positive or zero, and the series of f_i converges to a bounded function f.[9] Then we have, by noting that $A(f_i)$ and $A(-f_i)$ are equal and of opposite signs, and by setting $\theta_i = \pm 1$ according to the sign of $A(f_i)$, we have

$$\begin{aligned}\sum_1^p |A(f_i)| = \sum_1^p \theta_i A(f_i) &= A\left[\sum_1^p \theta_i f_i\right] \\ &\leq M \times \text{max. of} \left|\sum_1^p \theta_i f_i\right| \\ &\leq M \times \text{max. of} \sum_1^p f_i \leq M \times \text{max. of } f.\end{aligned}$$

Thus, when a non-decreasing or non-increasing sequence of functions s_p of C tends towards a bounded limit, the sequence $\sum A(s)$ converges.

Another property we will use: if f is a function of C and if k is a constant, we have

$$A(kf) = kA(f).$$

This property is obvious for integer k or inverse of an integer; we then arrive at commensurable (rational) k, then finally we attain any k by passage to the uniform limit.

This being said, let us set $\alpha(a) = 0$ and, for $a < x_i \leq b$, let us take $\alpha(x_i)$ equal to the limit of $A(f)$ for the decreasing sequence of continuous functions $f_{n,x_i}(x)$ equal to 1 from a to x_i, equal to 0 from $x_i + \frac{1}{n}$ to b, linear from x_i to $x_i + \frac{1}{n}$.

Any continuous function in (a, b) can be approximated as closely as desired using a linear combination of functions $f_{n,x_i}(x)$. Indeed, let us form the function

$$\begin{aligned}g_n(x) =& f(\xi_p) f_{n,x_p}(x) + [f(\xi_{p-1}) - f(\xi_p)] f_{n,x_{p-1}}(x) \\ &+ [f(\xi_{p-2}) - f(\xi_{p-1})] f_{n,x_{p-2}}(x) + \cdots + [f(x_1) - f(x_2)] f_{n,x_1}(x),\end{aligned}$$

for

[9] In the field C, a similar examination for the case of uniform convergence would be unnecessary; on the contrary, for other fields, it provides an interesting result, which, along with the one that will be given in the text, can be used to directly study the extension of the domain of definition of a functional, without resorting to the notion of Stieltjes integral.

$$a = x_0 < \xi_1 < x_1 < \xi_2 < x_2 < \cdots < x_{p-1} < \xi_p < x_p = b;$$

it is, in each (x_{i-1}, x_i), contained between $f(\xi_{i-1})$ and $f(\xi_i)$ when $\frac{1}{n}$ is less than the smallest difference $x_i - x_{i-1}$.

Therefore, if we chose x_i in such a manner that the oscillation of $f(x)$ is less than ε in each (x_{i-1}, x_i), then the difference $f(x) - g_n(x)$ is less than 2ε when n is large enough. And then the difference $A[f(x)] - A[g_n(x)]$ will be less than $2M\varepsilon$.

Now we know the limit S of $A[g_n(x)]$ for very large n; it is written as

$$S = f(\xi_p)\alpha(x_p) + [f(\xi_{p-1}) - f(\xi_p)]\alpha(x_{p-1}) + \cdots + [f(\xi_1) - f(\xi_2)]\alpha(x_1)$$

or again

$$S = f(\xi_1)[\alpha(x_1) - \alpha(x_0)] + f(\xi_2)[\alpha(x_2) - \alpha(x_1)] + \cdots + f(\xi_p)[\alpha(x_p) - \alpha(x_{p-1})]$$

And consequently, $A(f)$ is the limit of previous sum, meaning $A(f)$ is the Stieltjes integral of $f(x)$, taken with respect to $\alpha(x)$. The theorem of M. Riesz will be proved as soon as we have verified that $\alpha(x)$ is of bounded variation.

However, this is obvious; if $\alpha(x)$ were not of bounded variation, it would have a positive variation equal to $+\infty$. Therefore, we would be able to find intervals (a_1, b_1), (a_2, b_2), ... distinct from each other and finite in number such that $\sum[\alpha(b_i) - \alpha(a_i)]$ exceeds $2M$; where M is always the number which appears in the second property of the linear functional. Therefore, for the continuous functions φ_i equal to 1 in (a_k, b_k), zero in the intervals $\left(b_k + \frac{1}{i}, a_{k+1} - \frac{1}{i}\right)$ and linear in the intervals $\left(a_k - \frac{1}{i}, a_k\right)$, $\left(b_k, b_k + \frac{1}{i}\right)$, functions which decrease towards the function φ equal to 1 in (a_k, b_k) and to 0 outside, the numbers $A(\varphi)$ would tend towards

$$\sum[\alpha(b_i) - \alpha(a_i)] > 2M;$$

which is impossible, since $A(\varphi)$ cannot exceed M.

This theorem of M. Riesz associates with each linear functional, defined for the continuous functions $f(x)$ in an interval (a, b), a determinate function $\alpha(x)$. As we have seen regarding the extension of Stieltjes integral, *such a functional can be extended to the field of all functions which, through the change of variable from x to the total variation ν of $\alpha(x)$ in (a, x), transform into summable functions of ν.* This family varies with $\alpha(x)$, but it is important to note that it always contains all B measurable and bounded functions.

When we use the linear functionals of continuous functions $f(x)$, one of the most useful properties is this: the sum $A[f(x)] + B[f(x)]$ of two linear functionals $A[f(x)]$ and $B[f(x)]$ is itself a linear functional.

When we want to extend the field of functions $f(x)$ while still retaining the advantage of this property, it is necessary to focus on a functional field independent of the determinate function $\alpha(x)$. Therefore, it is very interesting to know that *all linear functionals defined on the field of continuous functions can be extended to the field of B measurable and bounded functions.*

It is clear that the functionals extended to the field of B measurable and bounded functions, as we have obtained, possesses the following property:

3. *If functions $f_i(x)$ increase towards a limit $f(x)$, we have*

$$A[f(x)] = \lim_{i \to \infty} A[f_i(x)].$$

We will verify that *the properties* 1, 2, *and* 3 *are sufficient to characterise the extension we have made to the field of B measurable and bounded functions of a given linear functional defined on continuous functions.*

Indeed, we will see how, from the properties 1, 2, and 3, can extend to a larger field, a linear functional $A(f)$, given in the field of continuous functions.

We have seen that, for n increasing indefinitely, the functions $f_{n,x_i}(x)$ decrease towards the function $f_{x \leq x_i}(x)$ which equals 1 for $a \leq x \leq x_i$ and 0 for $x_i < x \leq b$. Therefore, according to 1 and 3, $A[f_{x \leq x_i}(x)]$ is deduced.

Let us set, for $\alpha < \beta$,

$$f_{\alpha < x \leq \beta} = f_{x \leq \beta}(x) - f_{x \leq \alpha}(x).$$

From 1 we deduce the value of $A[f_{\alpha < x \leq \beta}(x)]$.

The sequence of functions $f_{\alpha - \frac{1}{n} < x \leq \beta}(x)$ decreases, when n increases indefinitely, towards a limit $f_{\alpha \leq x \leq \beta}(x)$; therefore, the value of $A[f_{\alpha \leq x \leq \beta}(x)]$ is determined by 1 and 3.

Thus, we have determined the value of $A[f_E(x)]$ for certain functions $f_E(x)$ that are zero outside a set E and equal to 1 on E. We have just arrived at the functions $f_{\alpha \leq x \leq \beta}(x)$ for which the set E is a closed interval. Now, from such intervals, we construct any B measurable set by repeating two operations: addition of disjoint sets and subtraction of one set from another which contains it; from first of these operations

$$E = E_1 + E_2 + E_3 + \cdots$$

the conditions 1 and 3 imply the equality

$$A[f_E] = A[f_{E_1}] + A[f_{E_2}] + A[f_{E_3}] + \cdots$$

for the second operation

$$E = E_1 - E_2,$$

the condition 1 corresponds to

$$A[f_E] = A[f_{E_1}] - A[f_{E_2}],$$

therefore, $A[f]$ is defined for each function $f_E(x)$ relative to a B measurable set E.

Now, if $f(x)$ is any B measurable bounded function, it differs by at most ε only, from the function

$$f^{\varepsilon}(x) = \sum i\varepsilon f_{E[i\varepsilon \leq f(x) < (i+1)\varepsilon]}(x);$$

therefore, the sequence of functions $f^{\varepsilon}(x)$ tends uniformly towards $f(x)$ when ε tends towards zero and we deduce from 1 and 2, as we did previously, that the numbers $A[f^{\varepsilon}(x)]$ converge to a limit that we must take as the value of $A[f(x)]$.

Thus, the extension to the entire field of B measurable bounded functions, if possible, is unique. But, we have seen that it was possible.[10]

The extension to this large functional field[11] is, therefore, well characterised by the conditions 1, 2 and 3.

Thanks to the concept of the Stieltjes integral, we will obtain another extension. With every definite integral, we have associated an indefinite integral providing a point function, an interval function, and a measurable set function. Thus, the notion of the Stieltjes integral leads to an indefinite Stieltjes integral as a point function, an indefinite Stieltjes integral as as an interval function, and an indefinite Stieltjes integral as a set function. This last one allows us to associate with the function $f(x)$ and a set E_x, a definite number

$$\int_{E_\nu} f[x(\nu)].'A(\nu)\,d\nu = \int_{E_x} f(x)\,d[\alpha(x)],$$

where E_ν is the set of values of ν for which $x(\nu)$ belongs to E_x. Thus, the *Stieltjes integral of $f(x)$ taken with respect to $\alpha(x)$, is now defined and extended to the set E_x.*

This integral is defined for the sets E_x for which E_ν is measurable; this family of sets contains all the B measurable sets E_x.

The integral is defined for every function having a definite value at the points of E_x and is equal on E_x to a function $f(x)$ for which $f[x(\nu)]$ is summable over $(0, V)$. Therefore, in particular *this definition is applicable to any function $f(x)$, that is bounded and B measurable on a B measurable set E_x.*

Now, $\int_{E_x} f(x)\,d[\alpha(x)]$ is obviously a linear functional in the field of functions $f(x)$ given on E_x. Therefore, *when a linear functional $A_1[f(x)]$ is defined on continuous functions in an interval I, we can deduce a family of linear functionals $A_E[f(x)]$, associated with each B measurable set E located in I and defied on B measurable functions on E, by the conditions*

1. $A_E[f_1(x) + f_2(x)] = A_E[f_1(x)] + A_E[f_2(x)]$;
2. $A_E[f(x)] \leq M \times \max_{x \in E} |f(x)|$

   where M is a fixed unknown number, independent of E and $f(x)$.
3. If f_i increases towards $f(x)$ on E, we have

---

[10] If we had not already extended it, thanks to the Stieltjes integral, we should, which would be easy, verify here that the values we have assigned to $A[f]$ are well-determined for each f of the new functional field and that they satisfy the conditions 1, 2 and 3. That is, in fact, what we will do shortly.

[11] We will see later how we can attain an even larger functional field, consisting of functions $f(x)$ which yield measurable functions $f[x(\nu)]$.

$$A_E[f(x)] = \lim_{i \to \infty} A_E[f_i(x)];$$

4. If $g(x)$ is equal to $f(x)$ on E and zero elsewhere, we have

$$A[f(x)] = A_I[g(x)].$$

We can be surprised by this result because given $A_I[f(x)]$, the determinate function $\alpha(x)$ is not unique. Indeed, it is quite clear that if we modify $\alpha(x)$ at a single point x_0 interior to I, it does not modify $A_I[f(x)]$ for the continuous functions $f(x)$ in I, because we have

$$A_I[f(x)] = \int_{a \le x < x_0} f(x)\, d[\alpha(x)] \\ + f(x_0)[\alpha(x_0 + 0) - \alpha(x_0 - 0)] + \int_{x_0 < x \le b} f(x)\, d[\alpha(x)];$$

and, however, this modifies in general

$$\int_{a \le x \le x_0} f(x)\, d[\alpha(x)] = \int_{a \le x < x_0} f(x)\, d[\alpha(x)] + f(x_0)[\alpha(x_0) - \alpha(x_0 - 0)].$$

This paradox arises solely from the fact that, for $x_0 > a$, the equality

$$A_{a \le x \le x_0}[f(x)] = \int_{a \le x \le x_0} f(x)\, d[\alpha(x)],$$

holds only with functions α which are continuous to the right of x_0. Such was the case for the function $\alpha(x)$ constructed in the course of proof of the theorem of M. Riesz, which can be verified easily. However, with this function, we do not have

$$A_{x_0 \le x \le b}[f(x)] = \int_{x_0 \le x \le b} f(x)\, d[\alpha(x)],$$

when x_0 is a point of discontinuity of $\alpha(x)$.

The difficulty that we encounter here is the same as the one that arose previously (Sect. 9.1). For each function $f(x)$, $A_E[f(x)]$ is a completely additive set function; therefore, if we want to have

$$A_{x_1 < x \le x_2}[f(x)] = \int_{x_1}^{x_2} f(x)\, d[\alpha(x)] = \mathfrak{F}(x_2) - \mathfrak{F}(x_1),$$

by denoting $\mathfrak{F}(x)$ as the indefinite integral $\int_a^x f(x)\, d[\alpha(x)]$, it is necessary that $\mathfrak{F}(x)$ serves as a generating function for an additive set function, and as a result, be, as we have seen in the indicated section, a right continuous function. However, the right jump of $\mathfrak{F}(x)$ at x_0 is equal to $f(x_0)[\alpha(x_0+0)-\alpha(x_0)]$; therefore, $\alpha(x)$ must be right continuous.

To circumvent the difficulty which thus arises, when we want to extend a known functional to measurable sets in an interval using a determinate function $\alpha(x)$ that is not right continuous, we can decompose $\alpha(x)$ into its jump function and its continuous part

$$\alpha(x) = S(x) + C(x);$$

$S(x)$ then gives rise to a functional of the form

$$\sum f(x_0)[\alpha(x_0+0)-\alpha(x_0-0)]$$

whose definition is applicable to a sequence of discontinuous functions $f(x)$ just as well as to continuous functions. As for the extension of the functional corresponding to $C(x)$, it does not pose any difficulty.[12]

12.3 Direct Definition of Integral of Stieltjes

This study of the linear functionals provides a better understanding of the meaning of conditions $1I, 2I, \ldots, 6I$ in the integration problem (Sect. 8.1). Let us compare these conditions with conditions $1F, 2F, 3F$ and $4F$ established for linear functionals (Sect. 12.2). $3I$ is identical to $1F$; $6I$ replaces $3F$; $2I$ replaces $4F$; as for $2F$, it turns out to be a consequence of $4I$ and $5I$. The conditions $1I, 4I, 5I$ serve only to characterise the functional $Af(x)$ relative to continuous functions that needs to be extended. In short, when the integral of a continuous function $f(x)$ is known in any interval where the function is given, in Chap. 7, we could have limited ourselves to setting the conditions $1F, 2F, 3F$ and deducing the extension of the integral by reasoning presented earlier. However, while for the extension of the general linear functional, we could limit ourselves to proving that the extension was unique because the case of the integral had been previously examined, we would now need to directly verify that the extension is possible. This is what we will do by placing ourselves in the case of most general Stieltjes integral. In this way, we will obtain a direct definition of this integral, from which the definition of the ordinary integral will be deduced by setting $\alpha(x) \equiv x$.

[12] If we decompose $C(x)$ in its function of singularities and its kernel

$$C(x) = C_s(x) + AC(x),$$

the functional relative to $AC(x)$ is written $\int_a^b f(x).'AC(x)\,dx$; only the functional relative to $C_s(x)$ requires the use of Stieltjes integral (FRÉCHET, *Comptes rendus du Congres des Societes savantes en* 1913).

But, before looking for a new form of the definition of the Stieltjes integral, it is necessary to determine the cases in which the initially established definition for continuous functions can be employed without modification.

Let $f(x)$ be a bounded function in (a, b), and let $\alpha(x)$ be a function of bounded variation. We will form, as was said, the sum

$$S = \sum_0^n f(\xi_i)[\alpha(x_{i+1}) - \alpha(x_i)] = \sum_0^n f(\xi_i)\delta_i\alpha.$$

If, in (x_i, x_{i+1}), $f(x)$ varies between L_i and l_i, we will denote by L_i' and l_i', two numbers defined by the conventions

$$\begin{aligned} L_i' &= L_i, & l_i' &= l_i, & &\text{if } \delta_i\alpha \geq 0, \\ L_i' &= l_i, & l_i' &= L_i, & &\text{if } \delta_i\alpha < 0; \end{aligned}$$

L_i' and l_i' will be the upper and lower bounds of $f(x)$ defined with respect to $\alpha(x)$. The oscillation of $f(x)$ in the given interval will be

$$\omega_i = L_i - l_i = |L_i' - l_i'|.$$

This being said, if we vary ξ_i without varying x_i, then S varies between

$$\overline{S} = \sum_0^n L_i'\delta_i\alpha \quad \text{and} \quad \underline{S} = \sum_0^n l_i'\delta_i\alpha.$$

We will show that these sums, for a sequence of divisions $D_1, D_2, \ldots$, into intervals whose maximum length tends towards zero, converge towards definite limits

$$\overline{\int_a^b} f(x)\, d[\alpha(x)], \quad \underline{\int_a^b} f(x)\, d[\alpha(x)].$$

—which we will call the upper and lower Stieltjes-Darboux integrals of f—provided that every point of discontinuity of $\alpha(x)$ is a point of division of D_k, from a certain value of k.

If $p(x), n(x), \nu(x)$ are three total variations of $\alpha(x)$

$$\alpha(x) = \alpha(a) + p(x) - n(x), \quad \nu(x) = p(x) + n(x),$$

we have

$$\sum_0^n l_i\delta_i p - \sum_0^n L_i\delta_i n \leq S = \sum_0^n f(\xi_i)\delta_i[p - n] \leq \sum_0^n L_i\delta_i p - \sum_0^n l_i\delta_i n.$$

Now, for the increasing functions p and n, the study of sums such as $\sum L_i\delta_i p$ is easy. Let $\Delta_1, \Delta_2, \ldots$ be a second sequence of division subject to the same conditions as the sequence of D_i; let s_i and σ_j be the numbers $\sum L_i \delta_i p$ provided by D_i and Δ_j; we want to compare the sequence of s_i with that of σ_j.

If j is taken sufficiently large, i being fixed, in each interval given by Δ_j at most one of the points of D_i will be found. So, if (x, y) is one of the intervals given by D_i, x will be in an interval (r, s) of Δ_j and y in (t, u); $r \leq x \leq s \leq t \leq y \leq u$. If x (or y) is a point of discontinuity of $\alpha(x)$, for j sufficiently large, r and s will coincide with x (or t and u with y); so we will assume $s - r$ (or $u - t$) is different from zero only if x (or y) is point of continuity of $\alpha(x)$. Then, let us replace the contribution of (r, s) in σ_j with $f(x)[p(s) - p(r)]$; as $f(x)$ differs from the upper bound of f in (r, s) by as little as we want when j is taken sufficiently large,—because of the smallness of (r, s) and of the continuity of f at point x,—we will modify σ_j by as little as we want. Let us do that for each point of division of D_i, and we will have a number (σ_j) different from σ_j by less than ε. If, between s and t the points of Δ_j are $z_1, z_2, \ldots$ we can say that the contribution of (x, y) to (σ_j) is of the form

$$[p(s) - p(x)]f(x) + L^1[p(z_1) - p(s)] + L^2[p(z_2) - p(z_1)] + \cdots + [p(y) - p(t)]f(y).$$

Now, $f(x), L^1, L^2, \ldots, f(y)$ are at most equal to the upper bound L of f in (α, β); therefore, the previous sum is at most equal to

$$L[p(y) - p(x)],$$

that is, the contribution of (α, β) in s_i. Therefore, we have

$$(\sigma_j) \leq s_i$$

and as a result

$$\sigma_j \leq s_i + \varepsilon,$$

as soon as j is sufficiently large. It follows that the s_i and the σ_j converge towards the same limit.

In short, we have just proved the theorem for a monotone function and the existence of the limits.

$$\overline{\int_a^b} f(x)\, d[p(x)], \quad \underline{\int_a^b} f(x)\, d[p(x)],$$
$$\overline{\int_a^b} f(x)\, d[n(x)], \quad \underline{\int_a^b} f(x)\, d[n(x)]$$

is proved.

By denoting lim S as any of the limits of the number S, we can write:

$$\begin{cases} \underline{\underline{\int_a^b}} f(x)\,d[p(x)] - \overline{\int_a^b} f(x)\,d[n(x)] \le \lim S \\ \overline{\int_a^b} f(x)\,d[p(x)] - \underline{\int_a^b} f(x)\,d[n(x)] \ge \lim S \end{cases} \quad \text{or} \quad D \le \lim S \le E.$$

The difference between the extreme members of this inequality is, according to the manner in which it was obtained, the limit of

$$\left[\sum_0^n L_i \delta_i p - \sum_0^n l_i \delta_i n\right] - \left[\sum_0^n l_i \delta_i p - \sum_0^n L_i \delta_i n\right]$$
$$= \sum_0^n (L_i - l_i)(\delta_i p + \delta_i n) = \sum_0^n \omega_i \delta_i \nu.$$

Now, we have

$$\overline{S} - \underline{S} = \sum_0^n L'_i \delta_i \alpha - \sum_0^n l'_i \delta_i \alpha = \sum_0^n (L'_i - l'_i)\delta_i \alpha = \sum_0^n \omega_i |\delta_i \alpha|.$$

By comparing these two differences, we obtain

$$0 \le \sum_0^n \omega_i \delta_i \nu - \sum_0^n \omega_i |\delta_i \alpha| = \sum_0^n \omega_i [\delta_i \nu - |\delta_i \alpha|];$$

all the ω_i are at most equal to the oscillation Ω of $f(x)$ in (a, b), since the $\delta_i \nu - |\delta_i \alpha|$ are positive, this quantity is at most equal to

$$\Omega \sum_0^n [\delta_i \nu - |\delta_i \alpha|] = \Omega V - \Omega \sum_0^n |\delta_i \alpha|.$$

But we know (Sect. 5.1) that in the considered conditions here $\sum_0^n \delta_i \alpha$ tends towards V. From this, the following relations follow:

$$D \le \lim \overline{S} \le E; \quad D \le \lim \underline{S} \le E;$$
$$\lim(\overline{S} - \underline{S}) = E - D;$$

Therefore, we have

$$\lim \overline{S} = E, \quad \lim \underline{S} = D.$$

The theorem is proved and we have for the integrals by upper and lower sums, the expressions

$$E = \overline{\int_a^b} f(x)\,d[\alpha(x)] = \overline{\int_a^b} f(x)\,d[p(x)] - \underline{\int_a^b} f(x)\,d[n(x)].$$

$$D = \underline{\int_a^b} f(x)\,d[\alpha(x)] = \underline{\int_a^b} f(x)\,d[p(x)] - \overline{\int_a^b} f(x)\,d[n(x)].$$

For our statements to reduce exactly to those of Chap. 2, when we set $\alpha(x) \equiv x$, let us define the mean oscillation of $f(x)$ in (a, b), taken with respect to $\alpha(x)$, the limit of the ratio

$$\frac{\sum_0^n \omega_i \delta_i \nu}{\sum_0^n \delta_i \nu},$$

that is, the number

$$\omega = \frac{E - D}{V};$$

then, *the necessary and sufficient condition for the sums S to converge,*[13] *towards a definite limit, which we will call the Stieltjes–Riemann integral of $f(x)$ taken with respect to $\alpha(x)$, is that $f(x)$ has a mean oscillation, taken with respect to $\alpha(x)$, equal to zero in the given interval.*

Since the convergence of $\sum_0^n \omega_i \delta_i \nu$ towards zero is the condition for integrability, it immediately follows, as in Chap. 2, that *the necessary and sufficient condition for $f(x)$ to be integrable, in the Stieltjes–Riemann sense, with respect to $\alpha(x)$, is that the set of points where $f(x)$ has a discontinuity of at least ε, for any $\varepsilon > 0$, should form an integrable group with respect to $\alpha(x)$.* The meaning of integrable group with respect to $\alpha(x)$ is, any set of points which can be enclosed *in the interior*[14] of a finite number of intervals yield a sum of increments $\sum \delta\nu$ of the total variation, as small as desired.

The word *interior* is indispensable if $\alpha(x)$ is discontinuous; then a set reduced to a single point P is not always an integrable group. Indeed, if P is a point of discontinuity of $\alpha(x)$, there does not exist an interval containing P in its interior and for which $\delta\nu$ is less than $\nu(P + 0) - \nu(P - 0)$. Thus, the integrable groups with respect to $\alpha(x)$ are formed from points of continuity of $\alpha(x)$; a function $f(x)$ can be integrable, in the Stieltjes–Riemann sense, with respect to $\alpha(x)$, only if it is continuous at all the points of discontinuity of $\alpha(x)$. On the other hand, a function can have a Stieltjes–Riemann integral and still admit all the points of interval as points of continuity. It is sufficient for this interval to be an integrable group with respect to $\alpha(x)$, which means that $\alpha(x)$ is a constant in this interval. It is evident that the nature of integrable groups varies as $\alpha(x)$ varies.

---

[13] We will note that here it is no longer necessary to consider only sequences of divisions D_i such that every point of discontinuity of $\alpha(x)$ belongs to all the D_i from a certain value of the index.

[14] That is, ≪ enclosed in the open intervals ≫.

By using the theory of measure that will be developed, the reader will easily show, as in Chap. 2, *the necessary and sufficient condition for a function to have a Stieltjes–Riemann integral is that the set of its points of discontinuity should have a measure zero with respect to the determinate function* $\alpha(x)$.

Let us not study the initial definition of the Stieltjes integral any longer and, to prepare a broader definition of this integral, let us define the measure of a set, taken with respect to a function $\alpha(x)$, of bounded variation.

We will agree that the measure of closed interval (l, m) is

$$m_{\alpha(x)}[l \le x \le m] = \alpha(m+0) - \alpha(l-0),$$

the measure of a point is

$$m_{\alpha(x)}(x_0) = \alpha(x_0+0) - \alpha(x_0-0);$$

from there we determine the measure of an open or a semi open interval by subtracting from the measure of a closed interval, the measure of either or both of its two extreme points.

It is obvious that this function of intervals is completely additive. Therefore, as mentioned in Sect. 9.3, it corresponds to a completely additive set function, defined particularly for all B measurable sets. We now elaborate on this result.

Let us recall that the sets E_x, for which we defined the function $\mathfrak{A}_{\alpha(x)}(E_x)$, denoted here as $m_{\alpha(x)}(E_x)$, are those which are transformed into measurable sets $\mathcal{E}_t$ with respect to t by means of the change of variable

$$t = x + V(x);$$

$V(x)$ always denotes the total variation of $\alpha(x)$ from a to x; this change of variable is to be interpreted as it has been explained in Sect. 9.3.

To say that $\mathcal{E}_t$ is measurable means that we can enclose it in a set e_t of open intervals and its complement $\mathfrak{F}_t$ in a set f_t of open interval, such that the length of the subsets common to e_t and f_t is at most ε, as small as we want. To e_t and f_t, there correspond sets e_x and f_x of intervals enclosing E_x and its complement, F_x. This is achieved by ensuring that the intervals constituting e_t and f_t are chosen in such a way that none of them has an extremity in the interior of an interval (t_1, t_2) corresponding to a singular point of $\alpha(x)$; which is possible since these intervals are completely in $\mathcal{E}_t$ or $\mathfrak{F}_t$.[15] In each interval of the x-axis we have

$$\delta x \le \delta t, \quad \delta V < \delta t.$$

From the first inequality, we have already deduced that E_x is measurable, because it implies $\sum \delta x \le \varepsilon$, where the summation is extended over the intervals common to e_x and f_x. The second inequality provides us $\sum \delta V < \varepsilon$ and shows us that the E_x are measurable with respect to $\alpha(x)$ by agreeing that: *a set* E_x *is said to be measurable with respect to* $\alpha(x)$ *if it can be enclosed in an infinite number of open intervals* e_x *and, if we can enclose the*

[15] Compare, Sect. 9.3.

complement F_x of E_x in an infinite number of open intervals, f_x, such that the sum $\sum \delta V$, extended over intervals common to e_x and f_x, can be made as small as we want.

Therefore, we have defined $m_{\alpha(x)}(E)$ for sets which are at the same time measurable in the ordinary sense and measurable with respect to $\alpha(x)$.[16] If we were to stop there, measurability in the ordinary sense would play a separate role. We completely generalise the theory of measure only by defining measure with respect to $\alpha(x)$ for all the measurable sets with respect to $\alpha(x)$, without additionally requiring that they are measurable in the ordinary sense. That is what we are going to do now.[17]

The problem of measure that we have solved in Chap. 7 can be stated as follows.

Find a set function which is:

1. *Positive or null;*
2. *Completely additive;*
3. *Which, for the open and closed intervals, reduces to the usual measure of these intervals.*

When it comes to the measure with respect to a non-decreasing function $\alpha(x)$, we can retain this statement exactly. Then, we define the exterior measure of E_x as the lower limit of the sums $\sum \delta\alpha$ over the sets of intervals enclosing E_x. The measure of (a, b) minus the exterior measure of the complement F_x of E_x gives the interior measure of E_x. Clearly, that the first of these measures is at least as large as the second; these two measures are equal only for sets measurable with respect to $\alpha(x)$. In short, for a non-decreasing $\alpha(x)$, the theory of the measure with respect to $\alpha(x)$ is constructed identically to that of ordinary measure, that is, for $\alpha(x) \equiv x$.

However, if $\alpha(x)$ is only of bounded variation, condition 1 cannot be retained, since it is no longer satisfied for all intervals. We will replace it with the following:

1′. *The measure of a set with respect to a non-decreasing function is positive or zero. The measure of a set with respect to $\alpha(x)$ is at most equal, in absolute value, to the measure of the same set with respect to the total variation $\nu(x)$ of $\alpha(x)$.*

If E_x is a set, let us enclose it in sequences $\mathcal{E}_x^1, \mathcal{E}_x^2, \ldots$ of sets of open intervals such that the corresponding sums $\sum^1 \delta\nu, \sum^2 \delta\nu, \ldots$ tend towards the smallest possible value. Let $\mathcal{E}_x^{ij}$ be the set common to $\mathcal{E}_x^i$ and $\mathcal{E}_x^j$; as $\mathcal{E}_x^{ij}$ encloses E_x, it provides a sum $\sum^{ij} \delta\nu$, such that $\sum^i \delta\nu - \sum^{ij} \delta\nu$, $\sum^j \delta\nu - \sum^{ij} \delta\nu$ tends towards zero when i and j both increase indefinitely. Now, the three sets $\mathcal{E}_x^i$, $\mathcal{E}_x^j$ and $\mathcal{E}_x^{ij}$ provide the sums of increment of $\alpha(x)$ such that we have

$$\left|\sum^i \delta\alpha - \sum^{ij} \delta\alpha\right| \le \sum^i \delta\nu - \sum^{ij} \delta\nu, \quad \left|\sum^j \delta\alpha - \sum^{ij} \delta\alpha\right| \le \sum^j \delta\nu - \sum^{ij} \delta\nu.$$

[16] We have defined $m_{\alpha(x)}(E)$ for *all* the sets satisfying these two conditions at the same time.

[17] It will be noted that the result we are going to obtain is the one that would be given by the change of variable $\nu = V(x)$ used instead of the change $t = x + V(x)$.

Therefore, the numbers $\sum^i \delta\alpha$ converge when i increases indefinitely; their limit is what we call the *exterior measure, with respect to* $\alpha(x)$ *of* E_x. *The interior measure of* E_x *is the measure of* (a, b) *minus that of the complement* F_x *of* E_x.

Now, suppose E_x is measurable with respect to $\alpha(x)$ and let us enclose F_x in a sequence of sets of intervals—the sets $\mathfrak{F}_x^i$, providing sums $\sigma^i \delta\nu$ and $\sigma^i \delta\alpha$—such that the subsets common to $\mathcal{E}_x^i$ and $\mathfrak{F}_x^i$ provide a sum $\tau^i \delta\nu$ tending towards zero when i increases. Then, the sum $\tau^i \delta\nu$ provided by these common subsets tends *a fortiori* towards zero and as we obviously have

$$m_{\alpha(x)}[\mathcal{E}_x^i] + m_{\alpha(x)}[\mathfrak{F}_x^i] = m_{\alpha(x)}[\alpha \leq x \leq b] + \tau^i \delta\alpha,$$

it follows

$$m(\text{ext.})_{\alpha(x)}[E_x] + m(\text{ext.})_{\alpha(x)}[F_x] = m_{\alpha(x)}[a \leq x \leq b],$$

that is

$$m(\text{ext.})_{\alpha(x)}[E_x] = m(\text{int.})_{\alpha(x)}[E_x].$$

Thus, the exterior and interior measures, with respect to $\alpha(x)$, *of a measurable set with respect to* $\alpha(x)$ *are equal. Their common value is the measure of the set, taken with respect to* $\alpha(x)$.

For proving this last point, let us note that the set $\mathcal{E}_x^i - E_x$, which is enclosed in the subsets common to $\mathcal{E}_x^i$ and $\mathfrak{F}_x^i$, has measure, with respect to $\nu(x)$, equal to $\tau^i \delta\nu$, at most. A fortiori, we have

$$|m_{\alpha(x)}[\mathcal{E}_x^i - E_x]| < \tau^i \delta\nu;$$

hence we conclude

$$m_{\alpha(x)}[E_x] = \lim_{i \to \infty} m_{\alpha(x)}[\mathcal{E}_x^i] = m(\text{ext.})_{\alpha(x)}[E_x].$$

After finding the unique number which can satisfy the conditions of the problem of measure, it remains to verify whether it actually satisfies them. To simplify, and to return to previous considerations, let us conclude this from the correspondence between the sets E_x located in (a, b) and sets E_ν in $(0, V)$ that has already been useful[18] to us; in this correspondence, every point x in (a, b) is associated with the interval $[\nu(x-0), \nu(x+0)]$ in $(0, V)$. To any interval E_x, corresponds an interval E_ν whose length is the measure of E_x with respect to $\nu(x)$. Consequently, the sets E_x measurable with respect to $\nu(x)$ are those which provide sets E_ν measurable in the ordinary sense, and for them, we have

$$m_{\nu(x)}[E_x] = m[E_\nu].$$

Since the sets measurable with respect to $\alpha(x)$ and with respect to $\nu(x)$ are, by definition, the same, we now know which sets are measurable with respect to $\alpha(x)$.

---

[18] *See* Sect. 12.1. For direct proofs, we can refer to the aforementioned Book of M. de la Vallée Poussin.

Moreover, the measure with respect to $\alpha(x)$ of an interval $E_x = (l, m)$ is $\alpha(m+0) - \alpha(l-0)$, in other words, the increment $A(\mu) - A(\lambda)$ undergone by the function $A(\nu)$, in the interval $E_\nu = (\lambda, \mu)$ transformed from (l, m). Therefore, the measure just been defined is identical to the completely additive function of the set E_ν determined on $(0, V)$ by the absolutely continuous function $A(\nu)$; the definitions of these two functions are identical. Thus we have

$$m_{\alpha(x)}[E_x] = \int_{E_\nu} {}'A(\nu)\, d\nu;$$

which allows us to derive all the properties of the measure from the known properties of integrals of the summable functions. Let us limit ourselves to this indication and pass on to the extension of the concept of the integral.

The problem of integration that we solved in Chap. 7 can be stated thus:

To attach to any function f defined on (a, b) a number $I(f)$ such that

1. $I(f_1 + f_2) = I(f_1) + I(f_2)$;
2. $I(f_1 + f_2 + \cdots) = I(f_1) + I(f_2) + \cdots$,
   when the series $f_1 + f_2$ is uniformly convergent;
3. and when this series is convergent and of positive terms;
4. $I(f)$ reduces to the usual integral of f when f is continuous;
5. $I(f) = I(g)$ if f differs from g only at the points of a set of measure zero.

We will retain this statement for the extension of the Stieltjes integral; only, the integral and the measure referred to in number 4 and 5 will now be the integral and the measure with respect to $\alpha(x)$.

To solve this problem, we only need to revisit, with slight modification due to condition 5, the reasonings used for the extension of a linear functional, Sect. 12.2 and we will fall back on the same considerations as in Chap. 7.

Let us consider a function ψ taking only the values 0 and 1. If $E[\psi = 1]$ is an interval (l, m), then ψ is the limit of decreasing sequence of continuous functions f_p equal to 1 in (l, m), zero outside of $\left(l - \frac{1}{p}, m + \frac{1}{p}\right)$, and linear in $\left(l - \frac{1}{p}, l\right)$, $\left(m, m + \frac{1}{p}\right)$. From 1 and 3 it follows that $I(\psi)$ is the limit of $I(f_p)$, a number which, being the Stieltjes integral of the continuous function f_p, differs less and less from $\alpha(m+0) - \alpha(l-0)$ that is, from $m_{\alpha(x)}[E(\psi = 1)]$. We therefore have

$$I(\psi) = m_{\alpha(x)}[E(\psi = 1)],$$

when $E(\psi = 1)$ is an interval. By the addition of such functions and the application of condition 3, we see that we have the same equality when $E(\psi = 1)$ is a set of intervals. Now, let us suppose that $E(\psi = 1)$ is a measurable set with respect to α and, let us enclose this set in sets of intervals yielding sums $\sum \delta\nu$ tending towards the smallest possible limit. Moreover, we can assume that each of these sets of intervals contain following intervals.

To these sets of intervals correspond the functions $\psi_1, \psi_2, \ldots$, taking only the values 0 and 1, and such that these sets are denoted $E[\psi_1 = 1]$, $E[\psi_2 = 1], \ldots$. Let ψ_0 be the function towards which the functions ψ_i converge in a decreasing manner. We have, according to 3,

$$I[\psi_0] = \text{limit of} \quad I[\psi_i] = \text{limit} \quad m_{\alpha(x)}[E(\psi_i = 1)] = m_{\alpha(x)}[E(\psi = 1)]$$

and, according to 5,

$$I[\psi_0] = I[\psi].$$

Hence, again

$$I[\psi] = m_{\alpha(x)}[E(\psi = 1)].$$

Now, let $f(x)$ be a *bounded and measurable function with respect to* $\alpha(x)$, meaning that all sets such as $E[l_i \leq f(x) < l_{i+1}]$ are measurable with respect to $\alpha(x)$.

Let (l, L) be an interval containing in its interior, the interval of variation of $f(x)$. Let us partition this interval using the numbers

$$l_0 = l < l_1 < l_2 < \cdots < l_n = L,$$

let us suppose that $l_{i+1} - l_i$ is never greater than ε.

Let us denote by ψ_i, $(i = 0, 1, 2, \ldots, n)$ the function equal to 1 when x belongs to $E[l_i \leq f(x) < l_{i+1}]$ and zero elsewhere. Since we know $I(\psi_i)$, conditions 1 and 2 yield $I(\varphi)$ for the function

$$\varphi(x) = \sum_0^n l_i \psi_i(x).$$

Now, $\varphi(x)$ tends towards $f(x)$ uniformly when ε tends towards zero, as we have

$$0 \leq f(x) - \varphi(x) < \varepsilon,$$

therefore, according to 2,

$$I(f) = \lim_{\varepsilon \to 0} I[\varphi(x)] = \lim_{\varepsilon \to 0} \sum_0^n l_i m_{\alpha(x)}[E(\psi_i = 1)].$$

Thus, the integral of a bounded functions is obtained.

If $f(x)$ is still assumed to be measurable with respect to $\alpha(x)$, but is no longer necessarily assumed to be bounded, the l_i will be taken staggered from $-\infty$ to $+\infty$ and separated from one another by ε at most. Then we see immediately that if the infinite series in both directions

$$\sum_0^n l_i m_{\nu(x)}[E(\psi_i = 1)]$$

is absolutely convergent for one choice of l_i, it will be so for any choice of l_i and for any ε. The function $f(x)$ is, then, said to be *summable with respect to* $\alpha(x)$. By condition 3, it follows that, for such a function, we again have

$$I(f) = \lim_{\varepsilon \to 0} \sum_{-\infty}^{+\infty} l_i m_{\alpha(x)}[E(\psi_i = 1)];$$

what can be written again as

$$I(f) = \lim_{\varepsilon \to 0} \sum_{-\infty}^{+\infty} l_i m_{\alpha(x)}[E(l_i \leq f(x) < l_{i+1})].$$

The integral, with respect to $\alpha(x)$ of a summable function with respect to $\alpha(x)$, is thus defined in all cases exactly as in the particular case of $\alpha(x) \equiv x$.

It is easy to show directly, using arguments completely similar to those of Chap. 7, that this definition provides a definite number, which indeed satisfies the conditions of our problem, allows us to derive its main properties. However, we are going to do all of this in one step by proving that $I(f)$ is identical to the integral $\int_a^b f(x)\, d[\alpha(x)]$ defined in Sect. 12.1. We would use the functions $x(\nu)$ and $A(\nu)$ which were employed at that time.

The previous expression derived from the measure of a set, with respect to $\alpha(x)$, gives us

$$I(f) = \lim_{\varepsilon \to 0} \sum l_i m_{\alpha(x)}[E(\psi_i = 1)] = \lim_{\varepsilon \to 0} \sum l_i \int_{E_\nu(\psi_i = 1)} {}'A(\nu)\, d\nu;$$

where $E_\nu(\psi_i = 1)$ is the transform on $(0, V)$, of $E(\psi_i = 1)$.

Hence, we deduce, by using the properties of the ordinary integral of the summable functions,

$$\begin{aligned} I(f) &= \lim_{\varepsilon \to 0} \sum l_i \int_0^V \psi_i[x(\nu)].{}'A(\nu)\, d\nu = \lim_{\varepsilon \to 0} \int_0^V \varphi[x(\nu)].{}'A(\nu)\, d\nu \\ &= \int_0^V \lim_{\varepsilon \to 0} \{\varphi[x(\nu)].{}'A(\nu)\}\, d\nu = \int_0^V f(x, \nu).{}'A(\nu)\, d\nu \\ &= \int_a^b f(x)\, d[\alpha(x)]. \end{aligned}$$

The definition we have just given, which is due to M. Radon, is therefore equivalent to that in Sect. 12.1 and this exempts us from any direct study of the properties of the Stieltjes integral. However, we will show, by way of example, how the generalisation of the notion of the absolutely continuous function[19] appears. However, first let us examine how it happens

[19] Among the exercises that the reader can tackle, I highlight the following. Apply Jordan's methods to the problem of the measure relative to $\alpha(x)$; define the range with respect to $\alpha(x)$; show that the J measurable functions with respect to $\alpha(x)$ are the integrable functions in the Stieltjes–Riemann sense, with respect to $\alpha(x)$; compare the field of extension of the definition of M. Radon to that of the various definitions given in Sect. 12.1 of this chapter. In particular, show that, just as the extension

that the notions of the B measurable sets, B measurable functions are independent of the function $\alpha(x)$ determining the problems of measure or integration we are dealing with, whereas the notion of measurable sets and measurable functions vary with $\alpha(x)$.

The reason is that these are notions of entirely different natures; M. Denjoy would say that the first ones are descriptive and the second metric. M. Borel introduced the B measurable sets in the context of measure theory, and that is where their names come from, but he did not characterise them by a metric property: he indicated what geometric operations, performed from intervals and the points, allow us to obtain these sets. The importance of B measurable functions in Analysis is primarily from the fact that these functions encompass all those which fall into the classification of M. Baire, all those which are amenable to an analytic representation.[20] Here, the importance of B measurable sets and functions come, as we might have noticed, from the fact that for them, the measure or integral is determined by the conditions of our problems which can be expressed as equalities, and without the need to use those involving inequalities; that is to say, with the help of the conditions 2, 3 (Sect. 12.3), conditions 1, 2, 3, 4 (Sect. 12.3). Now, the field of these sets is so vast that it took great efforts to construct some examples of the sets or the functions which are B non measurable; that is to say, there would be no practical inconvenience in limiting ourselves to the study of the B measurable sets and functions.

Let us propose to characterise the functions $F(x)$ which are indefinite integrals with respect to a known function of bounded variation $\alpha(x)$, these functions $F(x)$ are the ones which we could call *absolutely continuous functions with respect to* $\alpha(x)$.

A first condition is that $F(x)$ is of bounded variation and have left and right jumps at every point that are proportional to those of $\alpha(x)$, according to the expression of the jump function of a given integral (Sect. 12.1)

$$\frac{F(x) - F(x-0)}{\alpha(x) - \alpha(x-0)} = \frac{F(x+0) - F(x)}{\alpha(x+0) - \alpha(x)}.$$

Let us search the other conditions: we want to have

$$F(x) = C + \int_a^x f(x)\, d[\alpha(x)] = C + \int_0^{V(x)} f[x(\nu)].'A(\nu)\, d\nu$$

for an unknown function $f(x)$. The continuous functions

---

of the notion of measure obtained (Sect. 12.3) using a change of variable of the form $t = x + \nu(x)$, applies only to sets that are measurable in both the ordinary sense and with respect to $\alpha(x)$, the extension obtained (Sect. 12.1) using the same change of variable applies only to the functions that are both measurable and summable in the ordinary sense and with respect to $\alpha(x)$.

It is also clear that two of these definitions are always consistent when they both apply, since they define linear functionals satisfying the third condition in Sect. 12.2 and they are identical for the continuous functions.

[20] LEBESGUE, *Journ. de Math.*, 1905.

$$\Phi(\nu) = C + \int_0^{\nu} f[x(\nu)].'A(\nu)\, d\nu$$

is defined on entire $(0, V)$. For any value of ν, such that the equation $\nu = V(x)$ have at least one root, we have

$$\Phi(\nu) = F[x(\nu)];$$

the only values of ν at which we do not have this equality are, therefore, those for which we have an inequality of the form

$$V(x_0 - 0) = \nu_1 \le \nu < V(x_0) = \nu_0$$

or, of the form

$$V(x_0) = \nu_0 < \nu \le V(x_0 + 0) = \nu_2.$$

in (ν_1, ν_2), $x(\nu)$ is constant and equal to x_0; $A(\nu)$ is linear in (ν_1, ν_0) and (ν_0, ν_2) and takes the values $\alpha(x_0 - 0)$, $\alpha(x_0)$, $\alpha(x_0 + 0)$ for ν_1, ν_0, ν_2. Therefore, $\Phi(\nu)$ is linear in (ν_1, ν_0) and (ν_0, ν_2) and undergoes the increments

$$f(x_0)[\alpha(x_0) - \alpha(x_0 - 0)], \quad f(x_0)[\alpha(x_0 + 0) - \alpha(x_0)]$$

Hence, it follows that if we set $\Psi[V(x)] = F(x)$ and, if we agree to complete the definition of Ψ in such a manner that it is everywhere continuous and, linear in the intervals where it was not yet determined, we must have $\Psi(\nu) \equiv \Phi(\nu)$. In other words, $\Psi(\nu)$ must be absolutely continuous with respect to ν.

Let us show that this condition, in conjunction with the previous one, is sufficient. Therefore, let us assume that these conditions are satisfied. If x_0 is a point of discontinuity of $\alpha(x)$, we will take

$$f(x_0) = \frac{\Psi[V(x_0)] - \Psi[V(x_0 - 0)]}{\alpha(x_0) - \alpha(x_0 - 0)} = \frac{F(x_0) - F(x_0 - 0)}{\alpha(x_0) - \alpha(x_0 - 0)},$$
$$f(x_0) = \frac{\Psi[V(x_0 + 0)] - \Psi[V(x_0)]}{\alpha(x_0 + 0) - \alpha(x_0)} = \frac{F(x_0 + 0) - F(x_0)}{\alpha(x_0 + 0) - \alpha(x_0)};$$

these two expressions of $f(x_0)$ are quite in agreement because of the first condition.

If x_0 is a point of continuity of $\alpha(x)$, for which the equation $V(x) = V(x_0)$ admits only the root x_0, we will take

$$f(x_0) = \frac{'\Psi(\nu)}{'A(\nu)};$$

a finite quantity since $'A(\nu)$ has been taken equal to $+1$ or -1.

Finally, at the points where $f(x)$ is not yet defined, we will take $f(x)$ arbitrarily. Indeed, these points correspond to a countably infinite number of values of $\nu = V(x)$ having no influence on the integral taken from 0 to V.

With these choices, it is clear that we have

$$\Psi(\nu) = C + \int_0^\nu f[x(\nu)].'A(\nu)\,d\nu,$$

therefore,

$$F(x) = C + \int_a^x f(x)\,d[\alpha(x)].$$

Let us transform the second condition we found; for that let us note that, as soon as the first condition is met, we can calculate $f(x)$ at the points of discontinuity of $\alpha(x)$, and therefore $f[x(\nu)]$ in the various intervals (ν_1, ν_2). Let us denote by $g(\nu)$ the function equal to $f[x(\nu)]$ in the intervals (ν_1, ν_2) and zero elsewhere. The function $g(\nu).'A(\nu)$ is summable, because in (ν_1, ν_2) its integral is $F(x_0) - F(x_0 - 0)$ and in (ν_0, ν_2) it is $F(x_0 + 0) - F(x_0)$, and the sum of the absolute values of all these integrals is bounded since $F(x)$ was assumed to be of bounded variation.

The function

$$\Psi_1(\nu) = \int_0^\nu g(\nu).'A(\nu)\,d\nu$$

can therefore be calculated as soon as the first condition is met. Moreover, it is clear from the above that $\Psi_1(\nu)$ is the transform of the jump function $S(x)$ of $F(x)$.

$\Psi_1(\nu)$ being absolutely continuous, it is sufficient for us to express that $\Psi(\nu) - \Psi_1(\nu) = \Psi_2(\nu)$ is also absolutely continuous. This means that if we form the sum $\sum \Delta\Psi_2$, extended to a set I_ν of intervals—assuming that they do not have their origins or the extremities in the intervals (ν_1, ν_2), since $\Psi_2(\nu)$ is constant in such intervals, and we can assume them to be open since $\Psi_2(\nu)$ is continuous—and if the total measure of these intervals is ε, it tends towards zero with ε.

However, since the I_ν have neither their origins nor their extremities in the (ν_1, ν_2), and that these are the open intervals; they are indeed transformed from sets I_x of open intervals of x-axis and the sum to be considered is

$$\sum_{I_x} |\Delta[F(x) - S(x)]|.$$

$\Psi_1(\nu)$ is absolutely continuous,

$$\sum_{I_\nu} |\Delta\Psi_1| = \sum_{I_x} |\Delta S|$$

tends towards zero with ε; therefore it is necessary and sufficient that $\sum_{I_x} |\Delta F|$ tends towards zero.

For a function $F(x)$ to be an indefinite integral with respect to $\alpha(x)$, it is necessary and sufficient that:

1. *$F(x)$ is of bounded variation;*
2. *At every point, the left and right jumps of $F(x)$ are proportional to the jumps of $\alpha(x)$;*
3. *The sum $\Delta|F(x)|$, extended to a set of open intervals whose measure, with respect to the total variation $V(x)$ of $\alpha(x)$, is equal to ε, tends towards zero with ε.*

The answer is much simpler when it comes to determining how to recognise that a function of B measurable sets is an indefinite integral with respect to a given $\alpha(x)$. First, let us recall that for calculating such an indefinite integral, we can eliminate all *unnecessary singularities* of $\alpha(x)$, meaning we replace $\alpha(x)$ with a function that is equal to $\alpha(x)$ at a, b, and all points of continuity of $\alpha(x)$, and which does not exhibit two jumps of opposite signs at any point. Therefore, let us assume that $\alpha(x)$ has only useful singularities left.

An indefinite integral with respect to $\alpha(x)$ *is a function of* B *measurable set:*

1. *Which is completely additive;*
2. *Which is zero in every set of measure zero with respect to the total variation* $V(x)$ *of the function* $\alpha(x)$*, assuming there are no unnecessary singularities, and conversely.*

The stated property of the indefinite integrals is simply a translation of the fact that when considered as attached to sets of $(0, V)$, this set function is absolutely continuous. Let us examine the converse.

To a function φ on the B measurable set E_x, our change of variable associates a function Φ on the B measurable set $\mathcal{E}_\nu$, but is not defined for every B measurable set of $(0, V)$. Indeed, we know its value W in an interval (ν_1, ν_2) corresponding to a singular point x_0 of $\alpha(x)$, but we do not know its value U in a B measurable set e_ν contained in (ν_1, ν_2); let us agree that we will have

$$\frac{U}{m(e_\nu)} = \frac{W}{\nu_2 - \nu_1}$$

This convention is sufficient to determine Φ for all the B measurable sets of $(0, V)$, because we want that Φ to be completely additive.

Since φ is zero in any set of measure zero with respect to $V(x)$, Φ is zero in any set of measure zero. Therefore Φ is an indefinite integral, and according to how Φ has been chosen in the interior of each (ν_1, ν_2), the function that is summable with respect to ν, whose Φ is the indefinite integral, is constant in (ν_1, ν_2);

We do have

$$\Phi(\mathcal{E}_\nu) = \int_{\mathcal{E}_\nu} f[x(\nu)].'A(\nu)\,d\nu,$$

for a certain function $f(x)$; the almost derivative $'A(\nu)$ is constant in every (ν_1, ν_2) because $\alpha(x)$ does not have unnecessary singularities.

The final form we have given, (Sect. 9.3), of the condition of absolute continuity, can be generalised quite literally; from this, we could easily derive the generalisation of other forms of this condition.

12.4 Physical Significance of Integral of Stieltjes

We have just generalised one of the modes of analytic definition of the integral; the other modes of analytic definition are amenable to similar generalisations.[21] But is there not, for the Stieltjes integral, a definition similar to the geometric definition of the integral, one that appears as a simple refinement of an intuitive definition? This mode of definition does exist; it certainly guided Stieltjes' initial ideas, although Stieltjes does not emphasise it; his analytic presentation is completely satisfactory from a logical point of view, as a result, the intuitive significance of the Stieltjes integral was somewhat forgotten.

Nevertheless, Stieltjes does say: let us suppose that there is heavy matter spread over Ox. Let $u(x)$ be the mass located on $(0, x)$; let us calculate the moment of the total mass with respect to the origin. For that, let us partition the considered interval using increasing values ξ_i. We will have an approximate value of the moment in the form

$$\sum \xi_i [u(\xi_{i+1}) - u(\xi_i)];$$

from this, we obtain, for the exact value of the moment, an integral $\int x\, d[ux]$.

However, the meaning of the Stieltjes integral is much more clearly provided by Cauchy, who had considered integration with respect to a function before Stieltjes. He did so from a more extensive physical perspective, albeit in a much less precise form, from the logical point of view[22] than Stieltjes.

Cauchy's starting point is the notion of *coexisting quantities*, a broader notion than that of a function, with the later being just a particular case.

The quantities are said to be coexistent when they are determined by the same geometric or physical conditions. For instance, the surface and volume of a cylinder are coexisting quantities, determined simultaneously by the datum of the cylinder. In a gaseous expanse, we isolate the matter contained within a certain domain in our thoughts; we can consider the volume of the conceived body, its mass, and the amount of heat needed to raise its temperature by one degree at constant volume as three coexistent quantities.

The numbers that measure these quantities are not necessarily functions of each other, as the previous examples illustrate. They can be functions in some cases, though. For example, the radius, height, surface area, and volume of a cylinder of revolution are coexistent quantities and any two of them can determine the other two. In a more general sense, if one quantity is a function of another, then all these quantities are coexistent. We are accustomed to reason on the variables and functions, but there is just as much a reason to argue on the coexisting quantities. For example, we can establish relationships of inequality between

---

[21] It has already been said that the definition of M. W. H. Young was first generalised (Sect. 12.2, in note).

[22] *Sur le rapport differentiel de deux grandeurs qui varient simultanement* (*Ex. d'Analyse, t. II, pp.* 188–229; *OEuvres,* 2nd *serie, t.XII, pp.* 214–262*). See* also the *Traite de Mecanique analytique de l'abbe Moigno.*

the surface area and volume of a cylinder. To provide a solid base to the arguments on the coexistent quantities, we need to clarify this notion, which will also limit its scope.

In the previous examples, coexistent quantities appear to be associated with the same object, such as a cylinder or gaseous body, and they are functions of same domain. Physically measurable quantities also appear, always as functions of domains, although these domains are not always three dimensional. They can be domains on a straight line, meaning, the intervals, planar domains, or domains with more than three dimensions. In the later case, the domain is not directly perceptible, and its purely mathematical conception is somewhat artificial. For example, if we wanted to discuss the quantity of the heat required to raise the temperature of an isolated gaseous body C by δt degrees, and we wanted to vary both δt and the body itself, we would need to consider the quantity of heat associated with a domain of four dimensional space in x, y, z, t coordinates. This space would be obtained by translating the body C, traced at time $t = t_0$, by δt units parallel to the t-axis.

Therefore, we will admit that the quantities we are talking about are functions of domain and, we will replace the notion of coexistent quantities with the much more precise notion of functions of the same domain or, through an abstraction of mathematician, functions of the same set.

In physics, one also encounters functions of points or, if you will, of a certain number of variables. Some are still functions of a domain, but attached to special domains that depend only on a finite number of parameters: for instance, the mass m of the quantity of water contained in a container up to the height h is a function of h, as both m and h are functions of the same domain. The others are truly attached to points; these numbers generally serve to calibrate states, qualities, to distinguish, for example, movements of varying speeds (velocity), materials of varying densities (density).

If we consider the precise definition of these numbers, we will see that we obtain them as the limit of the quotient of two functions of the same domain:

$$\text{Speed} = \lim .\text{mean Speed} = \lim \frac{\text{arc length of trajectory}}{\text{time taken to travel this arc}},$$
$$\text{Density} = \lim .\text{mean Density} = \lim \frac{\text{mass of a body}}{\text{volume of this body}}.$$

As our goal is not to study the numbers of physics, we do not need to investigate if all these numbers fit well into one or the other of two categories mentioned, or if the distinction between these two categories of numbers is absolute. It suffices to have noted the importance, in physics, of functions of domains and the kind of derivation from one domain function with respect to another which provides the functions of points.

It should not surprise us that domain functions are introduced into physics and appear to be more directly suited to the physicist's needs than point functions. A point is merely the limiting concept of successively smaller bodies, and a point function can be introduced in physics only as the limit of a body function, a domain function. However, little is spoken about these functions because mathematicians have not yet created the Algebra and Analysis of domain functions. On the contrary, we possess a remarkably convenient notations for point functions. Therefore, through various techniques—which essentially reduce to reasonings about sufficiently special domains that depend on only a finite number of variables—we always replace the use of domain functions with that of point functions.

The operation of differentiation we encountered is one studied by Cauchy. It is defined as follows: $\varphi(D)$ and $\psi(D)$ being two domain functions, to obtain the derivative at a point P of φ with respect to ψ, we take the limit of ratio $\frac{\varphi(D_i)}{\psi(D_i)}$ for a sequence of increasingly smaller domains D_i that ultimately reduce to the limit at the single point P.

We can similarly define the derivative of a set function with respect to another. We may also need to impose additional restrictions on the sequence of D_i or E_i for the limit to exist, as we did (Sect. 10.2); let us leave these details aside.

Let us propose, following Cauchy, to calculate $\varphi(D)$ knowing $\psi(D)$ and the derivative $f(P)$ of $\varphi(D)$ with respect to $\psi(D)$. This problem would not be well defined, and we would hardly know how to study it, if we allowed the notion of domain functions in all its possible generality. *We will assume that we are dealing with additive domain functions.* This is an important restriction to the concept of Cauchy: the surface and the volume of a body are two coexistent quantities, these are certainly two domain functions, but only the second one is additive.

In reality, for addressing the problem at hand, Cauchy, like we are going to do, unconsciously restricts himself to the case of additive domain functions. Moreover, this restriction is practically justified by the fact that the numbers provided by physics, which we call measure of the quantities,[23] are indeed additive domain functions.

Every point P_0 is then interior to a domain $D(P_0)$ such that the ratio $\frac{\varphi}{\psi}$ differs from $F(P)$ by less than ε for the domain $D(P_0)$ and for all the interior domains, subject to the restrictions which could have been imposed in the definition of the derivative. Using a finite number of these domains $D(P_0)$, we can cover, according to the theorem of M. Borel (Sect. 8.2), the entire domain D under consideration. By restricting these domains $D(P_0)$, we can assume that they have no common interior points. Then we would have partitioned D into partial domains $D_1, D_2, \ldots, D_n$ and taken a particular point P_i in each of them or in neighbourhood of each of them, in such a manner that we have

$$\varphi(D_i) = \psi(D_i)[f(P_i) + \theta_i\varepsilon] \quad (-1 \leq \theta_i \leq +1).$$

If $f(P)$ is continuous, this will be only slightly modified by assuming P_i is taken inside D_i.

From this, it follows

$$\varphi(D) = \sum \varphi(D_i) = \sum \psi(D_i) f(P_i) + \theta\varepsilon \sum |\psi(D_i)|.$$

[23] In my opinion, quantities must be defined right *from the beginning* as numbers associated with domains, such that the quantities associated with domains resulting from the subdivision of another domain have their sum equal to the quantity associated with the latter domain.

If therefore, regardless of the fragmentation of D into the D_i and the choice of P_i, the first sum tends towards a definite limit for increasingly smaller D_i, and the second remains bounded, the latter fact expresses that ψ is of bounded variation - with this knowledge we know how to calculate $\varphi(D)$.

The precision of these insights leads very naturally to the Stieltjes integral. We just need to assume that it concerns one dimensional domains, intervals; that $f(P)$ is $f(x)$, and that $\psi(D)$ is the interval function associated with a function $\alpha(x)$ of bounded variation, to recover the definition posed by Stieltjes, for $\int f(x)\, d[\alpha(x)]$.

It is clear that from these insights, generalisations of this integral to functions of several variables can also be derived. However, we would not dwell on this because, in this Book, we are always limited to the integration of functions of one variable.

These integrals reduce to the ordinary integrals if the function $\psi(D)$ reduces to the measure, in the ordinary sense, of the domain D. Therefore, the extension of the notion of measure studied by M. de la Vallée Poussin, and the form given to the definition of the integral by M. Radon, are closely related to the considerations of physics which we have just developed.

The definition of M. Radon is particularly appropriate if, instead of examining, as Cauchy did, a generalisation of the problem of primitive functions, we study a generalisation of the problem of quadratures. Let us assume that we know that for any domain or set E, the product $\psi(E)\lambda$-the number λ is an intermediate number between the lower limit l and the upper limit L of the values taken over E by the function $f(P)$-is an approximate value of $\varphi(E)$, and this approximation becomes more accurate as $L - l$ gets smaller. Naturally, to calculate $\varphi(D)$, we will be led to examine the sum $\sum i\varepsilon\psi(E_i)$, where E_i is formed from points of D at which we have $i\varepsilon \leq f(P) < (i+1)\varepsilon$. Now, this is precisely the definition given by M. Radon.

It is worth noting that in classical Analysis, sums of the form $\sum \psi(D_i).f(P_i)$ are considered on various occasions. For example, when we calculated a curvilinear integral $\int_c f(x, y, z)\, dx$ we looked for the limit of

$$\sum [x(t_{i+1}) - x(t_i)] f[x(t_i), y(t_i), z(t_i)];$$

In this case, we are dealing with a function $\psi(D)$ equal to the measure of segment projection, on Ox, of the arc D.

Ordinarily, such integrals are considered collectively

$$\int P\, dx + Q\, dy + R\, dz.$$

If we compare this with the classical formula which gives the arc length of a curve in simple cases

$$\int \sqrt{dx^2 + dy^2 + dz^2},$$

a formula that we must treat as a curvilinear integral, by substituting for x, y, z, a function of same parameter t. We would be led to examine the integration with respect to several functions of sets, or if we want, with respect to several coexistent quantities. Let $f(P, u_1, u_2, \ldots, u_p)$ be a function of point P and p variables. Let us assume f to be homogeneous and of degree 1 with respect to the set of these variables. On the other hand, let $\psi_1(E)$, $\psi_2(E)$, $\ldots$ $\psi_p(E)$ be p additive functions of coexistent domain.

The sum

$$\sum f[P_i, \psi_1(E_i), \psi_2(E_i), \ldots, \psi_p(E_i)]$$

extended over a division of a domain or a set into partial sets E_i, where P_i denotes a point of E_i, would define a sort of Stieltjes integral of f with respect to functions $\psi_1, \psi_2, \ldots, \psi_p$, by its limit when it exists.

These summations have still not been studied; M. Hallinger[24], however used an integration which is denoted $\int f(x)\frac{dx\,d\chi}{d\rho}$ and which is the limit of sums

$$\sum f(x_i)\frac{\psi(E_i)\chi(E_i)}{\rho E_i};$$

the integral which was subsequently studied by M. Radon.

Let us conclude this section by noting that, according to Cauchy, the notion of coexistent quantities is of elementary nature and renders a great service if it were used from the beginning of Analysis or even of Geometry. In any case, it seems to us that there would be a significant advantage in initially presenting a notion of integration with respect to a function of domain. This would provide a synthetic view of all types of integrals of continuous functions, the theory of these integral would be obtained more rapidly, and yet it would better prepare for geometric and physical applications.

12.5 Function Primitives and Totalisation with Respect to a Function

For the case of functions of a single variable, the problem of function primitives that we have just encountered is stated as follows: *Being given a function of bounded variation $\alpha(x)$ and a function $f(x)$ in (a, b), find a function $F(x)$ which admits $f(x)$ as derivative with respect to $\alpha(x)$ at every point.*

To that f is the derivative of F, means that we have

$$f(x) = \lim_{\substack{h\to 0^+ \\ k\to 0^+}} \frac{F(x+k) - F(x-h)}{\alpha(x+k) - \alpha(x-h)}$$

if we translate exactly the considerations from the previous section. However, in that case, it is obvious that $F(x)$ would not be determined at any point since $F(x-0)$ and $F(x+0)$

[24] *Journal de Crelle, Bd* 136.

alone[25] would be involved. Therefore, we require that

$$f(x) = \lim_{h \to 0} \frac{F(x+h) - F(x)}{\alpha(x+h) - \alpha(x)},$$

to say that F admits f as derivative. In finding the limit of the right-hand side, we only consider the numbers h for which the right-hand side, has a definite, finite or infinite value. At an interior point of an interval in which both F and α are constant, the derivative would be indeterminate. For any $\varphi(x)$, it could be considered equal to $f(x)$.

We will still use the method of chains of intervals to study this problem. However, we need to proceed with caution, primarily because $F(x)$ is not continuous. Indeed, if we let, h tend towards zero with positive values in the derivative's definition formula, we see that $F(x+0)$ exists, and we obtain

$$\frac{F(x+0) - F(x)}{\alpha(x+0) - \alpha(x)} = f(x).$$

This formula and the similar formula for $x - 0$ reveals us the points of discontinuity and jumps of $F(x)$.

---

[25] This is completely natural, because $\alpha(x)$ appears only to define the function $\psi(D)$, and $F(x)$ is defined by a domain function $\varphi(D)$. Now, the functions $\alpha(x)$ and $F(x)$ correspond to $\psi(D)$ and $\varphi(D)$, respectively, for which

$$\alpha(x-0), \quad \alpha(x+0), \quad F(x-0), \quad F(x+0)$$

are determined up to an additive constant, while $\alpha(x)$ and $F(x)$ are not.

Stieltjes had already noted that for a given distribution of mass on Ox, in other words, for a function $\varphi(D)$, there does not correspond a unique function $\alpha(x)$. When $\alpha(x)$ is given directly, any point of discontinuity x_0 of $\alpha(x)$ corresponds to the concentration of a mass $\alpha(x_0+0) - \alpha(x_0-0)$ at point x_0. Stieltjes imagines that this concentration occurs at two geometrically coincident points at x_0; the first one, carrying the mass $\alpha(x_0) - \alpha(x_0-0)$, belongs to (a, x_0); and the second one, of mass $\alpha(x_0+0) - \alpha(x_0)$ belongs to (x_0, b).

This is equivalent to considering symbols like $x+0$, $x-0$ as numbers in the same way as the symbols x, and all these symbols can be classified in order of magnitude. It is also equivalent to treating sets of numbers $\alpha \leq x \leq \beta$, where α and β are two different numbers, with $\alpha < \beta$, as intervals, and to consider $\varphi(D)$ as a function of such intervals. The intervals would then fall into nine different categories depending on whether their origin and extremity are represented by the symbols $x-0$, x or $x+0$. There would be three kinds of null intervals,

$$(x-0, x), \quad (x, x+0), \quad (x-0, x+0).$$

These conventions would do away with need for the precautions that we previously had to take when dividing one interval into several others (Sect. 9.1); in such a division, any null interval of the form

$$(x-0, x) \quad \text{and} \quad (x, x+0),$$

should appear once and only once.

Because of this discontinuity, if $a, x_1, x_2, \ldots,$ are the division points of the chain, we apply the formula

$$\begin{aligned}F(b) - F(a) =& [F(x_1 - 0) - F(a)] + \cdots \\ &+ \sum [F(x_{i+1} - 0) - F(x_i - 0)] + \cdots + [F(b) - F(b - 0)];\end{aligned}$$

in this formula, the sum $\sum$ must be extended to all the indices i of points of the chain, finite or transfinite, and it must be calculated while taking into account the order of succession of these indices. To show that, under these conditions, the formula is exact, it is sufficient to prove that we have, for any index I,

$$\begin{aligned}F(x_I - 0) - F(a) =& [F(x_I - 0) - F(a)] \\ &+ \sum^{i<I} [F(x_{i+1} - 0) - F(x_i - 0)].\end{aligned}$$

This formula is obviously true for $I + 1$ if it is true for I. Moreover, it also holds for an index I that a transfinite number of the second kind, provided it holds for $I_0 < I$, since as I_0 increases towards I, $F(x_{I_0} - 0)$ tends towards $F(x_I - 0)$. In other words, whether I is of first or second kind, the formula is true for I because it is true for smaller indices. The formula is general. We will take the intervals of the chain in such a manner that we have

$$\left| \frac{F(x_{i+1} - 0) - F(x_i - 0)}{\alpha(x_{i+1} - 0) - \alpha(x_i - 0)} - f(x_i) \right| < \varepsilon,$$

and in a similar manner for (a, x_1). Then

$$\begin{aligned}&F(x_{i+1} - 0) - F(x_i - 0) \\ &= f(x_i)[\alpha(x_{i+1} - 0) - \alpha(x_i - 0)] + \theta_i \varepsilon |\alpha(x_{i+1} - 0) - \alpha(x_i - 0)|,\end{aligned}$$

where θ_i is included between -1 and $+1$; this is again written

$$\begin{aligned}&F(x_{i+1} - 0) - F(x_i - 0) \\ &= f(x_i) m_{\alpha(x)}[x_i \leq x < x_{i+1}] + \theta_i \varepsilon [V(x_{i+1} - 0) - V(x_i - 0)].\end{aligned}$$

$V(x)$ denotes, as always, the total variation of $\alpha(x)$ from a to x. Hence

$$\begin{aligned}f(b - 0) - f(a) =& f(a)[\alpha(x_1 - 0) - \alpha(a)] \\ &+ \sum f(x_i)[\alpha(x_{i+1} - 0) - \alpha(x_i - 0)] + \theta \varepsilon V.\end{aligned}$$

From this formula and the similar one for (a, x), we easily deduce that, whenever $f(x)$ has an integral of Stieltjes–Riemann, $F(x)$ is the indefinite integral of $f(x)$ taken with respect to $\alpha(x)$. Let us content ourselves by deducing that if $f(x)$ is identical to zero, then $F(x)$ is a constant, from which it follows that *the primitive function, with respect to a given function of* *bounded variation $\alpha(x)$, of a given function $f(x)$ is determined up to an additive constant*, since the difference of two primitive functions of $f(x)$ has an identically zero derivative with respect to $\alpha(x)$.

Let us return to the formula which we have just found

$$F(b-0) - F(a) = \lim_{\varepsilon \to 0} \sum_{i=0} f(x_i) m_{\alpha(x)}[x_i \leq x < x_{i+1}],$$

formula in which, for simplicity, we have included under the contribution of (a, x) under the $\sum$ sign, despite its special form. And let us specify, the choice of intervals of the chain, as in Sect. 10.1 and following. For this purpose, let us assume that $f(x)$ be summable with respect to $\alpha(x)$ and let E_l be the set $E[l\varepsilon \leq f(x) < (l+1)\varepsilon]$. Let us enclose E_l in a set A_l of non-intersecting intervals, where the measure of A_l with respect to $V(x)$, does not exceeds that of E_l by more that ε_l at most. Choose the numbers ε_l in such a way that the series $\sum \varepsilon_l$ and $\sum |l|\varepsilon_l$ are convergent and are of very small sums η and ζ.

Let us subject the intervals (x_i, x_{i+1}) whose origin x_i belongs to E_l to be enclosed in A_l and such that

$$\left| \frac{F(x_{i+1}-0) - F(x_i - 0)}{\alpha(x_{i+1}-0) - \alpha(x_i - 0)} - l\varepsilon \right| < \varepsilon;$$

our formula becomes

$$F(b-0) - F(a) = \lim_{\varepsilon \to 0} \sum_{i=0} l\varepsilon m_{\alpha(x)}[x_i \leq x < x_{i+1}].$$

Let us show that the series which appears there is absolutely convergent. Let Λ_l be the measure, with respect to $\alpha(x)$, of those intervals of the chain whose origins are points of E_l, and let Λ'_l be the measure of these same intervals with respect to $V(x)$. By summing the absolute values of the terms of the series, we see that their sum is at most

$$\sum |l|\varepsilon\Lambda'_l \leq \sum |l|\varepsilon m_{V(x)}[A_l] \leq \sum |l|\varepsilon[m_{V(x)}(E_l) + \varepsilon_l],$$

a quantity which is at most equal to

$$\int_a^b [|f(x)| + \varepsilon]\, d[V(x)] + \varepsilon\zeta.$$

Since the series is absolutely convergent, we can group the terms and write

$$F(b-0) - F(a) = \lim_{\varepsilon \to 0} \sum l\varepsilon\Lambda_l$$

Let us show that, as η and ζ tend towards zero, we can replace each Λ_l under the $\sum$ sign with its limit $m_{\alpha(x)}(E_l)$. The series of the right-hand side, having terms which vary less than those of the derivative $\sum |l|\varepsilon\Lambda'_l$, it suffices to justify the passage to the limit for this series. Now, we have

$$m_{V(x)}(E_l) - \eta \leq \Lambda_l' \leq m_{V(x)}(A_l) \leq m_{V(x)}(E_l) + \varepsilon_l,$$

hence

$$\sum^{p} |l|\varepsilon[m_{V(x)}(E_l) - \eta] \leq \sum |l|\varepsilon \Lambda_l' \leq \sum |l|\varepsilon[m_{V(x)}(E_l) + \varepsilon_l,]$$

the index p indicates that the first summation must be extended only to the positive terms.

When η tends towards zero, the left-hand side increases towards $\sum |l|\varepsilon m_{V(x)}(E_l)$; the last expression is $\sum |l|\varepsilon m_{V(x)}(E_l) + \varepsilon\zeta$. Consequently, the limit of $\sum |l|\varepsilon\Lambda_l'$, as η and ζ tend towards zero, is $\sum |l|\varepsilon m_{V(x)}(E_l)$. In other words, we can pass to the limit under $\sum$ sign. Therefore, we have

$$F(b-0) - F(a) = \lim_{\varepsilon \to 0} \sum l\varepsilon m_{\alpha(x)}(E_l) = \int_a^{b-0} f(x)\, d[\alpha(x)].$$

But we have seen that

$$\frac{F(b) - F(b-0)}{\alpha(b) - \alpha(b-0)} = f(b),$$

whence

$$F(b) - F(a) = \int_a^b f(x)\, d[\alpha(x)].$$

Similarly, the same reasoning applies to (a, x). Therefore, the *primitive function, with respect to $\alpha(x)$, of a summable function $f(x)$, is the indefinite integral function of one variable of $f(x)$ with respect to $\alpha(x)$.*

We have just revisited the reasoning from Chap. 9, but under particularly simple conditions. To follow the reasoning of this chapter more accurately, it is necessary to introduce the concept of derivative numbers with respect to a function $\alpha(x)$. The reader will easily see that all the results of Chap. 9 would then extend to integration and differentiation with respect to $\alpha(x)$, and very often quite literally. Let us limit ourselves to verifying this statement in connection with the preceding one: *The indefinite integral function of one variable of $f(x)$, with respect to a function of bounded variation $\alpha(x)$, admits $f(x)$ as its derivative with respect to $\alpha(x)$, except at most in a set of points of measure zero, with respect to $V(x)$.*

We have, by the very definition,

$$F(x) = \int_a^x f(x)\, d[\alpha(x)] = \int_0^{V(x)} f[x(\nu)].'A(\nu)\, d\nu.$$

Except possibly for a set E_ν of values of ν with measure zero, the function

$$\int_0^{\nu} f[x(\nu)].'A(\nu)\, d\nu$$

admits $f[x(\nu)].'A(\nu)$ as its derivative and $A(\nu)$ admits $'A(\nu)$ as its derivative.

In other words,

$$\frac{F[x(\nu+\delta\nu)]-F[x(\nu)]}{\delta\nu} \quad \text{and} \quad \frac{\alpha[x(\nu+\delta\nu)]-\alpha[x(\nu)]}{\delta\nu}$$

tends towards the two indicated limits as $\delta\nu$ tends towards zero. And since $'A(\nu)$ is different from zero, from this, it follows that

$$\frac{F[x(\nu+\delta\nu)]-F[x(\nu)]}{\alpha[x(\nu+\delta\nu)]-\alpha[x(\nu)]}$$

tends towards $F[x(\nu)]$. However, we cannot not conclude immediately, as the function $x(\nu)$ does not take all values in (a, b) and ν and $\nu+\delta\nu$ can give the same value of x. We can say only that almost at every point $x = x(\nu)$, given by a value of ν not belonging to E_ν, and for which $\alpha(x)$ is continuous, $F(x)$ admits $f(x)$ as its derivative with respect to $\alpha(x)$. The points which are not given by a value of ν through the formula $x = x(\nu)$ belong to an interval in which $\alpha(x)$ is constant; the set of these intervals give a finite or countable set of points $V(x)$, and therefore it has measure zero with respect to $V(x)$. Moreover, in these intervals, the derivative of $F(x)$ is indeterminate. On the other hand, if x is the point of discontinuity of $\alpha(x)$, at this point $F(x)$ indeed admits $f(x)$ as its derivative, according to the calculation that we have done for the jumps of $F(x)$. Therefore, the theorem is entirely proved.

Now let us try to solve the problem of primitive functions, without subjecting the function $f(x)$, so far assumed summable with respect to $\alpha(x)$, to any restrictive condition. We assume that the derivative of an unknown function $F(x)$, finite everywhere, is given by $f(x)$ with respect to a given function of bounded variation $\alpha(x)$.

It is clear that, for the determination of $F(x)$, integration with respect to $\alpha(x)$ will be insufficient, and we will need to turn to a generalisation of totalisation, since the operation of totalisation is necessary when $\alpha(x)$ reduces to the function x. But such a generalisation presented itself to us (Sect. 12.1) when we decided to take the formula as the definition of the Stieltjes integral:

$$\int_a^b f(x)\,d[\alpha(x)] = \int_0^V f[x(\nu)].'A(\nu)\,d\nu,$$

we have considered only the case where the theory of the summable functions gives meaning to the right-hand side, without invoking the theory of totalisation. Let us now agree to call the *definite total of* $f(x)$ *with respect to* $\alpha(x)$, the expression

$$T_a^b f(x)\,d[\alpha(x)] = T_0^V f[x(\nu)].'A(\nu)\,d\nu,$$

in which the symbol T_0^V on right hand side denotes the definite total, in sense of M. Denjoy, of the function $f[x(\nu)].'A(\nu)$ assumed to be totalisable.[26]

[26] By doing away with use of the word integral instead of total and the symbol $\int$ instead of the symbol T, I follow the example of M. Denjoy, who has always carefully distinguished integration from totalisation in both vocabulary and formulae.

This new approach to totalisation simultaneously defines *the indefinite total of* $f(x)$ *with respect to* $\alpha(x)$. *These two totals are obtained, as previously, by repeated use of operations similar to the operations A and B of Sect.* 11.4*: by transfinite induction*

A_1. *We assume indefinite totals* $F_k(x)$ *of* $f(x)$ *to be known, in intervals* (a_k, b_k) *tending towards* (l, m)

$$l < \cdots < a_2 < a_1 < b_1 < b_2 < \cdots < m.$$

The function $F(x)$, *equal to* $F_1(x)$ *in* (a_1, b_1), *equal to*

$$F_k(x) - F_k(a_{k-1}) + F_{k-1}(a_{k-1}) - F_{k-1}(a_{k-2}) + \ldots + F_1(a_1)$$

in (a_k, a_{k-1}), *and defined in a similar manner on* (b_{k-1}, b_k) *is taken as indefinite total of* $f(x)$ *in* $(l + 0, m - 0)$. *We achieve the determination of the indefinite total, in* (l, m), *by agreeing that* $F(x)$ *has, at* l, *a right jump equal to*

$$f(l)[\alpha(l + 0) - \alpha(l)]$$

and, at m, a left jump equal to

$$f(m)[\alpha(m) - \alpha(m - 0)].$$

B_1. *We have a closed set E contained in the interior of an interval* (l, m); *we assume totals of* $f(x)$ *to be known for the various intervals* (α, β) *contiguous to E and contained in* (l, m); *if the series* $\sum[F(\beta - 0) - F(\alpha + 0)]$ *provided by these totals is convergent and, if* $f(x)$ *is summable over E with respect to* $\alpha(x)$, *we take*

$$\int_E f(x)\, d[\alpha(x)] + \sum[F(\beta - 0) - F(\alpha + 0)]$$

as indefinite total of $f(x)$ *in* $(l + 0, m - 0)$.

For $f(x)$ to be totalisable with respect to $\alpha(x)$, it is necessary that: 1. *that the operation* A_1 *yields a function for which* $F(l + 0)$ *and* $F(m - 0)$ *exist;* 2. *for any closed set* $\mathcal{E}$, *there exists an interval* (l, m) *containing points of* $\mathcal{E}$ *in its interior, where the necessary conditions for operation* B_1 *are satisfied.*

From the definition, it also follows that: *a function* $F(x)$ *is an indefinite total with respect to* $\alpha(x)$ *if, and only if:*

1. *It admits only points of discontinuity of the first kind;*
2. *For any closed set E, the function* $G(x)$ *equal to* $F(x)$ *on E, linear in the intervals contiguous to E and admitting the same jumps* $F(x)$ *at each point of E, is absolutely*

Other authors, on the contrary, have used the word integral and the symbol $\int$ for all cases. Both approaches have their advantages and disadvantages.

continuous, in an interval containing points of E in its interior, with respect to the function deduced from $\alpha(x)$, just as $G(x)$ is derived from $F(x)$.

Finally, from the relationship between the indefinite integral with respect to $\alpha(x)$ and the integrated function $f(x)$, it follows that *the indefinite total, taken with respect to $\alpha(x)$, of a function $f(x)$ admits $f(x)$ as its approximate derivative with respect to $\alpha(x)$, except at points of a set of measure zero with respect to the total variation $V(x)$ of $\alpha(x)$.*

By approximate derivative of $F(x)$ with respect to $\alpha(x)$, at a point x_0, we mean the limit of ratio

$$\frac{F(x_0+h)-F(x_0)}{\alpha(x_0+h)-\alpha(x_0)}$$

taken for numbers x_0+h, forming a set E of density one, with respect to $V(x)$, at point x_0. This amounts to saying that, if we let x and y tend towards x_0, $x \leq x_0 \leq y$, and if E_{xy} denotes the subset of E located in xy, the ratio $\frac{m_{V(x)}E_{xy}}{V(y+0)-V(x-0)}$ tends towards one.

Let us limit to the theory of totalisation with respect to a function with these assertions, that the reader can immediately justify, and return to the search for primitive functions.

Therefore, our goal is to construct a function $F(x)$ knowing the finite value $f(x)$ of its derivative with respect to a given function $\alpha(x)$ of bounded variation. Let us go back to what we have already done (Sect. 12.3).

For any value ν_0 of ν such that we have $\nu_0 = V(x_0)$ for one or several values x_0, let us associate the number $\mathfrak{F}(\nu_0) = F(x_0)$. This convention does not lead to any ambiguity because, if there are several values x_0 corresponding to ν_0, it is because they all yield the same value for $V(x_0)$ and therefore to $\alpha(x_0)$ and as a result for $F(x_0)$. Indeed, if $F(x)$ was not constant in an interval where $\alpha(x)$ is constant, $F(x)$ would not have a finite derivative, with respect to $\alpha(x)$, at all the points of this interval.

For a value ν_0 such that we have:

$$\begin{aligned} &\text{either} \qquad V(x_0-0) \leq \nu_0 < V(x_0), \\ &\text{or} \qquad V(x_0) < \nu_0 \leq V(x_0+0), \end{aligned}$$

Let us set, respectively

$$\begin{aligned} &\text{either} \qquad \mathfrak{F}(\nu_0) = F(x_0-0) + \frac{F(x_0)-F(x_0-0)}{V(x_0)-V(x_0-0)}[\nu_0 - V(x_0-0)], \\ &\text{or} \qquad \mathfrak{F}(\nu_0) = F(x_0) + \frac{F(x_0+0)-F(x_0)}{V(x_0+0)-V(x_0)}[\nu_0 - V(x_0)]. \end{aligned}$$

It is clear that the function $\mathfrak{F}(\nu)$, defined on the interval $(0, V)$, is continuous. Let us set aside the countably infinite set D of points ν provided by the intervals where $\alpha(x)$ is constant and by the formulae

$$\nu = V(x_0), \quad \nu = V(x_0-0), \quad \nu = V(x_0+0),$$

in which we will assign to x_0, only the values corresponding to discontinuity of $\alpha(x)$.

Let us examine what the differentiation of $\mathfrak{F}(\nu)$ yields at a point not belonging to D. In an interval $V(x_0-0)<\nu<V(x_0+0)$ the function $\mathfrak{F}(\nu)$ is formed by two linear functions of slopes

$$\frac{F(x_0)-F(x_0-0)}{V(x_0)-V(x_0-0)}=\frac{F(x_0)-F(x_0-0)}{\alpha(x_0)-\alpha(x_0-0)}\frac{\alpha(x_0)-\alpha(x_0-0)}{V(x_0)-V(x_0-0)}=f(x_0).'A(\nu),$$
$$\frac{F(x_0+0)-F(x_0)}{V(x_0+0)-V(x_0)}=\frac{F(x_0+0)-F(x_0)}{\alpha(x_0+0)-\alpha(x_0)}\frac{\alpha(x_0+0)-\alpha(x_0)}{V(x_0+0)-V(x_0)}=f(x_0).'A(\nu),$$

in $V(x_0-0)<\nu<V(x_0)$ and $V(x_0)<\nu<V(x_0+0)$. Except at $\nu=V(x_0)$ the derivative of $\mathfrak{F}(\nu)$ exists and can be denoted by $f[x(\nu)].'A(\nu)$, by using the function $x(\nu)$ of Sect. 12.1.

Any point ν_0 not belonging to D and not located in the various intervals $[V(x_0-0), V(x_0+0)]$ corresponds to a unique value x_0 by the formula $\nu_0=V(x_0)$, and x_0 is not a value of discontinuity of $\alpha(x)$. If we consider a value ν_1 tending towards ν_0, the number $x_1=x(\nu_1)$ then tends towards $x(\nu_0)=x_0$. The number $F(x_1)=F[x(\nu_1)]$ is equal to $\mathfrak{F}(\nu_1)$ if this number $\mathfrak{F}(\nu_1)$ results from the first part of the definition of $\mathfrak{F}(\nu)$. In this case we have

$$\frac{\mathfrak{F}(\nu_1)-\mathfrak{F}(\nu_0)}{\nu_1-\nu_0}=\frac{F(x_1)-F(x_0)}{\alpha(x_1)-\alpha(x_0)}\frac{\alpha(x_1)-\alpha(x_0)}{\nu_1-\nu_0},$$

and when ν_1 tends towards zero the first ratio tends towards $f(x_0)$, the second remains contained between -1 and $+1$ and tends almost everywhere towards $'A(\nu_0)$.

If the value of $\mathfrak{F}(\nu_1)$ arises from the second part of the definition of $\mathfrak{F}(\nu)$, it means that $x(\nu_1)$ is the abscissa of a singular point of $\alpha(x)$ and ν_1 is included in $\{V[x(\nu_1)-0], V[x(\nu_1)+0]\}$.

Let us assume, for example, that we have

$$V[x(\nu_1)]<\nu_1\leq V[x(\nu_1)+0];$$

In this interval $\mathfrak{F}(\nu)$ is linear and the ratio $\frac{\mathfrak{F}(\nu_1)-\mathfrak{F}(\nu_0)}{\nu_1-\nu_0}$ is included between

$$\frac{\mathfrak{F}\{V[x(\nu_1)]\}-\mathfrak{F}(\nu_0)}{V[x(\nu_1)]-\nu_0}\quad\text{and}\quad\frac{\mathfrak{F}\{V[x(\nu_1)+0]+\varepsilon\}-\mathfrak{F}(\nu_0)}{V[x(\nu_1)+0]+\varepsilon-\nu_0}$$

or at least, differs as little as we want from the right-hand side, where ε denotes a very small positive number chosen in such a manner that for $V[x(\nu_1+0)+\varepsilon]$, the value of $\mathfrak{F}$ arises from the first part of the definition of $\mathfrak{F}(\nu)$.

Now, these last two ratios fall into the category of those studied at the beginning, therefore the largest and the smallest limits of these ratios are always included between $+f(x_0)$ and $-f(x_0)$ and they tend almost everywhere towards a definite limit.

$$f(x_0).'A(\nu_0)=f[x(\nu_0)].'A(\nu_0).$$

Therefore:

The function $\mathfrak{F}(\nu)$ has, except, possibly, at points of a countable set D, all its derivative numbers finite, and as a result $\mathfrak{F}(x)$ is an indefinite total;

The function $\mathfrak{F}(\nu)$ has almost everywhere a definite and finite derivative equal to $f[x(\nu)].'A(\nu)$, therefore $\mathfrak{F}(\nu)$ is the indefinite total

$$\mathfrak{F}(\nu) = \text{const.} + T_0^{\nu} f[x(\nu)].'A(\nu);$$

From this, it follows that

$$F(x) = \text{const.} + T_0^x f(x)\, d[\alpha(x)].$$

Thus, the search for a function $F(x)$, whose finite derivative $f(x)$ taken with respect to a given function $\alpha(x)$ of bounded variation is known, can always be carried out by the indefinite totalisation of $f(x)$ with respect to $\alpha(x)$.

This result is a direct generalization of M. Denjoy's result for the case where $\alpha(x) \equiv x$. It would be very interesting to systematically revisit the reasoning which served us in this particular case and extend it to the general case. The reader will not encounter any particular difficulty in this study. The reader can also show that the method developed in Sect. 11.2 and onwards, which allows for solving the problem of primitives functions without invoking the notion of integration, still applies to any function $\alpha(x)$ of bounded variation. He can also examine the problem of primitive functions of the derivative numbers to which the method of change of variable used here does not seem to provide a solution. We will not explore this generalisation of problem of primitive functions. However, there are others, much more elementary and immediate, which remain unsolved; we will see them.

The case where $\alpha(x)$ is of bounded variation is, as explained, probably the only one of physical interest. However, from a mathematical point of view, there is no reason to only consider the derivative of a function $F(x)$ with respect to a function $\alpha(x)$ under hypothesis where $\alpha(x)$ is of bounded variation.

Now, if we abandon this hypothesis, almost none of our conclusions would survive. For instance, let us show that if $\alpha(x)$ is of unbounded variation, there exist *continuous* functions $f(x)$ which do not have the Stieltjes integral with respect to $\alpha(x)$. In other words, for these functions, the sums $S = \sum f(\xi_i)[\alpha(x_{i+1}) - \alpha(x_i)]$ do not tend towards any definite and finite limit, when we vary the choices of x_i and ξ_i so that the maximum λ of $x_{i+1} - x_i$ tends towards zero.

Indeed, when $\alpha(x)$ is of unbounded variation, we can (Sect. 5.1), find an ordered sequence of points X_i, such that the series

$$\sum |\alpha(X_{i+1}) - \alpha(X_i)|$$

is divergent. Then we can find a sequence of numbers ρ_i tending towards zero and such that the series

$$\sum \rho_i[\alpha(X_{i+1}) - \alpha(X_i)]$$

is of positive terms and divergent.

Let us suppose, for fixing our ideas, that our points X_i succeed each other in the order

$$a \leq X_1 < X_2 < \cdots < b_1 \leq b;$$

where b_1 is the limit of X_i. Let us take for $f(x)$ a continuous function, that is zero from a to X_1, from b_1 to b and at the points X_i, and attains the value ρ_i in (X_i, X_{i+1}).

We claim that, for any given maximum length λ imposed on the intervals of subdivision of (a, b), we can choose these intervals and the points ξ_i so that the corresponding sum S exceeds any prescribed bound.

Let k be the value of index i from which the difference $X_{i+1} - X_i$ remains less than λ. Let us arbitrarily divide (a, X_k) into intervals of length λ at most and let us choose in each of them a point ξ; do the same for (b_1, b). It remains to divide (X_k, b_1). Let us take the subdivision points as $X_{k+1}, X_{k+2}, X_{k+l}$; where l is an arbitrary integer. Let us take points ξ in $(X_k, X_{k+1}), (X_{k+1}, X_{k+2}), \ldots, (X_{k+l-1}, X_{k+l})$, at which $f(x)$ has the values $\rho_k, \rho_{k+1}, \ldots, \rho_{k+l-1}$, respectively. In (X_{k+l}, b_l) we take ξ at X_{k+l-1}. Then we have

$$S = s + \sum_{k}^{k+l-1} \rho_i[\alpha(X_{i+1}) - \alpha(X_i)],$$

where s is the contribution of intervals (a, X_k), (b_1, b), which do not depend on choice of l. Now, for sufficiently large l, the second term in the right-hand side exceeds any given bound. Therefore, S is as large as we want. Hence, the definition of Stieltjes does not apply to $f(x)$ and $\alpha(x)$.

Thus, we can no longer associate with each continuous function $f(x)$ an integral, with respect to $\alpha(x)$, when $\alpha(x)$ is of unbounded variation.

Moreover, the problem of primitive functions will no longer arise for all continuous functions $f(x)$.

Let us take $\alpha(x) = x \sin \frac{1}{x}$; the function $\alpha(x)$ is increasing in the intervals

$$p_k = \left[\frac{1}{\left(2k + \frac{1}{2}\right)\pi}, \frac{1}{\left(2k + \frac{3}{2}\right)\pi}\right],$$

and decreasing in the intervals

$$n_k = \left[\frac{1}{\left(2k - \frac{1}{2}\right)\pi}, \frac{1}{\left(2k + \frac{1}{2}\right)\pi}\right].$$

Let us take $f(x)$ continuous in $(0, 1)$, zero in n_k, positive in p_k. That will be possible even if we require that the integral $\int_{p_k} f(x)\, d[\alpha(x)]$ has a value π_k, provided that π_k tends towards zero faster than the growth of $\alpha(x)$ in p_k.

Let us take

$$\pi_k = \frac{1}{\left(2k + \frac{1}{2}\right) \pi \mathrm{Log} k}.$$

Then, it is clear that, in $(\varepsilon, 1)$, however small $\varepsilon > 0$ may be, $f(x)$ is the derivative, with respect to $\alpha(x)$, of the function

$$\int_{\varepsilon}^{x} f(x)\, d[\alpha(x)].$$

But this integral increases indefinitely when ε tends towards zero, because the series of π_k is divergent. Therefore, in (0, 1), the continuous function $f(x)$ is not the derivative of a function $F(x)$ with respect to $\alpha(x)$.

Thus, when we no longer assume that $\alpha(x)$ is of bounded variation, the problem of primitive functions appears completely different from the one we have solved.

However, here is a category of functions $\alpha(x)$ to which the previous considerations immediately apply.

The function $\alpha(x)$ satisfies the following two conditions:

1. *$\alpha(x)$ has only points of discontinuity of first kind;*
2. *E being an arbitrary closed set, there exists an interval i containing points of E in its interior, where the function $\beta(x)$ is of bounded variation. $\beta(x)$ is equal to $\alpha(x)$ at points of E, linear in the intervals (l, m) contiguous to E, and such that $\alpha(l) = \beta(l)$, $\alpha(m) = \beta(m)$.*

Let us take for E as the interval (a, b) itself. $\beta(x)$ is identical to $\alpha(x)$ in i; in i we know the derivative $f(x)$ of $F(x)$, with respect to the function of bounded variation $\alpha(x)$. Therefore, $f(x)$ is totalisable, with respect to $\alpha(x)$. Consequently, an interval j exists in i, in which $f(x)$ is summable with respect to $\alpha(x)$; the Stieltjes integral of $f(x)$ will yield $F(x)$.

Thus, $F(x)$ is known in intervals which cover the interior of intervals contiguous to a certain closed set. From there, we deduce $F(x)$, first in the contiguous intervals considered as open sets; and then, as we know the jumps of $F(x)$ at every point, in the closed contiguous intervals.

Let us assume that, by this operation or any other, we have determined $F(x)$ in the close intervals contiguous to a closed set E. let $G(x)$ be the function derived from $F(x)$ as $\beta(x)$ is derived from $\alpha(x)$.

If (l, m) is an interval contiguous to E, $G(x)$ is linear in it, and admits, with respect to $\beta(x)$, a know derivative

$$\frac{G(m) - G(l)}{\beta(m) - \beta(l)} = \frac{F(m) - F(l)}{\alpha(m) - \alpha(l)}.$$

At the points of E, we have $F = G, \alpha = \beta$, hence it easily follows that G admits at these points $f(x)$ as derivative with respect to $\beta(x)$.

This is, however, only true for left-hand derivative at points l, right-hand derivative at points m; the right-hand derivative at l, and the left-hand derivative at m, have been calculated above.

Thus, in i, we know the derivative of $G(x)$ with respect to the function of bounded variation $\beta(x)$. This derivative is finite and definite, except at the points of a countable set D_x where there exists a finite and known right derivative and a finite and known left derivative.

The conditions we are in are therefore a little more general than the ones previously examined, but nothing essential will change. When the derivative $f(x)$ exists everywhere, we deduce that the function $\mathcal{G}(\nu)$ arising from $G(x)$ has a finite right-hand superior derivative number, except possibly at the points of a countable set D. It is necessary for us to say now, except at the points of $D + D_\nu$, where D_ν is the transform of D_x. If all the points of D_x are points of continuity of $\alpha(x)$, D_ν is countable and nothing changes in our previous conclusions. We would not even need to know that at points of D_x the derivative numbers of $G(x)$ are finite. If x_0 is a point of discontinuity of $\beta(x)$ belonging to D_x, to this point correspond, in D_ν, two intervals[27] $[V(x_0 - 0), V(x_0)]$, $[V(x_0), V(x_0 + 0)]$; but thanks to the right and left derivatives at x_0, $f_d(x_0)$, $f_g(x_0)$, which are known, we know $\mathcal{G}(\nu)$ in these two intervals, and we know that $\mathcal{G}(\nu)$ has at any point a finite and known right derivative. Thus, nothing essential changes; $\mathcal{G}(\nu)$ has finite right-hand superior derivative number at every point, except at most at countable set of points. $\mathcal{G}(\nu)$ is an indefinite total in i. However, to calculate $G(x)$, it would be necessary to slightly modify the method of calculation of Stieltjes integral $\int \varphi(x)\, d[\beta(x)]$; because now $\varphi(x)$ represents the known derivative of G with respect to $\beta(x)$, which is multi valued at the point of discontinuity x_0, if at this point G has a right derivative $f_d(x_0)$ and a left derivative $f_g(x_0)$. Then x_0 may appear in a set $E[l\varepsilon \leq \varphi < (l+1)\varepsilon]$, either only because of the value $f_d(x_0)$, or only because of the $f_g(x_0)$, or both; depending on these cases, x_0 will appear in $m_{\beta(x)}\{E[l\varepsilon \leq \varphi < (l+1)\varepsilon]\}$ for

$$\beta(x_0 + 0) - \beta(x_0), \quad \beta(x_0) - \beta(x_0 - 0), \quad \beta(x_0 + 0) - \beta(x_0 - 0)$$

[27] Here, $V(x)$ denotes the total variation of $\beta(x)$ and not $\alpha(x)$; for the particular function $\beta(x)$ in the text and the special set D_x, one of the two considered intervals does not exist. Indeed, $\beta(x)$ is continuous to the right at l and to the left at m.

By[28] such elementary modifications, we will therefore find *a function* $F(x)$ *knowing the finite value* $f_d(x)$ *of its right derivative with respect to a function of bounded variation* $\alpha(x)$ *at every point, and knowing, at the points of discontinuity of* $\alpha(x)$, *the finite value* $f_g(x)$ *of its left derivative.*[29]

But these modifications are unnecessary here because the points of D_x are origins l or extremities m of intervals contiguous to E; $\beta(x)$ is continuous to the right at l, and to left at m, so it is sufficient to take $\varphi = f$ at the points of D_x without taking into account the values $f_d(l)$ and $f_g(m)$.

As a result, in i, we can find an interval j containing points of E in its interior and in which φ is summable with respect to $\beta(x)$. The integral of φ, in j, is the sum of the contributions from the intervals (l, m), which is known, and the contribution from the subsets E_j of E located in j. Therefore, we have

$$\mathfrak{A}_{G(x)}(j) = \int_{E_j} f(x)\, d[\beta(x)] + \sum_{j}^{E_j} [G(m) - G(l)].$$

This could be replaced by

$$\mathfrak{A}_{G(x)}(j) = \int_{E_j} f(x)\, d[\alpha(x)] + \sum_{j}^{E_j} [F(m) - F(l)],$$

the extremities of j being assumed to be taken in E, but only if, in the calculation of the integral with respect to $\alpha(x)$, we consider the points l only for their left measure, and the points m for their right measure.

In one or the other form, we see that there is a possibility of determining F in intervals containing points of E; this is enough to be sure of being able to obtain F in entire (a, b) by transfinite induction. As a result, F is determined up to an additive constant.

The new category of functions $\alpha(x)$ is very vast, however *we do not always know how to find the continuous function* $F(x)$ *which admits, with respect to a given continuous function* $\alpha(x)$, *a given continuous derivative* $f(x)$. It is sufficient to take $\alpha(x)$ to be continuous and of unbounded variation in any interval and to take $f(x) \equiv 1$ for us to be in this case. Without doubt, in this example, we know one of the primitive functions of $f(x)$, namely the function

---

[28] This corresponds to considering, in the evaluation of the physical quantity $m_{\beta(x)}\{E[l\varepsilon \leq \varphi < (l+1)\varepsilon]\}$, either one or both of the material points that Stieltjes imagines to coincide at point x_0 (Sect. 12.5).

In a more abstract sense, we can say that we replace the notion of a set of points with that of set of null intervals. However, there are three types of null intervals, and one must take into account the nature of the null intervals which constitute it, in the evaluation of the measure of a set.

[29] The earlier considerations clearly prove that a function is not defined solely by the knowledge of one or several of its right-hand derivative numbers, with respect to a function $\alpha(x)$. It is necessary to have additional information about the behaviour of the sought function in the left neighbourhoods of the points of discontinuity of $\alpha(x)$.

$\alpha(x)$ itself; but we do not know whether there exist the primitive functions which are not of the form

$$\alpha(x) + \text{const.}$$

Indeed, we have not given here any method, either for finding the primitive functions of $f(x) \equiv 1$, or for limiting the extent of their indetermination.

Note. On Transfinite Numbers

A

A.1 The Derived Sets

We had to resolve, at the end of Chap. 1, the following question:

A continuous function is known up to an additive constant, varying from one interval to another, and in any interval that does not contain any points from a set E; what must be the nature of set E for the function to be uniquely determined?[1]

This problem was solved by M. G. Cantor, who used it in the theory of the trigonometric series. We will study the properties of sets which have been used in Chap. 1 for the solution of this question.

Let us consider a bounded set of points[2] E. The set of its limit points is *its first derivative,* denoted by E' or E^1. The derivative of E^1 is the *second derivative*, denoted by E^2; and so on.

I. *For any infinite set* (that is, it contains an infinite number of points), E^1 *exists*, and it is Bozano–Weierstrass principle. To show this, let us group all the numbers into two classes, A and B. Class A contains all numbers that are greater than only a finite number of elements of E, while class B contains the remaining numbers. The cut A, B defines a number that is obviously a limit point of E, and in fact, it is the smallest of its limit points.
   E^1 is obviously closed, meaning it contains its limit points, therefore it contains its derivative E^2; E^2 is closed, it contains E^3; and so on.

[1] We can always assume that the set E which appears in this statement is closed. Therefore, it would be sufficient to study only closed sets, but this limitation would not lead to any notable simplification.
[2] He deals with points on a straight line, therefore with numbers. There would only be a little change if he dealt with the set of points in space of several dimensions. Furthermore, the use of the curves, such as the curve of Peano, allows to limit ourselves to study the case of straight line.

R. Jain, *Lebesgue's Theory of Integration*,
https://doi.org/10.1007/978-981-96-1169-0

These sets E^1, E^2, $E^3 \cdots$ exist in certain cases. One straightforward case where their existence is obvious is when E^1 is perfect, because then E^1, E^2, $E^3 \cdots$ are all identical. In this case, the definition of E^2, $E^3 \cdots$ does not present any interest, and we will agree never to discuss the derivative of any perfect set. However, these sets can all be distinct. Here is the construction method we will use to see this:

Let there be sets $e_1, e_2, \ldots$ located in (0, 1). Let us divide (0, 1) into partial intervals $(\frac{1}{2}, 1), (\frac{1}{2^2}, \frac{1}{2}), (\frac{1}{2^3}, \frac{1}{2^2}), \ldots$. Let us perform on e_i the homothetic transformation which replaces (0, 1) by $(\frac{1}{2^i}, \frac{1}{2^{i-1}})$; e_i becomes $\mathcal{E}_i$. The sum of these sets $\mathcal{E}_i$ and of point 0 will be denoted by $A(e_1, e_2, \ldots)$.

If $e_1, e_2, \ldots$ reduce to the point 0,

$$A_1 = A(e_1, e_2, \ldots)$$

is a set for which E^1 reduces to point 0. If $e_1, e_2, \ldots$ are identical to A_1, we obtain $A_2 = A(A_1, A_1, \ldots)$ for which E^2 reduces to point 0, and so on.

If $e_1 = A_1$, $e_2 = A_2, \ldots,$ for $A(A_1, A_2, \ldots)$, the derivatives E^1, $E^2, \ldots$ all contain points.

II. *When the derivatives E^1, E^2, $E^3, \ldots$ all contain points, there exist points common to all these derivatives.* Indeed, let M_i be a point of E^i; M_i is also a point of $E^{i-1}, E^{i-2}, \ldots, E^1$. The set $M_1, M_2, \ldots$ has at least one limit point. Since it is a limit of points $M_i, M_{i+1}, \ldots$ of E^i, it must be a point E^{i+1}. Therefore, this point belongs to all the E^i.[3]

The set of points whose existence is thus shown is called the *derivative* E^ω.

For $A_\omega = A(A_1, A_2, \ldots)$, E^ω contains only the point 0. For

$$A_{\omega+1} = A(A_\omega, A_\omega, A_\omega, \ldots)$$

the derivative of E^ω reduces to the point 0. The derivative of E^ω is denoted by $E^{\omega+1}$, and more generally, the successive derivatives of E^ω, are denoted $E^{\omega+1}$, $E^{\omega+2}, \ldots$. It is not necessary to attach any importance to the particular form of the indices used here; it is sufficient to imagine that different symbols are used to denote the different derivatives.

We will say that one derivative of a set comes *after* another if it is contained within the other. With this convention, the words *before* and *after* can be used as in ordinary language.

So far, we have used the following definition: *When a derivative contains an infinite number of points and is not perfect, its derivative is by definition the first derivative which comes after it.* A second definition is necessary; to formulate it, let us first note

[3] In summary, we have just proven that there are always points common to the sets $E_1, E_2, \ldots$ when these sets are closed and each of them contains all those that follow in the sequence $E_1, E_2, \ldots$. The set of these points is obviously closed.

that if E^α, $E^\beta, \ldots$ are derivatives, countable or finite in number, if they all contain points and are pairwise different, there exist points which are common to all of them. Indeed, let us arrange E^α, $E^\beta, \ldots$ in a finite or possibly infinite sequence S, say $E_1, E_2, \ldots$. We can compare two sets $E_i = E^\lambda$ and $E_j = E^\mu$, from two perspectives: either as derivatives, considering the higher indices λ, μ, and we will use the terms *before* and *after* to state the result of this comparison; or as terms of the sequence $E_1, E_2, \ldots$, considering the lower indices i, j, and we will use the terms *in front of* and *behind.*

Within the sequence S let us mark the terms that are both before E_1 and behind E_1, we obtain $E_1, F_2, F_3, \ldots$. Within this sequence, mark the terms that are both before and behind F_2, and so on. Finally, we will obtain a sequence $E_1, F_2, G_3, \ldots$ for which there is an agreement between the terms *before* and *front*, as well as the terms *after* and *behind.* This implies that each set within this sequence Σ contains all those that are behind it. As these sets are closed, there exists a closed set of points that is common to all the sets in Σ.

Every point in this set $\mathcal{E}$ is obviously common to all the sets in S, and conversely, since every set in S is either a part of Σ or comes before a set in Σ. For the same reason, $\mathcal{E}$ would not change if we replace the sequence S with any other sequence S' formed by derivatives of E, and such that every term in one of these sequences belongs to the other or comes before a set in the other.

Because of these facts, and by analogy with the definition of E^ω, we state the following proposition and definition:

III. *When derivatives of a set, whether countable or finite in number, all contain points, there exist points common to all these derivatives. The set of these points, which is obviously closed, is by definition, the first derivative which does not come before the given ones.*[4]

For this definition to hold, if there exists a set e in sequence S that comes after all others, our definition must lead to that set. Indeed, this is the case because e, coming after the others, is contained in all the others, and it indeed constitutes the set of points common to all the sets of S. In the case we are currently examining, the sequence Σ is bounded and conversely; e is the last term of Σ. In other cases, the defined derivative is called the first derivative that comes after the given derivatives.

By definition, there are at most countably infinite number of derivatives before each derivative.

[4] The reasoning which has served us provides a general result which can be stated, using a notion defined in the following section, under the form: For a countable well ordered set of closed sets, such that each of them contain all those which follow it, there exist points common to all these sets and, these points form a closed set.

A.2 The Well Ordered Sets. The Transfinite Numbers

Cantor states that a set of elements is ordered if a relationship established between any two elements of that set, which can be expressed using the words *before and after*; following the customary rules. In other words, if we say α is before β, it implies: β is after α. Similarly, if we say α is before β, which is before γ, then α is before γ.

Cantor further defines a set E as well ordered, if it is ordered in such a manner that, in this set E, and in each partial set P obtained by removing elements from E, there exists an element that comes before all the others.

If we agree that *before* means *smaller than*, every set of real numbers is ordered, but it is not necessarily well ordered. The set of real numbers contained in (0, 1), as well as the set of positive and negative integers are not well ordered. On the contrary, the set of positive numbers, that can be written in the decimal system with a non-increasing sequence of decimal digits is well ordered. That is, the set of numbers

$$n + \frac{a}{10} + \frac{b}{100} + \frac{c}{1000} + \cdots$$

where n is an integer, and the digits $a, b, c, \ldots$ satisfy the inequalities $9 \geq a \geq b \geq c \geq \cdots$, is well ordered.

The set of sets $A_1, A_2, \ldots, A_\omega, A_{\omega+1}$ is well ordered when arranged in the order in which they were obtained, but merely ordered when arranged in the reverse order.

The set $\mathcal{E}$ of derivatives of a set is well ordered, with the word *before* having the indicated meaning, that is, the derivative Δ_1 is before the derivative Δ_2, if Δ_1 contains Δ_2. Indeed, $\mathcal{E}$ is ordered, containing an element before all the others, the derivative E^1. Let P be a partial set obtained from $\mathcal{E}$. Let α be a derivative located in P; then there are at most countably many derivatives before α, the set $\mathcal{E}_1$ of derivatives located before all those in P, contains at most countably infinite number of elements. If E^1 is in P, $\mathcal{E}_1$ does not exist. In this case P includes the element E^1 that comes before all the others. If E^1 is not in P, $\mathcal{E}_1$ exists and there exists a derivative of P which is the first one, coming after all those in $\mathcal{E}_1$. This derivative λ belongs to P, and there cannot be a derivative in P that comes before this one; λ is, therefore an element of P that comes before all the others. Therefore, $\mathcal{E}$ is well ordered.

The sets of derivatives are such that before each of their elements, there are at most a countably infinite number of elements. We will deal only with well ordered sets that possess this property and reserve the term *sequence* for these sets.[5] The transfinite numbers that we are going to define are those which are used to represent the order of the elements in these sequences. Let us consider one of these sequences S; it contains an element that comes before all others, which we will call the first and can be denote with index 1, u_1. Then, if S is not reduced to u_1, the sequence $S - u_1$, will have a first element that we will call the

[5] We will reserve the term *simply infinite sequence* for the sequences whose elements are denoted using successive integers.

second element of S and can be denoted by u_2. If $S - u_1 - u_2$ exists, we will deduce the existence of a third element of S, etc.

The position in the sequence, of the elements encountered in this way is characterised using the usual ordinal adjectives or integers playing the role of ordinal numbers.

However, if the sequence S is not exhausted at the same time as the sequence of integers, it would be necessary to use the new symbols to represent the position of remaining elements. We will agree to denote u_ω as the first element of the sequence $S - u_1 - u_2 \ldots$, $u_{\omega+1}$ as the first element of $S - u_1 - u_2 - \cdots - u_\omega$. These new symbols are called transfinite numbers of the first numeric class of transfinite numbers. For brevity, we will refer to them simply as transfinite numbers. In short, transfinite numbers are the different indices that we imagined earlier for distinguishing the successive derivatives; they are used for all sequences.[6] The elements of the sequences will therefore be denoted by u_α, with α being either a finite or transfinite number. The numbers α themselves form a sequence, when we agree that α is before β, when α is used to denote an element which comes before the one denoted by β. By analogy with the case where the numbers are finite—and thus these numbers have both an ordinal and a cardinal meaning,—we also say that α is less than β, and we write $\alpha < \beta$, when α is before β.

To constitute the sequence of transfinite numbers, it would be sufficient to provide a procedure for generating the number which immediately follows a given finite or countable number of given finite or countable transfinite numbers. For example, one could agree that the number following $\alpha, \beta, \ldots$ will be denoted as $(\alpha, \beta, \ldots)$. With this convention, there would be several notations for the same number, but that is not important. It is also quite clear that writing the symbols with this convention will be impossible since it would require an infinite of strokes. Let us leave aside the question of enumeration for the transfinite numbers for now, as it is incidental for our purposes; we will return to it later.

IV. *The set of transfinite numbers is uncountable.*—In fact, if it were countable, we could simply add a new symbol to the countable sequence S_0 of finite and transfinite numbers, placing it after all the elements of S_0, to obtain a sequence S_1 where the position of last element cannot be denoted using the symbols from S_0. Thus, S_0 is uncountable. However, before any element of S_0, there are only countably infinite numbers, at most. This fact is completely analogous to the following: the sequence s_0 of integer is such that before any element of s_0, there are only finitely many integers. However, as adding an element to any finite sequence results in another finite sequence, the sequence s_0 is not finite.

Throughout these considerations, we have repeatedly assumed as obvious, that the numbering of elements of an ordered sequence using the successive symbols of S_0, could only be done in one way. For example, we have regarded it as clear that if elements of S_0 are enumerated as terms of a sequence with the elements of S_0 as numbers, each

[6] The transfinite numbers conceived in this way are known as ordinal transfinite numbers. Cantor also considers transfinite cardinal numbers, which he uses to represent the cardinality of sets.

element is identical to the number which determines its position. We can, if not make these fact clearer, at least present them in a form in that we are more accustomed to, by saying:

Let us establish a correspondence between the elements of two sequences S and T in such a way that the first two elements of S and T correspond to each other, and if the first elements in a finite or countable number of elements of S, say $s, s', \ldots$ correspond to the first elements $t, t', \ldots$ of T, then the first element of S after $s, s', \ldots$ corresponds to the first element of T, after $t, t', \ldots$; if these two elements exist.

Let us show that this correspondence is well-determined. Indeed, there are elements of S for which the correspondence is determined; the first element of S, for example. Let us consider the set Σ of all the elements of S for which the correspondence is determined, as well as for all the preceding elements. Then, if there were still elements remaining in S and T, the correspondence, according to its very definition, would still be determined for the first element of S after Σ. This would contradict the definition of Σ. Therefore, the correspondence is entirely determined and exhausts S or T, or both. *In this later case, we say that the sequences are similar.*

In the case where S is the sequence S_0 of transfinite numbers, it would be contradictory to assume that the correspondence exhaust S without exhausting T: every sequence is similar either to S_0 or to a segment of S_0. (By a segment of a sequence, we mean all the elements which precede a given element).

All the non-finite sequences we encounter, except for S_0, will be countable; thus, they will be similar to a segment of S_0. Furthermore, a segment of S_0 is entirely determined by the number which follow it. For each countable sequence S we can attach a transfinite number α that informs us of the finite and transfinite numbers necessary to enumerate the elements of S. This number α, thus give us an extremely important information on S, but naturally does not allow us to distinguish between S and similar sequences since, by definition, α is the same for all similar sequences. It immediately follows from the fact that the correspondence between the successive elements of two sequences is determined, that if S is similar to T and T to U, S is similar to U and the correspondence established directly between the elements of S and U is consistent with what one could established through the intermediary of T.

This number α is called the *order type* of the countable sequence S. It provides us with information on S that is exactly analogous to what we know about a finite sequence when we know that it contains 10 elements.[7] G. Cantor introduced transfinite numbers as order types.[8]

---

[7] We would note that if we applied the previous definition of order type to finite sequences, it would be the number 11 which would be the order type of a sequence of 10 elements.

[8] His principal Memoirs on this subject can be found in the Math. Ann., Bd 46 and Bd 49. These Memoirs were translated by F.Marotte (*Sur les fondements de la theorie des ensembles transfinis, Hermann,* 1899).

It is necessary to make a fundamental distinction between the various transfinite numbers. Some of them, known as *first kind,* immediately succeed a transfinite number. In other words, they belong to order types characterised by sequences having a last term. The rest are referred to as *second kind.* If α belongs to first kind, there is a transfinite number that can be denoted by $\alpha - 1$. This notation does not apply if α is of second kind. ω is of second kind, whereas $\omega + 1$ is of the first kind.

A.3 The Sets of Points

We have just logically conceived a set S_0 of finite and transfinite numbers which is countable. However, it is not certain that this set is entirely useful. Let us, therefore, show by examples, that a segment of S_0 cannot suffice, either to enumerate all the derivatives of sets of points, or to provide the order types of all sequences of points. These examples will not actually be constructed; we will only show that assuming we must stop during their construction would be absurd. G. Cantor frequently used this method of proof.

Therefore, let us suppose that we have constructed sets $A_1, A_2, \ldots, A_\omega, A_{\omega+1}, \ldots$, for which the derivatives are enumerated respectively using segments of S_0 where the last term is the number 1, the number 2, ..., the number ω, the number $\omega + 1, \ldots$. And this for all numbers of a and certain segment Σ of S_0. If α is the first term of the sequence S_0 after Σ, we aim to construct a set A_α whose enumeration of derivatives requires all the numbers of S_0 up to and including α.

If α is of first kind, that is, if $\alpha - 1$ exists, we will take $A_\alpha = A[A_{\alpha-1}, A_{\alpha-1}, \ldots]$; if α is of second kind, that is, if $\alpha - 1$ does not exist, we arrange all the numbers of Σ into a simply infinite sequence, let $a_1, a_2, a_3, \ldots$ be this sequence, and we will take

$$A_\alpha = A(A_{a_1}, A_{a_2}, A_{a_3}, \ldots).$$

We immediately verify that the set thus constructed answers the question. Indeed, in the first case the derivative of order $\alpha - 1$ is composed of points $\frac{1}{2}, \frac{1}{4}, \frac{1}{8}, \ldots$ and of point 0. In the second, if α_p is the largest of the numbers $a_1, a_2, a_3, \ldots, a_p$, there are no more points in the derivative of order $\alpha_p + 1$ in $\left(\frac{1}{2^{p+1}}, 1\right)$ while some points remain in $\left(0, \frac{1}{2^{p+1}}\right)$; where α is the first transfinite number which comes after all the numbers $\alpha_p + 1$, the derivative of order α will exist and reduces to point 0.[9]

Let us agree that ≪ before ≫ means smaller than and show that we can obtain a sequence of points, located in (0, 1), which has a given ordinal type. We are familiar with finite sequences and simply infinite sequences, that is, of type ω. As an example of the later sequences, we can consider the sequence S_ω composed of points whose abscissa are the

[9] Note that we intend to construct a set whose last derivative has a given rank α, rather than a set whose first derivative does not exist at a rank α. According to proposition *III*, such a set could not exist when α is of second kind.

successive sums of the series

$$\frac{1}{2}+\frac{1}{4}+\frac{1}{8}+\cdots$$

If we know a sequence S_α of type α, we will take a sequence $S_{\alpha+1}$ of type $\alpha+1$ that is obtained by taking, with respect to 0, the homothetic, in the ratio $\frac{1}{2}$, of the sequence constituted by S_α plus the point 1.

If we know the sequences $S_{a_1}, S_{a_2}, S_{a_3}, \ldots,$ of order types $a_1, a_2, a_3, \ldots,$ and if the simply infinite sequence $a_1, a_2, a_3, \ldots,$ does not contain a term greater than the others, and if α is the first transfinite number after $a_1, a_2, a_3, \ldots,$ we can form a sequence S using the sequences $s_{a_1}, s_{a_2}, s_{a_3}, \ldots$. Since the sequence s_{a_p} is directly similar to S_{a_p} and contained in the interior of pth interval in (0, 1) determined by points of S_ω.

It is clear that this sequence S has an order type greater than $a_1, a_2, a_3, \ldots$. Therefore, if we enumerate the successive elements using the numbers of S_0, we will encounter a segment of S—possibly S itself—whose enumeration requires all the numbers less than α. It is this segment which is the sequence S_α, of order type α, that we wanted to obtain.

Thus, the sequence S_0 is indispensable for either of the uses we have mentioned. However, we will see that a segment of S_0 always suffices for the enumeration of the elements of a well ordered set of *determined* points and also for enumeration of the successive derivatives of the *determined* set. The first fact is obvious, as it follows from the fact that any well ordered set of points on a line is countable. Indeed, if we count the abscissa x of these points starting from the first one and in the order in which the points succeed one another, the transformation $X=\frac{1}{x+1}$ replaces the given well ordered set with a well ordered set E of points of (0, 1) in decreasing order. Each point A of E is the extremity of an interval whose origin is the point B of E with an abscissa immediately lower than that of A. Therefore, we have as many of these intervals, (A, B) as there are points A. These intervals are non-intersecting, therefore there are at most p of them with a length not less than $\frac{1}{p}$ and this is true for any p. Therefore, the set of these intervals is finite or countable; so the proposed set is at most countable.

The other fact is much more hidden. It is the result of the Cantor's analysis of the properties of derivatives. In this section, without following Cantor's considerations step by step, we will not digress significantly from the ideas that guided him.

V. *The points of* E^1 *which do not belong to* E^α, $(\alpha > 1)$, *form a countable set.*—Indeed, the points of E^1 which do not belong to E^2 are isolated in E^1; each of them can therefore be enclosed in an interval that does not contain any other points of E^1, and these intervals with no common points can be taken. Therefore, these intervals form at most a countable infinity; thus, points of E^1 which do not belong to E^2 therefore form a finite or countable set B_1.

Similarly, points of E^β that do not belong to $E^{\beta+1}$ form a finite or countable set B_β. Now, the set mentioned in the statement is the sum of B_β for values of β less than α, which are finite or countable. Therefore, this set is itself finite or countable.

It can also be said that the points of E which do not belong to E^α form a finite or countable set, because the same information which showed that the sets B_β are at most countable proves the same for the set of points of E that do not belong to E^1. Hence, it follows that *any set of points whose one of the derivatives does not contain any points is at most countable.* These sets are called *reducible sets.* These are specific countable sets: for example, the set of rational numbers is countable, but not reducible.

When a set is not reducible, two cases are possible a priori. Either, in the course of derivation operation, we arrive at a derived set E^α which is perfect and then, following our conventions, we would stop at our operations of derivation at this set. In other words, we only consider the different derivative $E^1, E^2, \ldots, E^\alpha$. Or, the operations of derivation will always yield the different derivative, resulting in an uncountable set similar to S_0 of such derivatives. We will see that this case does arise.

Let us consider the set S of derivatives of E, which actually exist, meaning they all contain points, and they are not perfect, except perhaps the last. They are all different. Let (a_1, b_1), $(a_2, b_2), \ldots$ be the intervals contiguous to E^1, arranged in a simply infinite sequence. Let (a_i^α, b_i^α) be those intervals contiguous to E^α which contain (a_i, b_i). Let l_i^α be the increment in the length when we pass from (a_i^α, b_i^α) to $(a_i^{\alpha+1}, b_i^{\alpha+1})$. Some of l_i^α are zero. However, for each value of α, some of l_i^α are non zero, since $E^{\alpha+1}$ is always different from E^α. If $E^{\alpha+1}$ does not exist, we will take (a, b) as the interval $(\alpha_i^{\alpha+1}, \beta_i^{\alpha+1})$.

Now for a given i, there can be at most a countable number of values of l_i that are non-zero, because the length of (a_i^α, b_i^α) cannot exceed beyond the length of the interval (a, b) containing the given set E.[13] Since this is true for each integer value of i, the set of all the l_i^α different from zero is at most countable. The set of terms of S is therefore countable,[14] therefore

$VI.$ *Every set of points either has a derivative which does not contain any point, or has a perfect derivative.*

$VII.$ *Any closed set is the sum of a finite or countable set and a perfect set; one or the other of these sets may not exist.*

Indeed, a closed set E is contained in its derivative E^1, and the latter is the sum of a set that is at most countable and its last derivative, which is either empty or perfect.

These two theorems constitute the property known by the name of Cantor–Bendixson theorem.[15] To give the decomposition indicated by this theorem its full value, it is

---

[13] Therefore, we suppose that E is entirely at a finite distance; this assumption is not at all essential. We would treat the general case either by a transformation such as $x = \tan \frac{\pi x}{2}$, or by decomposing the set E into sets E_i, where E_i is formed of those points of E which are in $(i, i+1)$. In the case where E is not entirely at a finite distance, it is convenient to agree that the derivative of E contains a point at infinity.

[14] In essence, we have just proved a general property of sequences of closed sets, which will be stated in the text shortly.

[15] See *Acta mathematica, t.* 2.

I mentioned that we would not deviate from the ideas of Cantor, but the form of the proof in the text is very different from that of Cantor in the following respect. Cantor uses the concept of the total set of different of non different derivatives, that can be associated with each number from S_0. He

obviously necessary to note that *every perfect set is either uncountable or, more precisely, has the cardinality of the continuum.* This is obvious if the set contains an interval; therefore, let us assume that E is perfect set that is non-dense everywhere, with a and b as its extreme points. Now we have associated (Sect. 2.2) with such a set E, a continuous function $\varphi(x)$, which is constant in every interval contiguous to E, increasing in every interval containing points of E, and varies from 0 to 1 when x is contained between a and b. The equation $\varphi(x) = m$ admits, for each value of m between 0 and 1, either only one solution x_0, in which case x_0 is the abscissa of a point of E, or an infinite number of solutions given by a double inequality $x_1 \leq x \leq x_2$, in which case x_1 and x_2 are the abscissa of the two extremities of an interval contiguous to E. Thus, for each value m of $(0, 1)$, we associate either a point x_0 in E, or two points x_1 and x_2 in E; therefore, E has the cardinality of the continuum.
Also note that the reasoning from Sect. A.3 proves that: *Any sequence of closed sets, that are all different, each of which contains those that come after it in the sequence, must necessarily be finite or countable.*

A.4 Is a Notation for Transfinite Numbers Necessary for Us?

Throughout this book, we have made several references to the notion of the transfinite numbers. Therefore, it is important to clarify this notion and if possible, to convince the reader that they need only familiarise themselves with the sequence of transfinite numbers by using it frequently to gain as clear a view of this sequence as they have of the sequence of integers.[16]

When we first heard about the sequence S_0 of finite and transfinite numbers, we readily believed that everything would become clear if we had a notation for transfinite numbers. However, it is clear that we cannot have one; we can only manipulate a finite number of symbols or sounds and combine them in a finite number of ways, which would always yield only a finite number of notations. This number could be large, depending on the number of symbols, their more or less fortunate choices, and the ingenuity of the prescribed combination

---

names E_Ω as the set of points common to all these derivatives. This set E_Ω is a sort of derivative that comes after all those, related to the transfinite numbers of S_0. For Cantor, the symbol Ω is the first transfinite number of second class of transfinite numbers, and it comes after all the elements of S_0.

On the contrary, we have avoided reasoning about the set S_0 as it is *currently* conceived in its entirety. Instead, we have only reasoned about the procedure of formation of S_0 and any of its segments obtained in the course of the application of this procedure. In this way, we do not use an uncountable well ordered set at any point.

[16] My goal is thus clearly defined; I will therefore not address the questions, which are more philosophical than mathematical, raised by the concept of the sequence of integers. I consider this notion to be established and perfectly clear.

rules, but it will always be finite. Any enumeration applied to an infinite set of numbers is, in certain respects, fictitious.

Let us examine, for example, the decimal enumeration applied to the numbers included between 0 and 1; it claims to allow us to represent any number. For this purpose, it uses an indefinite sequence of digits $0, a, b, c, d, \ldots$. However, such a sequence can neither be written, nor pronounced. There are a few fortunate cases in which we could state the law of succession of decimal digits $a, b, c, \ldots$ of x and this law determines x; most of the times, all we could say is this: the sequence $0, a, b, c, \ldots$ is determined by the number x. It is obvious how illusory the decimal notation of numbers in the interval (0, 1) is.

Let us again examine the enumeration of integers. It is highly practical and easy for everyday use and represents a significant progress over the methods of Greeks, although those methods were already very powerful and somewhat akin to our current methods.

But, just like the methods of the Greeks, it does not allow us to denote all numbers. It only convinces us, as the Greeks did, that no matter how large a number is, we will succeed in establishing conventions ingenious enough to represent this number and smaller numbers.

The enumeration would only allow us to represent all the numbers if we could repeat the same act (to write a digit, to state a digit) an arbitrarily large number of times. But then, we could draw as many bars, or pronounce the sound 'one', as many times as there are objects in the finite collection to be counted. Therefore, we have only solved the problem of the representing numbers by placing ourselves in an assumption where the problem does not actually arise.[17]

It is quite clear that, similarly, the notation of transfinite numbers would be immediate for those who could repeat the same action a well ordered and countably infinite number of times of any type of order.

Despite everything, our habits are such that we desire a semblance of notation; even if it is merely conceptual and practically unattainable—meaning entirely illusory. Being able to denote the first transfinite numbers $\omega, \omega + 1, \ldots, 2\omega, 2\omega + 1, \ldots, 3\omega, \ldots, \omega^2, \ldots$ has already helped us.

Let x be a number between 0 and 1, let us write it in the decimal system. We assume this expansion to be infinite, let $0, a_1, a_2, a_3, a_4, \ldots$. Let us set

$$
\begin{array}{lllll}
x_1 = & 0, & a_1, & a_3, & a_5, \ldots, \\
x_2 = & 0, & a_{2\times 1}, & a_{2\times 3}, & a_{2\times 5}, \ldots, \\
x_3 = & 0, & a_{2^2\times 1}, & a_{2^2\times 3}, & a_{2^2\times 5}, \ldots, \\
\ldots & & & &
\end{array}
$$

If all the numbers x_i are different and if the set of these numbers is ordered in the direction of increasing x, for example, let us agree to say that x denotes the order type of this set.

[17] We note with interest what M. Borel says about a conversation he had with M. Baire (*Leçons sur la théorie des fonctions* 2nd *edition, p.*178).

What we did in the previous section regarding the construction of sequences S_α of order α shows that every transfinite numbers is denoted by some points x. However, this notation is very imprecise: the same number α corresponds to an infinite number of points x. For this *notation* to get close to the properties of ordinary notation, it would be necessary for each transfinite number α to be associated with well-determined[18] sequence S_α and therefore to a well-determined[19] number x.

We are thus led to ask whether it is possible to name a set E contained in (0, 1) such that there exist a bijective correspondence between the points of E and the numbers of S_0. This problem has preoccupied a lot many mathematicians; we have asserted several times that we could take the interval (0, 1) itself as E . However, the question is far from being resolved.[20] Besides, for us, all this is secondary, because it is clear that this mode of representation of transfinite numbers would hardly help us in understanding them better.

Yet, this pseudo-notation fulfils one of the essential conditions of notation. After all, what is the purpose of enumeration of integers? It never appears in any reasonings, except of course, in the matters which deal with the enumeration itself. Still, it allows us to characterise the number we are discussing, provided it is not very large. In other words, it enables us to construct and determine, a well ordered finite set similar to the one under consideration. When someone talks to us about the number 3, we know that it is just about the sets similar to this one: *I I I*, and when we argue about the number three, we make arguments that apply to all well ordered sets similar to set of such features. Whenever we argue about an integer, given in any manner whatsoever, we are in the similar circumstances.

Therefore, it is clear that we could do away with the notion and the term 'number'; the more tangible notion of ordered and finite sequence of objects would suffice for us. Exactly in the same manner, to assume a given transfinite number is to assume a well-odered countable set as determined. To reason about this number, is equivalent to making reasons that apply indiscriminately to all sets similar to this one. The notion and the term transfinite number are therefore unnecessary for us; the notion of a well ordered countable set suffices for us.

Basically, the obscurities we believed to arise in the notion of transfinite number might already be present in the notion of finite integer, when we want to see it as a metaphysical entity, and therefore somewhat unclear.

G. Cantor says[21]: We call the power or the cardinal number of a set M, the general notion that we deduce from M using our faculty of thought, abstracting from the nature of different elements of M and their order.

---

[18] The sequence we have called S_α is determined only if $S_{\alpha-1}$ is, when α is of the first kind. When α is of the second kind, the definition we provided for S_α assumes that the numbers less than α have been arranged in a determined manner, in a simply infinite sequence.

[19] In order for x to be well-determined when S_α is, it is necessary to take some elementary precautions, to account for the fact that decimal numbers have two decimal expansions.

[20] In a very recent Memoire, to which I can only refer the reader, M. Hilbert revisited the question (*Sur l'infini, Acta mathematica, t.*48).

[21] These citations are made according to the translations of M. Marotte.

And, for him, a type of order is the general notion which results from M when we abstract from the nature of the elements of M, but not from their order of succession.

It appears difficult to find anything else in these philosophical definitions other than the previous remarks, and as a result, these definitions reduce to a matter of language convention: when we use the word 'number', we are merely reminding ourselves that we are reasoning about a set, but we are doing so using reasonings that apply equally well to all sets similar to the one under consideration.

Thus, the definition of the finite and transfinite numbers is void of its metaphysical content and no longer presents any obscurity. It is true that we have shown simultaneously that the use of the word 'number' is unnecessary; but we are habitual of using the words 'whole numbers', so it will be easy for us to use the word 'transfinite numbers' as well, which we know we can do without any inconvenience.[22]

In summary, when we speak of a transfinite number, we mean that we are dealing with a sequence—that is, a well ordered countable set—defined only up to similarity.

A.5 Reasoning by Transfinite Induction

How can we obtain a property of transfinite numbers?[23] To answer this question, let us examine how we obtain a property valid for all integers. This is always done by using the method of mathematical induction at some point.

This method of reasoning is perfectly convincing[24]; it is constantly used by mathematicians, and there cannot be any question for us to raise the slightest doubt on its validity. Due to the fact that we will recall, the method of mathematical induction is not reducible to syllogistic reasoning, we can, therefore, simply conclude that the syllogistic reasoning is not the only one that can be used in mathematics.

---

[22] What is happening to us here always occurs when one clarifies a notion which was earlier obscure. Once it was understood that reasoning about the imaginary numbers was, in fact, reasoning about pairs of real numbers, the use of imaginary numbers became both clear and unnecessary, at least from the logical point of view. However, practically, there were significant advantages to using the imaginary numbers, as their use became at both legitimate and advantageous.

[23] We are only concerned with the reasonings applicable to any transfinite number; reasonings related to a particular transfinite number, however, may be totally different from those we are going to study:

Let us consider the function $F(x) = \frac{1}{f(x)-1}\Gamma\left[\frac{1}{f(x)}\right]$ in the expression of which, $f(x)$ represents any continuous monotonically increasing function which varies from $-\infty$ to $+\infty$ when x varies from $-\infty$ to $+\infty$. Let us reason about the set of values of x in the neighbourhood of which $F(x)$ is infinite, is reasoning about the number $\omega + 1$. However, it is clear that in doing so, we are not using the general notion of a well ordered set.

[24] I purposely used the word 'convince' to emphasise that in my opinion, the reasons to declare oneself satisfied with an argument are of a psychological nature, in mathematics as elsewhere. The logic gives us reasons for rejecting some of the reasonings; it cannot make us believe in a reasoning.

Let us assume that we have demonstrated: 1. A property P holds for number one, 2. if P holds for a number, then it also holds for its successor. Given this, let us demonstrate the property P for the number 3. We will say:

The property P holds for one. Since P is true for a number whenever it is true for the previous one, it must also hold for two, which follows one.

The property P is true for two; and since the property P is true for a number when it is true for the previous one, therefore, the property P is true for the number three which follows two.

Thus, to reach the number three, we had to repeat a line of reasoning reducible to syllogisms, twice.

To obtain the property P for all the numbers, we would need to perform infinitely many syllogisms. It is true that we can reason by contradiction: if the property was not true for all the numbers, we could find a number N for which it is not satisfied. Then, by descending the ladder of numbers. From N, where P does not hold, down to a number where P holds, we would encounter two numbers, n and $n + 1$, such that P holds for n but not for $n + 1$, which is contradictory.

But, if we examine this reasoning, we see that the search for N will proceed in the following way: if the property P is not true for all the numbers, it is either false for one or for a number greater than one. Then, if it is true for one, it is false for two or for a number greater than two, and so on. The search for N would proceed by induction and with the help of a proposition whose correctness is proved only by contradiction which would arise from assuming that we have exhausted the sequences of integers without encountering N. The assertion of the existence of N, on which we based our proof by contradiction, is thus legitimised in the same way, by reductio ad absurdum and through an actual infinity of syllogisms.[25]

Therefore, we have not legitimised the method of proof by induction using syllogistic reasoning. It is quite clear that any legitimisation of this nature is impossible, because we have constructed the sequence of integer only by assuming that we could envision the repetition of the same operation, namely the addition of an element, an arbitrarily large number of times.

---

[25] The question of so called ≪ law of excluded middle ≫ arose in connection with the difficulties posed by certain reductio ad absurdum arguments. In this regard, one can refer to the work of M. Brouwer and of M. Weyl. I may be allowed to take advantage of this opportunity to recall that, long before the question assumed this philosophical form, I wrote (*Soc, math, de France,* 1904) ≪ ... I do not attribute any more value to the method by which we show that a non finite set contain a countable set. Although I strongly doubt that we will ever name a set which is neither finite nor infinite, the impossibility of such a set does not appear demonstrated to me. ≫

By this I wanted to say that given definitions had been established for the *two* words finite and infinite, it was not certain that non finite meant infinite. This observation is not unrelated to some of the ideas of M. Brouwer. However, my purpose in making it was quite the opposite of his. I, on my part, do not by any means, contest the value of the traditional logic and the method of reasoning by reductio ad absurdum; I merely remind that it must be used correctly.

The contemptuous manner in which M. Brouwer speaks of the Paris School highlights our disagreement very strongly.

We can hope to explore the domain thus constructed only by considering the possibility of repeating the same reasoning an arbitrarily large number of times, which precisely means reasoning by induction.

Our initial attempt to legitimise the reasoning by induction using syllogisms, therefore, gives us a result as favourable as we could hope for, since it provides us with a means of verifying P, for each particular number using a finite number of syllogisms.[26] Without doubt, however fast we may go churning out these syllogisms, there will be numbers large enough for which we cannot provide a demonstrative proof. However, it is easy for us to conceive that, each time we manage to name a number M, which means that each time we have succeeded in performing the addition of an element, $M - 1$ times in a row to form a set of M considered objects, we can find a way to repeat the induction reasoning $M - 1$ times, as is necessary to verify the property P for the number M. Moreover, for a mathematician, it is not the demonstration that creates the truth of a proposition; it merely allows one to ascertain that truth. Hence,we may not have the same requirements here as we did when dealing with notation, and we can be satisfied with a demonstration constructed using the same method by which the sequence of integers itself is constructed.

In short, the construction of a finite segment of the sequence of integers only requires the use of what can be called finite induction. The demonstration of a property for all the numbers of this segment requires only a reasoning by finite induction, reducible to syllogisms. The construction of the complete sequence of the integers requires a process of infinite induction, and demonstrating a property for all integers requires reasoning by infinite induction, which could only be replaced by an indefinite sequence of syllogisms.

Consequently, it is clear that, to prove a proposition for the sequence of transfinite numbers, it is necessary to use a method of reasoning which allows us to follow step by step the procedure of forming this sequence. This reasoning is called transfinite induction and it requires proving: 1. that a property P is true for the number one (or for the number ω, depending on whether we consider all finite and transfinite numbers or only transfinite numbers); 2. that if the property P is true for all the numbers less than a number α, it is true for the number α.

This second part of reasoning by transfinite induction is most often replaced by two distinct demonstrations. We only prove the previous statement of number 2 above for α of the second kind and, for α of the first kind, we show that if the property is true for $\alpha - 1$, it is true for α.

In either form, the transfinite induction reasoning would lead to the syllogistic verification of the proposition for any number α, if we were to assume that we can repeat a reasoning a well ordered and countable number of times.

---

[26] The legitimisation through reductio ad absurdum of the procedure by induction is far less perfect in this respect, since it requires an *actual infinity* of syllogisms.

The links between this mode of reasoning and syllogistic reasoning remain close; however, we are still very far from ordinary reasoning since, at least for α of second kind, the second part of our reasoning is based on the knowledge of actual infinity of premises.[27]

We can, as we have already done it, explain why we regard a property as demonstrated for all the transfinite numbers by transfinite induction reasoning by saying: if the property P was not true for all the transfinite numbers, there would exist numbers for which it would not be true, the sequence of these numbers would comprise an element α smaller than all the others; property P would be true for all the numbers smaller than α and false for α, which would be a contradiction.

However, we would not consider this as a justification for transfinite induction reasoning. This mode of reasoning is new, irreducible to others, and it is precisely why it is powerful and useful.

But is one obliged to accept this mode of demonstration? Absolutely not; it is up to each individual to decide whether he is entirely satisfied with reasoning by transfinite induction. Nevertheless, since almost everything in mathematics has been written only for those who admit the reasoning by ordinary induction, some passages of this book are written only for those who admit the reasoning by transfinite induction.

Even if we feel convinced by the demonstrations involving transfinite induction, we may not love them, desire to do without them or want to define what the domain of mathematics would be without this mode of demonstration. This is why we would examine some means of avoiding the use of the transfinite induction.

A.6 Examination of Some Reasonings by Transfinite Induction

Throughout the contents of this book, we have used transfinite numbers in three different ways. First, by using Cantor–Bendixson theorem; then, by using the chains of intervals; and finally, in connection with the work of MM. Baire and Denjoy. Let us examine these applications, beginning with the use of chains.

Let us start by assuming that a well-ordered set of points is given in (a, b), ordered in the direction of increasing x, for example. We will label these points as $x_1, x_2, x_3, \ldots, x_\omega$, $x_{\omega+1}, \ldots$, and assume that each point with an index of the second kind is a limit point of the set of points with smaller indices. Then we say that the division of (a, b) using the considered points is a chain of intervals.

We would use such a chain to evaluate the increment $F(b) - F(a)$ of a function $F(x)$ by the formula

$$F(b) - F(a) = \sum [F(x_i) - F(x_{i-1})],$$

[27] This infinity is countable; it is to avoid having to consider an uncountable infinity of premises that I previously refrained from discussing the uncountable sequence of distinct or non distinct derivatives of a given set.

but it is necessary to explain the meaning of right-hand side.[28]

Let us suppose that the numbers $u_0, u_1, \ldots$ are known, and they are associated as indices with all the numbers from the sequence S_0 that are less than a certain number α. Then, the ordered sequence of these numbers u_i is of order type α.[29] The symbol $S = \sum^{\lambda<\alpha} u_\lambda$ represent a transfinite series of order type α. Let us set

$$S = \sum_{\lambda=0}^{\lambda<i} u_\lambda$$

The S_i are called the partial series of S. The S_i, as numbers, hold meaning for finite i; for $i = \omega$ we agree that S_i, as number, exists only if the ordinary series $\sum_{\lambda=1}^{\lambda<\omega} u_\lambda$ converges, and in such a case, it is equal to the sum of this series. The numbers S_i thus defined, for $i < \omega$, are partial sums of S. For extending the definition of these numbers to any index equal to α at most, let us agree that we have, for any λ of first kind

$$S_\lambda = S_{\lambda-1} + u_{\lambda-1},$$

and, for any λ of second kind, S_λ will denote the limit, if it exists, towards which any simply infinite sequence $S_{\lambda_1}, S_{\lambda_2}, \ldots$ relative to increasing numbers $\lambda_1, \lambda_2, \ldots$ converges, defining λ as the smallest number greater than them.

When these definition apply to all values of λ, such that $\lambda \leq \alpha$, the transfinite series is said to be convergent, and its sum S is the number S_α, which has just been defined. The convergence of a series of order type α thus requires the existence of a definite limit for each number of second kind, at most equal to α.

For the series

$$\sum_{\lambda=1}^{\lambda<\alpha} [F(x_\lambda) - F(x_{\lambda-1})],$$

relative to a continuous function $F(x)$, in which we have assumed $x_0 = a$, $x_\alpha = b$, and where we replace $F(x_\lambda) - F(x_{\lambda-1})$ with zero when λ is of second kind, it is clear that:

1. We have

$$S_1 = F(x_1) - F(a);$$

2. If we have

$$S_{\lambda-1} = F(x_{\lambda-1}) - F(a),$$

then we have

$$S_\lambda = S_{\lambda-1} + u_{\lambda-1} = [F(x_{\lambda-1}) - F(a)] + [F(x_\lambda) - F(x_{\lambda-1})] = F(x_\lambda) - F(a).$$

[28] This explanation could have been omitted in the main body of this work and postponed until now, which indicates how natural and, in a way, necessary the following definitions are.

[29] The presence of a term u_0 restores the agreement between the definition of order types for finite and transfinite α. *See* the footnote 7 of this chapter.

3. If we have

$$S_\lambda = F(x_\lambda) - F(a)$$

for all the numbers λ less than a number μ of second kind, and if we take an increasing sequence of numbers λ_i greater than λ_i, then, since the point x_{λ_i} increases towards x_μ, we have, by the continuity of F,

$$S_\mu = \lim_{i \to \infty} S_{\lambda_i} = \lim[F(x_{\lambda_i}) - F(a)] = F(x_\mu) - F(a).$$

Hence, by transfinite induction, we see that the series is convergent and has the sum

$$S = F(b) - F(a).$$

Here are two propositions concerning transfinite series, some particular cases of which are frequently used in classical analysis, for example, in transforming an ordinary series into a double-indexed series.

In a convergent transfinite series, we can group consecutive terms in any manner and consider the resulting partial sums as performed. In other words, if we have

$$0 = \lambda_0 < \lambda_1 < \lambda_2 < \cdots < \lambda_\mu < \cdots < \alpha.$$

the μ's constitute a finite, simply infinite or transfinite sequence of order type β, if the first series is convergent, then

$$\sum_{\lambda=0}^{\lambda<\alpha} u_\lambda = \sum_{\mu=0}^{\mu<\beta} \left(\sum_{\lambda=\lambda_\mu}^{\lambda<\lambda_{\mu+1}} u_\lambda \right).$$

This theorem follows immediately from the equality

$$\sum_{\lambda=0}^{\lambda<\alpha} = \sum_{\lambda=0}^{\lambda<\alpha_1} + \sum_{\lambda=\alpha_1}^{\lambda<\alpha} \quad \text{for} \quad \alpha_1 < \alpha;$$

which we will easily verify.

If, by replacing the terms of a transfinite series with their absolute values, we obtain partial sums bounded as a whole, the series is convergent. It remains convergent in any order we may arrange its terms (well ordered), and its sum does not depend on this order.

We can prove this property in many ways using transfinite induction. However, I emphasise here that we can always obfuscate the use of this induction by employing a proof by contradiction.

It would be obviously sufficient to show that if we arrange in a series (simply infinite) $V_\alpha = \sum \nu_n$ the terms u_λ of a transfinite series $\sum_{\lambda=1}^{\lambda<\alpha} u_\lambda$ of the nature indicated in the statement, that is, absolutely convergent, each partial sum

$$S_\mu = \sum_{\lambda=1}^{\lambda<\mu} u_\lambda$$

exists and is equal to the sum of the series V_μ obtained by removing from V_α, the terms which do not appear in S_μ. This certainly holds for $\mu = 1$; if this did not hold for any value of μ, there would be a first value μ_0 for which the proposition would not be true. Now we suppose that μ_0 is of first kind. Then, we have

$$S_{\mu_0 - 1} = V_{\mu_0 - 1}$$

But

$$S_{\mu_0} = S_{\mu_0-1} + u_{\mu_0-1}, \quad V_{\mu_0} = V_{\mu_0-1} + u_{\mu_0-1},$$

hence

$$S_{\mu_0} = V_{\mu_0},$$

which is in contradiction with the definition of μ_0.

Therefore, let us assume μ_0 is of second kind. Then for $\mu < \mu_0$, we have

$$S_\mu = V_\mu;$$

furthermore, V_μ is a partial series deduced from V_{μ_0} by removing some terms, and every term of V_{μ_0} belongs to V_μ as long as $\mu < \mu_0$ is sufficiently large. Therefore, if the simply infinite sequence

$$\mu_1 < \mu_2 < \cdots < \mu_0$$

defined μ_0 as the first number greater than all the μ_i, then V_{μ_i} tend towards V_μ and as a result we have

$$S_{\mu_0} = \lim S_{\mu_i} = \lim V_{\mu_i} = V_{\mu_0};$$

This is the same contradiction as the previous one.

The proof of the theorem shows that when a transfinite sequence is assumed to be formed, we can reason about it using a proof by contradiction, without apparently needing to resort again to a transfinite induction.

But, in the case of transfinite chains, we can go further. To say that it is absurd for there to be a first transfinite number μ_0 from which a property does not hold, is equivalent to saying that, among the intervals (a, x_λ), there is a first interval (a, x_{μ_0}) for which some property does not hold.

And if we have succeeded in putting this property in such a form that it belongs to (a, x) as soon as it belongs to any (a, x_λ), with $x_\lambda > x$, we are brought back to proving that a certain property holds for an interval (a, X) as soon as it holds for all (a, x), with $x < X$. We no longer need to use either the chain or transfinite induction; we reason on the continuum.

The first transformation of this nature that was applied to a proof involving the transfinite was undoubtedly the one that allowed us to deduce the reasoning in Sect. 8.2 from a reasoning of M. Borel, which can be restated as: We have an infinite[30] number of intervals (a_i, b_i) such that every point of $a \leq x \leq b$ is strictly interior to at least one of them. Let us choose an interval (a_{i_1}, b_{i_1}) containing a in its interior, an interval (a_{i_2}, b_{i_2}) containing b_{i_1} in its interior, an interval (a_{i_3}, b_{i_3}) containing b_{i_2}, etc. If we do not reach b in this way, there is a point, let us call it b_{i_ω}, which is the limit point of b_i. Let us choose an interval $(a_{i_{\omega+1}}, b_{i_{\omega+1}})$ containing b_{i_ω}, etc. It is clear that by using b_λ, a chain of intervals covering (a, b) is defined. Moreover, it is clear that (a, b_{i_n}) can be covered with a finite number of intervals (a_i, b_i); and the same holds for $(a, b_{i_{\omega+1}})$ since this interval can be covered with $(a_{i_{\omega+1}}, b_{i_{\omega+1}})$ and some others, in finite number of the (a_{i_n}, b_{i_n}). By continuing in this transfinite manner, we see that (a, b) can be covered using a finite number of intervals (a_i, b_i).

The transformation of this reasoning yields the conclusion that the interval (a, b) can be covered using a finite number of intervals (a_i, b_i); otherwise, there would exist a value of x such that $(a, x + h)$ cannot be covered in this way, although $(a, x - h)$ could be, for any small $h > 0$. However, this is impossible; because there exists an interval (a_{i_0}, b_{i_0}) containing x, and consequently, (a, b_{i_0}) can be covered by this interval along with those which allow for the covering of (a, a_{i_0}); the existence of the value x is impossible. This is the reasoning from Sect. 8.2.

Examining this transformation of the reasoning highlights the advantages and disadvantages of each form: the second form is more rapid and appears familiar. It allows us to better visualise the possible generalisations, and restrictive hypothesis that can be eliminated. However, it provides only an existence theorem, while the first form offers an operational procedure for choosing a finite number of (a_i, b_i), fit to cover the entire (a, b).

M. Borel, right from the beginning, emphasised that his reasoning provided a ≪ regular procedure ≫ to ≪ effectively determine ≫ the intervals (a_i, b_i). Certainly, we could quibble on the word effectively; point out that one cannot ≪ effectively ≫ carry out an operation which comprises infinitely many stages.[31] However, following this line of thinking, we would not effectively determine a function when only a series expansion for it were found. We can, certainly, refuse to use the term ‘effectively’ in this case; but is this not already a significant achievement to have solved a problem as well as the one of determining a function when we have provided the law governing the successive terms in its series expansion?

The chains of intervals are introduced in the search for primitive functions in the following way: Let $f'(x)$ be the derivative of a continuous function $f(x)$; there exist intervals $(a_0, a_1), (a_1, a_2), \ldots$ in each of which we have an inequality of the form

$$|f(x) - f(a_i) - (x - a_i) f'(a_i)| \leq \varepsilon(x - a_i).$$

---

[30] M. Borel assumed the countable infinity, but that is unnecessary.

[31] I would like to add that, in some respects, there is no operation that can always be carried out in reality. Calculating the value of a known function, $\sin x$, for example, is one such operation. When say that $f(x)$ is given, if $f(x)$ known when x is known, we are not providing an explanation of what we mean by a given function. Rather, we make this convention: when we said that $f(x)$ is given, we reason *as if* we can calculate f for each value of x.

These intervals, and these inequalities, are the ones which we have considered in the proof of existence of solutions of the equation $y' = f'(x)$. However, we generally put ourselves in conditions where we can cover the entire considered interval using a finite number of these intervals (a_i, a_{i+1}). If we assume nothing about $f'(x)$, then the intervals (a_i, a_{i+1}) are infinite in number and they form a chain of intervals which can always be used for the approximation of $f(x)$ through the exact formula

$$f(x) - f(a) = f(x) - f(a_\alpha) + \sum_{\lambda=0}^{\lambda<\alpha}[f(a_{\lambda+1}) - f(a_\lambda)].$$

Where, x is assumed to be contained in $(a_\alpha, a_{\alpha+1})$. From this approximation of $f(x)$, we then deduce that, with some generalisation of integral, we have

$$f(x) - f(a_0) = \int_{a_0}^{x} f'(x)\,dx,$$

as a consequence of the inequality,

$$\left| f(x) - f(a_0) - \int_{a_0}^{x} f'(x)\,dx \right| \leq \varepsilon(x - a_0).$$

This result is obtained in various chapters of this book using arguments of transfinite induction. However, it is clear, according to what has been discussed earlier, that we can deduce it from a reductio ad absurdum argument whenever it is shown that no point x could be the last for which the previous inequality holds.

The reader could check the modifications that are required to be brought to our exposition to rid it from any use of the transfinite numbers and transfinite induction. We will perform this transformation of reasoning only for the first of the propositions related to derivatives obtained with using the method of chains, namely[32]:

A function is determined up to an additive constant when its derivative is known at every point. Indeed, let $f(x) = f_1(x) - f_2(x)$ be the difference of two functions f_1 and f_2, which have the same finite derivative; thus, $f(x)$ has a zero derivative at any point. Our method of chains yields

$$|f(x) - f(a_0)| \leq \varepsilon(x - a_0),$$

for any small $\varepsilon > 0$; therefore $f(x)$ is constant.

The reasoning by the contradiction is as following: there exist points $x > a_0$ for which the previous inequality holds, let x_0 be the last of those points for which this inequality is satisfied. Let us take, what is always possible, h sufficiently small and positive so that in

---

[32] We refer to Sect. 6.3; at this point, the conditions envisaged are a little more general than those stated in the text. Here, we are placing ourselves in the simpler conditions.

$(x_0, x_0 + h)$ we have

$$|f(x) - f(x_0)| \leq \varepsilon(x - x_0);$$

then in this entire interval, we have

$$|f(x) - f(a_0)| \leq |f(x_0) - f(a_0)| + |f(x) - f(x_0)| \leq \varepsilon(x_0 - a_0 + x - x_0),$$

which shows the contradiction in assuming that the considered inequality does not hold for $x > x_0$.

This version of the proof was given by M. Denjoy.[33]

The study of functions possessing a certain property at each of their points, leads very naturally to the use of chains of intervals,[34] and, therefore, transfinite induction. Each of these arguments can be transformed, as just mentioned, to rely solely on the properties of the continuum.

It seems much more difficult to eliminate the transfinite methods from other applications we have given here: such as the theorem of Cantor–Bendixson, results of M. Baire, and the theory of totalisation. However, a distinction is necessary. We have broken down the theorem of Cantor–Bendixson in two statements VI and VII. The statement VI refers to the concept of the derivatives, which was established only through transfinite induction. Therefore, we cannot justify this statement without the use of transfinite; the statement VII, on the contrary, is a property of closed sets that we can hope to obtain without the use of transfinite methods.

---

[33] (*Jour. de Math.*, 1915, *p.* 176). M. Denjoy, however, does not obtain this proof in the same way as we do here, by transforming a reasoning previously given in the first edition of this book. On the contrary, he contrasts his method with those presented in the first edition. However, he only refers to proofs using the theorem of finite increments. He obviously missed that the method of chains, since it allowed for the search for primitive functions of a given derivative, thereby provided the proof that these functions were determined up to an additive constant, and that I had used this method of proof on page 79 of the first edition. However, I had not bothered to show, as I have done in this new edition (Sect. 6.3), that the method of chains yielded *previously established* statements; I had primarily used it to establish *new* results.

I take this opportunity to mention that I disagree with M. Denjoy and some other authors who consider reasonings based on the derivatives as entirely different, with some using the notion of integral, while the others do not, but instead use the Mean Value theorem:

$$|f(b) - f(a)| < M|b - a| \quad \text{when} \quad |f'(x)| < M.$$

Any theorem based on the integration can be expressed using only the Mean Value theorem. Moreover, aiming to give more uniformity to this work, I wanted, in Chap. 10, to deduce some theorems obtained by M. Denjoy using totalisation without resorting to the totalisation, just as I had deduced, some theorems of integration in Chap. 9. This was quite easy for me and I did not digress significantly from the proofs of M. Denjoy, as there is very little contradictions between the ideas used in the reasonings which involve the integration and those which do not.

That, of course, does not mean that I ignore the interest of various forms of proofs, the grand elegance of some of them, and the progress that comes from using the simplest methods. On the contrary, it is the comparison of these forms of demonstrations which has led me to the reflections that I express in the text.

[34] *See* the footnote 14 of the Chap. 8.

One can argue that VII is *existence theorem of Cantor - Bendixson*, while VI provides a transfinite operational method for solving *Cantor–Bendixson problem*: ≪ given a closed set F, decompose it into a countable set D and a perfect set P ≫.

Similarly, in M. Baire's research, a distinction must be made between *M. Baire's theorem*: ≪ any function of class one is point-wise discontinuous on every perfect set and conversely ≫ *M. Baire's problem*: ≪ given a point-wise discontinuous function on any perfect set, find a series of continuous functions whose sum it is ≫. Nothing stops us from hoping that we can prove M. Baire's theorem without using the transfinite methods. However, the operational method given by M. Baire for solving his problem is inherently transfinite and it cannot be justified without transfinite methods.

In the M. Denjoy's research, the statements refer to the operation of totalisation, the very definition of which is transfinite; we cannot hope to avoid the transfinite methods here.

However, nothing stops us from hoping that we can eventually manage to replace the operational method of Cantor–Bendixson (statement VI), the operational method of M. Baire, and the totalisation of M. Denjoy with non transfinite methods, while still being able to solve the problem of Cantor–Bendixson, problem of M. Baire and the problem of primitive functions.

Indeed, it has been possible to demonstrate the Cantor–Bendixson theorem and Baire's theorem without transfinite methods, and the Cantor–Bendixson problem can be solved without transfinite numbers. Let us see how.

In three fields of researches currently under consideration, transfinite numbers were used to denote the successive elements of the sequences. These elements are closed sets, all different from each other, each of them containing those that come after it in the sequence. We know (Sect. A.4), that such a sequence is either finite or countable. Now, in the questions under study, the sequence can stop only when we reach a set that contains no isolated points or is perfect, and proof of the theorem boils down to showing that, under some conditions, the sequence does not terminate at a perfect set. Thus everything comes down to characterising the perfect set which could arise through a property of its points.

Here, I will focus only on the Cantor–Bendixson theorem[35]; in this case, it is obvious that the points of the perfect set P are characterised by being those for which there are an uncountably infinite number of points form the set in every neighbourhood.

---

[35] I have given several proofs of M. Baire's theorem without using the transfinite (*Bull. de la Soc. math. de France,* 1904.—Note II of the Lessons of M. Borel, *Sur les fonctions de variable reelle.- Journ, de Math.,* 1905). These proofs mainly relied, in accordance with what is mentioned in the text, on the definition of a perfect set.

M. E. Lindelof (*Acta mathematica, t.*29) and myself (in the first edition of this book) almost simultaneously showed that Cantor–Bendixson's theorem could be derived from the notion that M. Lindelof called a condensation point. I believe that it is more commonly referred to as accumulation point now, and I adopt this terminology.

In my presentation, as I arrived at theorems VI and VII simultaneously, it became unclear that the demonstration of theorem VII was independent of transfinite numbers. Here, I am adopting the exposition of M. Lindelof, which is preferable over mine in many respects, meaning that I focus only on theorem VII.

With this insight, let us define an *accumulation point* of a set E as any point x such that in every interval containing x in its interior, there exist uncountably many points of E.[36]

The Bolzano–Weierstrass theorem suggests the following statement: *Every set E, that is uncountable in* (a, b) *admits accumulation points.* Indeed, let x be the smallest value in (a, b) such that there are only at most a countably infinite number of points of E in $(a, x - h)$ and there are uncountably infinite number of points in $(a, x + h)$ for any $h > 0$. Such a point indeed exists. Now, in $(x - h, x + h)$ there are an uncountably infinite number of points of E; therefore, x is an accumulation point of E.

From this proof, it also follows that in (a, x), there are at most, only a countably infinite number of points of E since there are at most only a countably infinite number of points in $\left(a, x - \frac{1}{n}\right)$, for any n. Therefore, x cannot be at b. There is at least one accumulation point in the interior of (a, b).

According to this, an accumulation point cannot be isolated; for if x was the only accumulation point of E in (a, b), there would be at most only a countably infinite number of points of E in (a, x) and in (x, b), and hence in (a, b); and x would not be an accumulation point. Now, the set of accumulation points is obviously closed, therefore *the set of accumulation points of an uncountable set is perfect.*

Let us apply this to a closed set F. It contains its derivative F' and, *consequently*, the perfect set P of its accumulation points. Therefore, F is the sum of P and the set of those points which are contained in the various intervals (l, m) contiguous to P. However, in each (l, m) there are only at most a countably infinite number of points of F, and the (l, m) intervals are finite or countable in number. Hence, the set $F - P$ is countable. The Cantor–Bendixson's theorem is proved:

$$F = P + D.$$

If E had not been assumed to be closed, we would similarly observe that the set D of points of E that are not accumulation points of E, is countable. As for the points of $E - D$, these are those accumulation points of E that belong to E. They form a set which is not only everywhere dense over the set P of accumulation points, but also everywhere accumulated on P. By this, we mean that, in every interval (α, β) containing points of P in its interior, there are an uncountably infinite number of points that are common to E and P. Otherwise, E would indeed be countable in (α, β), which is absurd.

Therefore, this new method elegantly proves the Cantor—Bendixson theorem and even generalises it. Moreover, it solves the Cantor - Bendixson's problem if we adopt a convention

---

When I provided these methods, I indicate the interest of proofs derived from the transfinite because they offer regular operational procedures, not only for proving the theorems, but also for solving the problems (*Journ. de Math.*, 1905, *p.*183; *C.R.Acad. Sc.*, 1903).

The terms *theorem* and *problem*, which so aptly concretise the necessary distinctions, are due to M. de la Vallée Poussin.

[36] I am, therefore, considering the case of sets of points on a straight line, as we have always done so far. However, the reasoning extends to the n-dimensional space.

that determining the accumulation points of a set is one of the operations, we always consider to be feasible in practice. And this is because we actually know how to perform it frequently.

MIX
Papier aus verantwortungsvollen Quellen
Paper from responsible sources
FSC® C105338

If you have any concerns about our products,
you can contact us on
ProductSafety@springernature.com

In case Publisher is established outside the EU,
the EU authorized representative is:
Springer Nature Customer Service Center GmbH
Europaplatz 3, 69115 Heidelberg, Germany

Printed by Libri Plureos GmbH
in Hamburg, Germany